THE FRACTAL PHYSICAL CHEMISTRY OF POLYMER SOLUTIONS AND MELTS

THE FRACTAL PHYSICAL CHEMISTRY OF POLYMER SOLUTIONS AND MELTS

G. V. Kozlov, DSc, I. V. Dolbin, DSc,
and Gennady E. Zaikov, DSc

Apple Academic Press
TORONTO NEW JERSEY

Apple Academic Press Inc.
3333 Mistwell Crescent
Oakville, ON L6L 0A2
Canada

Apple Academic Press Inc.
9 Spinnaker Way
Waretown, NJ 08758
USA

Printed in the United States of America on acid-free paper

International Standard Book Number-13: 978-1-926895-81-9 (Hardcover)

Library of Congress Control Number: 2013954454

Library and Archives Canada Cataloguing in Publication

Kozlov, G. V., author
The fractal physical chemistry of polymer solutions and melts/G. V. Kozlov, DSc, I.V. Dolbin, DSc, and Gennady E. Zaikov, DSc.

Includes bibliographical references and index.
ISBN 978-1-926895-81-9
1. Polymers--Structure. 2. Polymer solutions. 3. Polymer melting. 4. Fractal analysis. I. Dolbin, Igor V., author II. Zaikov, G. E. (GennadiTMi Efremovich), 1935-, author III. Title.

QD381.9.S87K69 2013 547'.7 C2013-906807-4

ABOUT THE AUTHORS

Georgy V. Kozlov, DSc

Georgy V. Kozlov, DSc, is a Senior Scientist at UNIID of Kabardino-Balkarian State University in Nal'chik, Russian Federation. His scientific interests include the structural grounds of properties of polymeric materials of all classes and states: physics of polymers, polymer solutions and melts, and composites and nanocomposites. He proposed to consider polymers as natural nanocomposites. He is the author of more than 1500 scientific publications, including 30 books, published in the Russian Federation, Ukraine, Great Britain, German Federal Republic, Holland, and USA.

Igor V. Dolbin

Igor V. Dolbin was graduated Kabardino-Balkarian State University in 1999. He is a candidate of chemical sciences and has 208 scientific works and 2 monographs published in Russia and Germany.

Gennady E. Zaikov, DSc

Gennady E. Zaikov, DSc, is Head of the Polymer Division at the N. M. Emanuel Institute of Biochemical Physics, Russian Academy of Sciences, Moscow, Russia, and professor at Moscow State Academy of Fine Chemical Technology, Russia, as well as professor at Kazan National Research Technological University, Kazan, Russia. He is also a prolific author, researcher, and lecturer. He has received several awards for his work, including the Russian Federation Scholarship for Outstanding Scientists. He has been a member of many professional organizations and on the editorial boards of many international science journals.

ABOUT THE AUTHORS

CONTENTS

List of Abbreviations .... *ix*
List of Symbols .... *xi*
Summary .... *xiii*
Preface .... *xv*
Introduction .... *xvii*

1. The Theoretical Fundamentals of Macromolecules Fractal Analysis 1
2. Fractal Physics of Polymer Solutions 25
3. The Fractal Analysis of Polymer Melts 255
Index 309

LIST OF ABBREVIATIONS

APESF	aromatic polyethersulfonoformals
CLA	chemically limited aggregation
DLA	diffusion-limited aggregation
DMDAAC	dimethyldiallylammonium chloride
DMSO	dimethyl sulfoxide
EP	epoxy polymer
HDPE	high-density polyethylene
MCh	methylene chloride
MFI	melt flow index
MMA	methyl methacrylate
MWD	molecular weight distribution
P-Cl	particle-cluster
PAASO	polyarylatearylenesulfoxide
PAr	polyarylate
PC	polycarbonate
PDMDAAC	polydimethyl diallyl ammonium chloride
PDMS	polydimethylsiloxane
PESF	polyethersulfoneformals
PHE	polyhydroxiether
PMMA	poly(methyl methacrylate)
PPQX	polyphenylquinoxalines
PPTA	poly n-phenyleneterefthalamide
PS	polystirene
PSF	polysulfone
PVC	poly(vinyl chloride)
SAN	stirene with acrylonitrile
SANS	small angle neutron scattering

SAW	self-avoiding walks
THF	tetrahydrofuran
TMAC	trimethylammonium methacrylate chloride
TSAW	true self-avoiding walk
UCD	upper critical dimension

LIST OF SYMBOLS

A	Kuhn segment length
B	the proportionality coefficient
b_B	Burgers vector
c	concentration of particles
d	the dimension of Euclidean space
D_{ch}	fractal dimension of macromolecule part between clusters
E	elasticity modulus
E_{vap}	vaporization energy
f	times number
g	branching factor
G	shear modulus
K	Boltzmann constant
l	monomer link length
m	mass of the cluster
m_0	mean weight of polymer elementary link
MM	polymer molecular weight
$M\eta$	the viscosity-average molecular weight
N	the degree of polymerization
N_A	Avogadro number
n_{st}	equivalent statistical links between macromolecular entanglements nodes
R	universal gas constant
R_g	gyration radius of diffusible particle
s	a monomer links number per segment
S	the cross-sectional area of a macromolecule
T	temperature
t	copolymerization reaction duration
V_{mol}	molar volume

Greek Symbols

α	the coil swelling coefficient
β	an exponent, characteristic for each solvent
$\Gamma(\nu)$	Euler's gamma function
d_c	energy of interaction between atoms with deficiency (acceptor) and surplus (donor) of electrons
ε	the volume interactions parameter
η	intrinsic viscosity
h_0	initial viscosity of reactionary medium
ν	Poisson's ratio
ν_F	Flory exponent
r_d	the structure linear defects density
r_p	polymer density
σ_Y	yield stress
t_0	the characteristic time of the given process
Φ	a form factor of a macromolecular coil in the conditions of rotatory friction
c_s	empirical parameter

SUMMARY

The structural analysis of polymer solutions and melts properties was performed in this book. For this purpose, the modern physical conceptions have been used—fractal analysis, irreversible aggregation models, and cluster model of polymers amorphous state structure. The statistical flexibility of a polymer chain was used as the main characteristic of polymer chemical structure. Such approach allows determining the structure of the main polymer element in solution and melt (macromolecular coil) as the function of the polymer chain flexibility. In its turn, the knowledge of macromolecular coil structure, characterized by its fractal dimension, allows determining polymer solutions and melts main characteristics (e.g., their viscosity). Also the offered approach allows obtaining the intercommunication of polymer structure in different states (in solution, melt, and solid-phase state). This allows predicting solid-phase polymers properties that are already at the stage of synthesis. The fractal model of polymer nanocomposites melt viscosity, principally differing from the description of polymer microcomposites melt viscosity, should be distinguished specially.

PREFACE

Physical chemistry, which combines physics and chemistry, has been developing quite radically in the recent times. This is also the case for pure and applied science as well.

In spite of an enormous number of papers dealing with the influence of the medium on the rate of chemical reactions, no strict quantitative theory capable of "universal" application has been put forward up to now. Thus, it is not yet possible to describe the relationship between the reactions rate constants and the equilibrium constants with the nature of the medium in which the reactions take place by means of a single equation.

The absence of a general theory on the influence of the medium on the kinetics of a chemical reaction can be explained by the fact that the change of solvent can not only influence on the rate of the process but also frequently results in a complication (or change) of the reaction mechanism. The calculation of the individual contributions made by each factor is thus, in most cases, rather complicated and requires a deep and comprehensive study of the properties of the medium and of the reacting particles. This is because a quantitative evaluation of all types of interactions between the reacting particles with medium can only arrive at the basis of full knowledge of these properties.

One of these situations shows the absence of fundamental relationships structure properties for polymer solutions and melts. In the present monograph, this deficiency is removed by using modern physical conceptions—fractal analysis and irreversible aggregation models.

— G. V. Kozlov, DSc, I. V. Dolbin, DSc,
and Gennady E. Zaikov, DSc

PREFACE

[illegible]

INTRODUCTION

As professor G. P. Gladyshev justly pointed out [1], the developed in physical chemistry general representations were the basis of our understanding of polymers' particular properties. Any researcher who is engaged in polymers study uses the representations and results obtained in this field. Therefore, an enormous number of scientific works is dedicated to polymer solutions studies, which are summarized in the monographs [1–7]. In these works, a large number of approaches are used, both physically strict and empirical ones, that have defined a large number of equations used for the description of macromolecules behavior in solution. Suffice it to say that in the monograph [1], more than 900 different kinds of relationships were adduced. In the long run the situation is created, wherein each author chooses a description, that is most suitable and clear for the model (or its modification). Different models were obtained for polymers of different classes (flexible-chain, rigid-chain, branched ones) and solutions (diluted, concentrated ones). This reason obvious: all available equations of polymer solutions (melts) of physical chemistry describe relationships property-property, but not structure property. Let us remember that within the framework of modern representations, the description of an object structure means the description of the distribution in space of elements, forming an object (in the considered case—a polymer macromolecule) [8]. All those widely used in present polymer macromolecule large characteristics (Kuhn segment length, distance between macromolecule ends and so on) are not structural parameters. This situation was changed in principle with the appearance of the fractal analysis conception, the main characteristic of which is fractal dimension—the structural parameter in the strict sense of this term. Let us note that a polymer macromolecule, simulated by the self-intersecting random walk, is the ideal object for fractal analysis application that immediately gave its own result. So, Muthukumar [9] has shown that polymer solution viscosity within the framework of fractal approximation is described by one model, taking into account all hydrodynamical interactions, and excluded volume effect, frequency, concentration, and so on. The transition from the polymer solutions properties description to the description of similar melts characteristics is realized by an absolutely natural mode—by replacement of dimension of low-molecular solvent, surrounding macromolecular coil, by dimension of the same coils for melt [10].

The considered circumstances served as a cause for the development of the fractal analysis, particularly practical applications for polymer solutions (melts) properties description, which will be considered in the present monograph.

REFERENCES

1. Budtov, V.P. Physical Chemistry of Polymer Solutions. Sankt-Peterburg, Chemistry, **1992**, 384 p.
2. Tsvetkov, V.N.; Frenkel, S.Ya.; Eskin, V.E. Structure of macromolecules in Solution. Moscow, Science, **1964**, 720 p.
3. Tenford, Ch. Physical Chemistry of Polymers. Moscow, Chemistry, **1965**, 772 p.
4. Moravets, G. Macromolecules in Solutions. Moscow, World, **1967**, 364 p.
5. Rafikov, S.R.; Budtov, V.P.; Monakov, Yu.B. Introduction in Physics-Chemistry of Polymers Solutions. Moscow, Science, **1978**, 388 p.
6. Eskin, V.E. Light Scattering by Polymer Solutions. Leningrad, Science, **1986**, 288 p.
7. Tsvetkov, V.N. Rigid-Chain Polymer Molecules. Moscow, Science, **1986**, 380 p.
8. Ebeling, V. Structures Formation at Irreversible Processes. Moscow, World, **1979**, 275 p.
9. Muthukumar, M. Dynamics of polymeric fractals. *J. Chem. Phys.,* **1985**, *83(6)*, 3161–3168.
10. Vilgis, T.A. Flory theory of polymeric fractals—intersection, saturation and condensation. *Physica A,* **1988**, *153(2)*, 341–354.

CHAPTER 1

THE THEORETICAL FUNDAMENTALS OF MACROMOLECULES FRACTAL ANALYSIS

CONTENTS

Keywords 20
References 21

The study of polymeric fractals in different states started directly after the publication of Mandelbrot's fractal conception [1], that is, about at the end of 70s of the last century. The cause of the indicated approach application for a polymer macromolecules analysis is obvious enough: it is difficult to find a more suitable object for fractal models application. One from the first research in this field with the experimental method of electrons spin-relaxation for proteins [2] found, out that the protein macromolecule dimension was equal to 1.65±0.04, that coincides practically exactly with the dimension of 5/3 of random walk without self-intersections, by which a polymeric macromolecule is usually simulated [3].

The authors [4, 5] considered random walks without self-intersections or self-avoiding walks (SAW) as the critical phenomenon. Monte-Carlo's method within the framework of computer simulation confirmed SAW or polymer chain self-similarity, which is an obligatory condition of its factuality. Besides, in Ref. [4], the main relationship for polymeric fractals at their treatment within the framework of Flory conception was obtained:

$$D_f = \frac{1}{\nu_F}, \tag{1}$$

where D_f is a macromolecular coil dimension, ν_F is Flory exponent.

Family [6] divided fractals on two main types: deterministic and statistical (or random) ones. Self-similar objects, which are constructed exactly on the basis of some basic rules, are deterministic fractals. The best examples of these fractals are Koch curve, Sierpenski gaskets, Havlin gaskets and Vicskek snowflake. The most important property of deterministic fractals is that their fractal dimension is exactly known. Since the given line, plane, or volume can be divided in an infinite number of different ways, and then it is possible to construct infinitely many different deterministic fractals with different fractal dimensions. Therefore, deterministic fractals cannot be classified into a finite number of classes without introducing, in addition to their fractal dimension, other parameters to characterize them.

In contrast to deterministic fractals, statistical fractals are constructed by random processes. The element of randomness makes them a more realistic representation of physical phenomena. The fact that randomness alone, that is, without any spatial correlations, is sufficient to produce fractals was first pointed out by Mandelbrot [1]. The best example of such a fractal is the path of a random walker. Nevertheless, purely random models are inadequate for most applications to real physical systems [6]. The reason is that the excluded volume effect, which is the geometrical constraint that prohibits two different elements from occupying the same spatial point, is not taken into account in these models. For this reason, a wide variety of random models with excluded volume have been introduced, which are the main topic of study in the given field. The best-known examples of these models are SAW, lattice animals and random percolation.

Fractals have distinct topological structures depending on the maximum number of elements that can be joined to a given element of the system. If each element can be connected at most to two other elements, the resulting structure has no branches. In analogy with linear polymers, these types of structures were called linear fractals. In contrast, when branching can occur, that is, three or more elements can be joined to the given element, the resulting fractal has a network-like structure. Such type of fractals was called a branched one [6]. Figures 1 and 2 give the examples of the indicated fractals types.

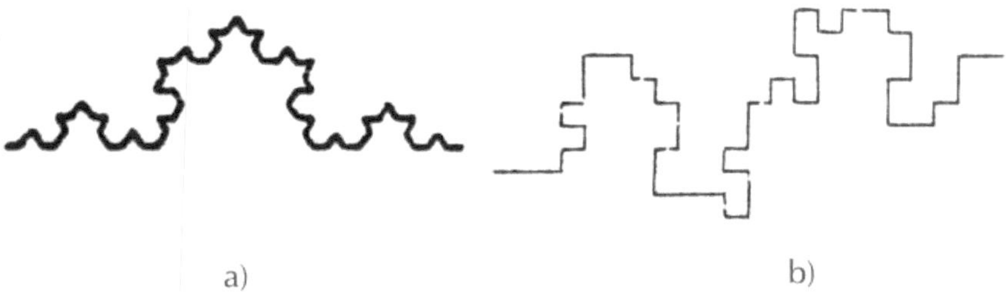

FIGURE 1 Two examples of linear fractals: Koch curve (a) and self-avoiding random walk (b), representing a deterministic fractal and statistical fractal, respectively.

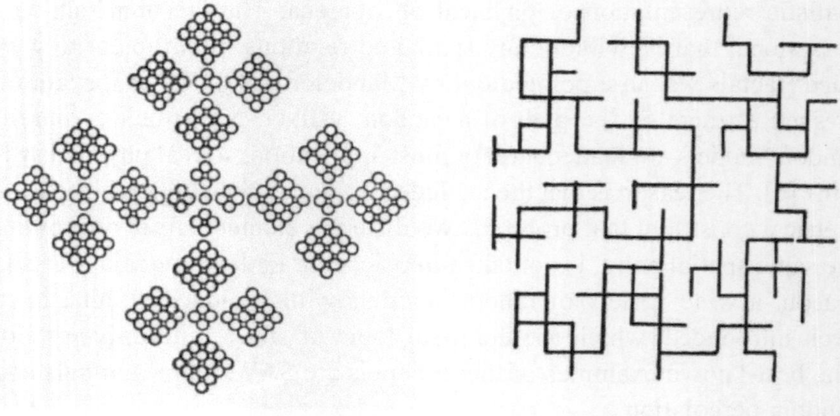

a) b)

FIGURE 2 Two examples of linear fractals: Vicsek snowflake (a) and percolation cluster (b), representing a deterministic fractal and random fractal, respectively.

Family [6] defined three states of polymeric statistical fractals depending on the system statistics: extended, compensated and collapsed ones. The two main factors, influencing fractal branching degree, were pointed out. The first from them is cluster concentration in system—if there are many clusters in the system, and then they occupy the entire volume. Therefore, other clusters availability restricts a fractal branching degree: it is more branched in isolation (very diluted solution), than in concentrated solution.

The second factor, which influences a cluster shape, is availability or absence of the attraction interactions. For example, the attraction interactions are less effective at high temperatures, than at low ones. Therefore the isolated statistical fractal at high temperatures will possess a more branched structure, than at low ones. The best examples of this type of fractals are isolated macromolecular coils in good solvents at high temperatures [7]. Family called such state of a statistical fractal extended, since in this case it has its least fractal dimension D_f.

The calculation within Flory theory [8] allows determining D_f value for the extended state of branched statistical fractal according to the equation [6]:

$$D_f = \frac{2(d+2)}{5} \tag{2}$$

where d is the dimension of Euclidean space, in which a fractal is considered.

For d=3 the value D_f=2, that coincides exactly with the results of Ref. [9] for branched polymers in diluted solution (ν_F=0.5).

Two variants of the dimension D_f increase exist for a branched fractal: the availability in its environment of the similar fractals and the attraction interactions introduction between this fractal elements, compensating the repulsion interactions [6]. Let us note, that the two indicated methods give different D_f values and as a matter of fact different states of a macromolecule. So, in case of the fractal cluster environment by similar clusters the following equation was received [6]:

$$D_f = \frac{d+2}{2} \tag{3}$$

that for d=3 is given D_f=2.5.

It is easy to see, that the indicated mode corresponds to the gelation point, where the value D_f≈2.5 was confirmed experimentally [10]. Besides, it is necessary to note the critical value D_f≈2.5 in the gelation point universality, which is the same for both linear and branched polymers. The same value of fractal dimension was obtained practically for all known critical phenomena (percolation, Ising and Potts models and so on), which correspond to the well-known Suzuki rule [6]. The indicated community of the critical dimension of 2.5 is defined by the common cause of its reaching—the excluded volume effect compensation by the screening effect.

The compensation of repulsion interactions by attraction interactions of fractal cluster elements allows obtaining the following equation for the dimension D_f determination in this point for a branched cluster [6]:

$$D_f = \frac{4(d+1)}{7} \tag{4}$$

that for d=3 gives D_f=2.286.

The indicated dimension corresponds to the state of branched polymer macromolecule in θ-solvent [8].

And at last, for the collapsed state the model of branched polymer in monodisperse melt was offered, within the framework of which the following identity was obtained [6]:

$$D_f = d \tag{5}$$

As it is known [7, 11], linear macromolecules depending on a number and sizes of monomer links (molecular weight), solvent quality, temperature, concentration and also a type and degree of the external power field influence can be found in different conformational and/or phase states. Most characteristic states are the following ones: 1—compact globule; 2—statistically turned coil in ideal (θ) solvent; nontransparent coil in good solvent; 4—leaking (transparent) coil (the state, which is typical for rigid-chain polymers) and, at last, 5—fully straightened rod-like molecule. The situation 1 is realized in the globular albumins or at precipitation (lower θ-point) of linear macromolecules from very diluted solutions. The conformation 5 is typical for some albumins or for linear chains at certain external influences; the conformations 2–4 are more trivial ones and are studied well enough [11]. The authors [12] determined the value of fractal dimension D_f for each from the indicated above states. In case of poor solvent for a globule the value D_f=d=3 was obtained, for a macromolecular coil in θ-solvent D_f=2.0. For the good solvent D_f=5/3, the leaking coil has the dimension D_f=1.5 and straightening chain—D_f=1.0 [12]. Let us note, that for linear and branched polymers do not coincide only two fractal dimensions—in good (5/3 and 2.0, accordingly) and θ-solvents (2.0 and 2.286, respectively).

In diluted polymer solutions the reaching of dynamical equilibrium between the macromolecules aggregation and degradation processes (which are realized by the way of formation of new bonds between monomers or decay of former ones) is possible, when macromolecules are, as a rule, in the fractal state. Moreover, the processes of synthesis and degradation can be realized simultaneously. In Ref. [13], the analytical expression was obtained for the mass spectra of macromolecular fractals in diluted polymer solutions.

The dynamical equilibrium of the macromolecules aggregation and degradation processes can be expressed mathematically by the equation for a monomers number in macromolecular fractal evolution (the fractal mass is directly proportional to monomers number) [13]:

$$\dot{N} = cN^a - mN^b, \quad 0<a<b, \tag{6}$$

where N is the monomers number, that is, the same individual subunits of a macromolecular structure.

The first item in the Eq. (6) right-hand part describes the aggregation process, the second—the degradation one. The type and character of the aggregation and degradation processes are defined by the parameters a and b and the intensity of these processes is defined by the coefficients c and m values. The condition $a<b$ ensures the dynamical equilibrium reaching, when monomers number N_0 in the stable cluster is equal to [13]:

$$N_0 = (c/m)^{1/(b-a)} \tag{7}$$

Let us note, that the Eq. (6) can be considered as the generalization of the relationship for dense inclusion growth (when a=1/3 and b=4.3) in case of fractal aggregation.

In diluted polymer solution macromolecules are in random environment, influencing on the process of either aggregation or degradation (more rarely on both of them) that results in the stochastic equation instead of the Eq. (6) [13]:

$$\dot{N} = \lambda N^a - N^b + \xi_t N_a, \quad \lambda=\langle c\rangle/\mathrm{m}. \tag{8}$$

In the Eq. (8) $\xi_t=c_t/m$, where c_t is stochastic contribution to aggregation process intensity, which is due to an environment influence, $\langle c\rangle$ is its mean value. For clarity the author [13] considered random influence on a macromolecules fractal aggregation process (case of influence on degradation process may be considered similarly). As a rule, the characteristic times of such influence are much smaller than aggregation times of macromolecules in diluted solution [13] and therefore ξ_t can be presented as the intensity white noise $\langle \xi_t^2 \rangle=\sigma^2$. The detailed elaboration and application of the considered general model to real polymer system will be considered in Chapter 2.

Let us consider some experimental data about determination of the polymer macromolecule dimension D_f in different conditions. Colvin and Stapleton [14] performed the value D_f calculation according to the data of electrons spin-relaxation and found this dimension variation within the wide enough limits: 1.19–1.82. It is easy to see, that the indicated D_f values differ essentially from the dimension of the typical states of polymer macromolecule, cited above. This circumstance assumes a factors number availability, changing continuously D_f value. Chapter 2 is dedicated to the indicated factors study.

The authors [15–17] measured the fractal dimension of cross-linked polydimethylsiloxane (PDMS) at the gelation point with the aid of rheological measurements and obtained the value D_f=2. By the confession of the authors themselves, this result is unexpected, since the indicated dimension corresponds to a linear macromolecule in θ-solvent. The similar measurements of the branched chains structure of the epoxy polymer (EP) with the aid of X-raying gave the value D_f=2.17±0.03 [16]. In connection with these measurements several comments should be made.

First of all, the dimension of a growing macromolecule is defined not only by polymer-solvent and macromolecule elements interactions between one another, but also by its formation mechanism. The two main models of fractal clusters formation by diffusion-limited aggregation (DLA) exist. The first from them, particle-cluster (P-Cl) was offered by Witten and Sander [18, 19]. Let us represent that a particle is placed in space. The second particle, accomplishing Brownian motion, appears remote from it. Meeting with the first particle, it sticks to it, moreover the sticking is irreversible. Then in the distance the next particle appears, which is also diffused and, coming to contact to the first particles, sticks to them. This process reiterates repeatedly. Some particles during walking are not contiguous to growing cluster and are diffused infinitesimally far from it. These particles are not taken into account afterwards.

Such cluster is presented in Fig. 3. For its construction Monte-Carlo method was used, realized during the computer simulation process. At this method all events totality are imitated in detail (Brownian motion is assigned by the generator of random numbers). In Ref. [20], it was found out that for such cluster D_f=2.51±0.06 for d=3.

FIGURE 3 The model of cluster (10,000 particles), forming during DLA process of type particle-cluster (P-Cl).

As it has been shown in Ref. [21], the model DLA P-Cl does not allow explaining colloid and aerosols formation. The aggregation model DLA of the type cluster–cluster explains their growth. It is necessary to note, that the same model describes polymers synthesis process [13, 22]. According to this model a clusters number in the system reduces and their sizes increase. Sutherland simulated the aggregation of type cluster–cluster for the first time [23]. He found the formation of branched loose aggregates, similar to sols, observed in electron microscope.

The author cannot characterize this looseness, since he does not use scaling approach (the fractal analysis was not still developed at that time). Meakin [24] and the authors group (Kolb, Botet and Jullien) [25] also performed independently aggregation cluster–cluster simulation. In two-dimensional space at the initial particles concentration c_0=0.03–0.15 both particles and growing clusters move over Brownian trajectories, that is, diffusion-limited aggregation of type cluster–cluster (DLA Cl–Cl) occurs. After all the sole cluster was obtained, which soaks up all particles (Fig. 4). In different works the somewhat differing values D_f were obtained, that is explained by sizes finiteness of simulating systems. The authors [26] assume the most well founded values D_f=1.75–1.80 for d=3 (for more detail in the review [21]). Clusters of the model DLA Cl–Cl are more loose than clusters of the model DLA P-Cl, that allows them to screen their internal regions to a larger extent [24, 25].

The authors of Ref. [27] noted that in many physical processes the situation, at which only rare clusters collisions resulted in their joint, is the most real one. Such situation can be created at a chemical bond formation. This results in this process name—chemically limited aggregation of type cluster–cluster (CLA Cl–Cl) (in other works—reactionally limited aggregation). At infinitesimal probability of clusters joining at collision all variants of two clusters joint are equally probable. At imitation in the process CLA Cl–Cl of physically real situation, when joining clusters have different sizes, the values D_f=2.11±0.03 for d=3 were obtained [28].

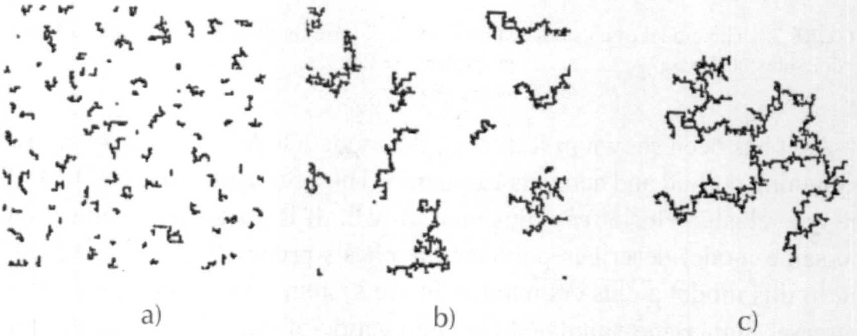

FIGURE 4 The stages of diffusion-limited aggregation process of type cluster–cluster (c_0=0.06): (a) 86 clusters; (b) 8 clusters; (c) one cluster.

The authors [16] explained the obtained by them value D_f=2.17±0.03 at branched EP formation kinetics study by just macromolecules formation mechanism.

The influence of macromolecules formation mechanism on their dimension has been shown particularly clearly on the example of lignin polymers [29, 30]. The biosynthesis realization in vitro dehydropolymers by two methods (end-wise-polymerization and bulk-polymerization) gives lignin's with dimension D_f=2.62±0.27 and 1.66±0.16, respectively. Thus, the adduced above D_f values indicate unequivocally on synthesis realization by the mechanisms P-Cl and Cl–Cl, respectively, for the same polymer at the synthesis conditions variation.

The authors [31] introduced the notion of true self-avoiding walk (TSAW), which describes a path of random walk, restricted so as to avoid the given point visit in space with the probability, proportional to a times number, which this point was visited already. This restriction results in an excluded volume interaction reduction in comparison with SAW. The large chain compactness is a resulting effect. If in Flory approximation the exponent ν_F for SAW is given as follows [32]:

$$\nu_F = \frac{3}{d+2}, \tag{9}$$

then in TSAW approximation this relationship changes up to [3]:

$$\nu_F = \frac{2}{d+2} \tag{10}$$

According to the Eq. (1) this gives D_f=1.67 for SAW and D_f=2.5 for TSAW. Family [32] proposed to consider two different cases: a typical chain with molecular weight of order of average weight and a very long chain with much higher molecular weight than average one. A very long chain senses other monomers availability, but on large length scales this chain interacts itself, creating the excluded volume effect. Other chains act already as good solvent. Therefore polydispersity does not influence on very long chains. For typical chains the excluded volume interaction is screened by other chains, therefore, these chains in condensate have TSAW statistics. Hence, in practice D_f experimental measurements can give some mean dimension value for SAW and TSAW at approaching to the gelation point.

Kaufman and other [33, 34] performed the electroconducting polymer polypyrrole polymerization and found out for it the value D_f=1.74±0.03. It is easy to see, that the indicated D_f value corresponds well to both a macromolecular coil in good solvent dimension and to fractal cluster dimension, formed by the mechanism DLA Cl–Cl.

Let us note in conclusion, that the authors [35] performed the study of dielectric spectroscopy methods of polymers dynamics in diluted and semidiluted solutions and found their results correspondence to Muthukumar fractal model [36].

Vilgis proposed [37–39] the most complete fractal treatment of polymers behavior in solution and melt. Since this very treatment is assumed as a basis of subsequent description of real polymer solutions and melts, then here its complete enough interpretation will be given. As the indicated treatment basis the polymeric fractal notion, proposed by Catas [40] and Muthukumar [36] was accepted. The term "polymeric fractal" describes the fractal objects class, which consists of flexible polymer chains on small length scales, but has arbitrary self-similar connectivity on large scales. This model is suitable particularly for the description of the macromolecules, formed by random curing (by gelation or aggregation) of flexible chains-precursors, but it can also be used for other fractals description, which have no necessary rigidity and therefore possessing diffusive (but not vibrational one) dynamics [40].

The first stage of the considered treatment is the formulation of "polymeric fractal" dimension D_f and its spectral (fracton) dimension d_s, which characterizes fractal object connectivity degree [41], intercommunication. In this case the value d_s=1.0 for linear polymer chain and d_s=1.33 for very branched (cross-linked) chain. Using Flory-de Gennes mean field approximation, Vilgis obtained the following equation for D_f calculation [39]:

$$D_f^{ph} = \frac{2d_s}{2-d_s} \tag{11}$$

for the phantom (which does not take into account excluded volume interactions) polymeric fractal and

$$D_f = d_s \frac{(d+2)}{(d_s+2)} \tag{12}$$

for swollen (with excluded volume effect accounting) polymeric fractal.

In general case the intersections number N_{int} of two arbitrary fractals with dimensions D_{f_1} and D_{f_2} can be estimated as follows [39]:

$$N_{\text{int}} \sim R_g^{D_{f1}+D_{f2}-d} \tag{13}$$

where R_g is a gyration radius of polymeric fractal.

Fractals are opaque (nonleaking) one for another if N_{int} increases at R_g growth, that is, at $D_{f_1}+D_{f_2}>d$ and transparent (leaking), if an intersections number decreases at R_g enhancement [39]. In other words, for the case d=3 and two fractals with the same dimension D_f the Eq. (13) predicts transparent polymers fractal at $D_f \leq 1.5$, that corresponds completely to the data of Ref. [12].

It is recognized, first of all for the case of linear chains, that the excluded volume becomes a screened one if small solvent molecules are replaced by linear chains of comparable sizes [7]. In case of solutions or melts of fractals with the same fractal dimensions the dimension Δ_f for such melts can be calculated according to the equation [39]:

$$\Delta_f = \frac{d_s(d+2)}{2} \tag{14}$$

and the upper critical dimension (UCD) d_{UCD} for polymeric fractal in melt is given as follows [39]:

$$d_{UCD} = \frac{2d_s}{2-d_s} \tag{15}$$

From the Eqs. (14) and (15) it follows, that for melts of linear polymer chains (d_s=1.0 [41]) in case d=3 the condition $\Delta_f > d_{UCD}$ is fulfilled, that is, in melt of linear chains the excluded volume interactions are screened.

The situation for the branched polymer chains (d_s=1.33) [41] differs essentially from the described above. In three-dimensional Euclidean space for such chains the Eq. (14) predicts Δ_f=10/3, that is physically incorrect, since the fractal dimension Δ_f cannot be larger than surrounding Euclidean space dimension d [1]. Moreover, in case d_s=4/3 Δ_f is always larger than d, that corresponds to the saturation effect and cluster loses its fractal properties. It becomes introduced into the space, filled with three-dimensional densely packed melt, although space Euclidean dimension d is less than d_{UCD}, equal to 4 for branched polymer chains [39].

It is obvious, that the cause of the saturation effect is a large fractal dimension of phantom fractal (compare the Eqs. (11) and (12)) or, in other words, higher spectral dimension d_s>1 in comparison with linear polymer chains. The upper limit D_s for spectral dimension allowing to avoid saturation can be determined from the condition Δ_f=d, that is given [39]:

$$D_s = \frac{2d}{(2+d)} \tag{16}$$

Hence, it can be concluded, that the polymeric fractals at the condition d_s>D_s (D_s=1.20 for d=3) become saturated ones, that is, they lose their structure fractal nature and become compact ones. Such effect takes place at globules formation in cross-linked polymers [22, 42].

Let us consider further the situation, when a solvent consists of fractal clusters with Gaussian (phantom) dimension δ_f. In this case the excluded volume interactions screening is observed again and the value d_{UCD} is determined according to the equation [39]:

$$d_{UCD} = 2D_f^{ph} - \delta_f \tag{17}$$

In its turn, the value D_f depends on δ_f as follows [39]:

$$D_f = \frac{d+2}{2-(\delta_f-2)/D_f^{ph}} \tag{18}$$

The Eq. (18) contains all cases, described above. In case of δ_f=0, that is, low-molecular solvent (molecules in the shape of points), let us obtain usual swelling (the Eq. (12)). If solvent molecules and polymeric fractal have the same phantom dimension, that is, $D_f^{ph}=\delta_f=2d_s/(2-d_s)$, then let us obtain again the fractal dimension of clusters in melt (the Eq. (14)).

Let us consider in this context the saturation effect realization conditions, that is, the transition from fractal behavior to nonfractal one. The conditions are defined by the criterion $D_f=d$ and predict the upper boundary for solvent phantom fractal border according to the Eq. (18) [39]:

$$\delta_f < 2 + D_f^{ph}\left(\frac{d-2}{d}\right) \tag{19}$$

The inequality Eq. (19) demonstrates, that if phantom fractal dimension of solvent molecules exceeds its right-hand part magnitude, then the cluster loses its fractal properties and saturates up to a densely filling space object [39].

As Alexandrowich [43, 44] has shown, branched polymers formation kinetics also influences strongly enough on their topological structures distribution and a mean macromolecules shape. Similarly one can expect deviation from the condition of equal accessibility of growing macromolecules ends at the expense of reactions diffusive control.

Up to now we considered polymeric fractals behavior in Euclidean spaces only (for the most often realized in practice case d=3). However, the indicated fractals structure formation can occur in fractal spaces as well (fractal lattices in case of computer simulation), that influences essentially on polymeric fractals dimension value. This problem represents not only purely theoretical interest, but gives important practical applications. So, in case of polymer composites it has been shown [45] that particles (aggregates of particles) of filler form bulk network, having fractal dimension, changing within the wide enough limits. In its turn, this network defines composite polymer matrix structure, characterized by its fractal dimension d_f and, hence, this polymer material properties. And on the contrary, the absence in particulate-filled polymer nanocomposites of such network results in polymer matrix structure invariability at nanofiller contents variation and its fractal dimension remains constant and equal to this parameter for matrix polymer [46].

In Euclidean *d*-dimensional spaces Flory exponent ν_F depends only on *d*. A good (although not exact) ν_F estimation was given by Flory formula, Eq. (9) [7, 8]. It is well known [47] that the critical phenomena depend by the decisive mode on various fractal characteristics of basic structure. It becomes obvious, that excepting the fractal (Hausdorff) dimension D_f physical phenomena on fractals depend on many other dimensions, including skeleton fractal dimension D_B [48], dimension of minimum (or chemical) distance d_{min} [29] and so on. It also becomes clear, that regular random walks on fractals have anomalous fractal dimension d_w [49] and that the vibrational excitations spectrum is characterized by spectral (fracton) dimension $d_s=2D_f/d_w$ [41, 50].

From the point of view of dimensions abundance the necessity of Flory modified formula for SAW on fractals obtaining arises. It was noted more than once [51, 52], that for ν_F determination the fractal dimension D_f was insufficient. Nevertheless, Kremer found that at *d* replacement by D_f in the Eq. (9) the formula:

$$\nu_F = \frac{3}{2 + D_f} \tag{20}$$

corresponds well to Monte-Carlo simulation of walks on percolation cluster at percolation threshold. The authors [45] have shown that the good description of polymer matrix structure of composites, modified by filler particles (aggregates of particles) network with dimension D_n can be obtained with the aid of the equation:

$$\nu_F = \frac{2.5}{2 + D_n} \tag{21}$$

The authors [51] criticized the Eq. (20), since they proved, that if SAW moved over fractal skeleton (otherwise it will be captured by handling or "dead" ends, that is important particularly for branched polymer chains), then Flory exponent ν_F should depend on the skeleton properties only, but not on D_f. Let us remind, that the skeleton notion is defined as follows [48]. The two cluster end points P_1 and P_2 are considered, which are divided by the distance, comparable with the correlation length of system.

If to force liquid to flow from point P_1 to point P_2 it can do this by some paths (bonds) set; this bonds set is called skeleton.

The authors [51] obtained the following relationship for Flory exponent determination in case of SAW on fractal lattices:

$$\nu_F = \frac{1}{D_B}\frac{3d_{s,B}}{d_{s,B}+2} = \frac{3}{d_{w,B}+D_B} \tag{22}$$

where $d_{s,B}$ and $d_{w,B}$ are skeleton spectral dimension and dimension of random walk on skeleton, respectively.

The Eq. (22) gives unsatisfactory correspondence to numerical calculations for precise fractals and the authors [51] expressed doubt about the general Flory approximation derivation possibility for fractal spaces (lattices), which would be both simple and exact at the same time.

As it is known [52], for the "lattice animals" case in d-dimensional Euclidean spaces the value D_f is given by the equation:

$$D_f = \frac{2d+4}{5} \tag{23}$$

which gives the precise results for d=3, 4 and 8.

In analogy with the Eq. (23) the following formula for D_B was proposed [52]:

$$D_B = \frac{2d+14}{15} \tag{24}$$

The authors [52] obtained the following Flory approximant for fractal spaces:

$$\frac{3}{2+D_f} = \frac{2D_B + 2\tilde{\xi}_R - d_{\min}}{d_{\min}\tilde{\xi}_R + D_B^2 + D_B\tilde{\xi}_R}, \tag{25}$$

where $\tilde{\xi}_R$ is the exponent of resistance between fractal two points scaling, $d_{\min}$ is minimum dimension, defined as follows [29]:

$$d_{\min} = \frac{D_f}{d_l} \quad (26)$$

where d_l is the chemical dimension or "spreading" dimension, which characterizes the distance between fractal object two points not as geometrical distance (the straight line, connecting these points), but as "chemical" distance, that is, the shortest way over object particles between these two points [53].

As the authors [51] assumed, the approximant Eq. (25) is not very simple one and includes three geometrical dimensions: D_B, $\tilde{\xi}_R$ and $d_{\min}$. In Ref. [54], one more variant of Flory approximant for fractal spaces with dimension D_{sp} was proposed:

$$\nu_F = \frac{1}{D_{sp}} \cdot \frac{4d_l - d_s}{2 + 2d_l - d_s} \quad (27)$$

For the dimension d_l value estimation the authors [54, 55] used the simplest variant based on Flory approximation and proposed in Ref. [56]. For the dimension $4/3<d<4$ the value of parameter z can be calculated:

$$z = \frac{4 + 3D_n}{8} \quad (28)$$

and then the value d_l can be determined as follows [56]:

$$d_l = \frac{D_n}{z} \quad (29)$$

The Eq. (27) application gave an exact enough (within the limits of 6%) description of composites polymer matrix fractal dimension change in fractal space, created by filler particles (aggregates of particles) network [45].

And in conclusion of the Chapter 1, let us consider the branched polymers behavior in restricted configurations, such as poses and slits. If for the linear polymers this situation is simple enough and is studied well [7], then for the branched polymers several restrictions exist, due to their larger connectivity degree [57]. The main difficulty in branched polymers swelling and conformational behavior treatment consists in the fact that they (contrary to linear polymers) are never capable of tension up to the size $R \sim M$, where M is total molecular weight. Only the chemical path can be stretched completely, which contains N monomers. From this assertion it follows directly that for the size of polymeric fractals with higher than for linear polymers connectivity degree (d_{s-}dimensional connectivity) two limits exist: firstly, a branched polymer can be stretched as $R \sim N$. This limit gives directly the minimum fractal dimension $D_f = d_s$, that is, the maximum size $R_{max}^{d_s} \sim M$ and corresponds to the greatest possible expansion for all physical processes, such as swelling and mechanical deformation. The other obvious restriction is given by a condensed or saturated phase, that is, the minimum size is defined by the greatest density: $R_{max}^{d} \sim M$. From the said above the conclusion can be made that $d_s \leq D_f \leq d$ [57].

For d_{s-}dimensionally connected polymer the following result was obtained (a case of good solvent) [57]:

$$\nu_F = \frac{d_s + 2}{d + 2} \tag{30}$$

Let us remind that these exponents take into account linear (chemical) size N, but not total molecular weight M. In this case the corresponding fractal dimension D_f is given as follows (compare with the Eq. (1)) [57]:

$$D_f = \frac{d_s}{\nu_F} \tag{31}$$

Further for d_{s-}dimensionally connected polymer in good solvent, which is placed between two parallel plates, the following result was obtained [57]:

$$R_{//} = a\left(\frac{a}{h}\right)^{1/4} N^{(2+d_s)/4} \tag{32}$$

where $R_{//}$ is the polymer size that is parallel to plates, a is the monomer size, h is the distance between two parallel plates, N is monomers number in one given direction in spectral vector space.

The fraction filling of slit between two plates by polymeric fractal $\Im$ is given as follows [57]:

$$\Im = \left(\frac{a}{h}\right)^{7/4} N^{(d_s-2)/2} \tag{33}$$

For the case of d_{s}- dimensionally connected polymer fractals (good solvent) in a cylindrical pore the most important parameter is the minimum pore diameter $D^*_{\min}$, through which a branched polymer still pass. This size is found from the conditions that for parallel direction $R \sim N$ is given by the equation as well [57]:

$$D^*_{\min} \sim N^{(d_s-1)/2} \equiv M^{(d_s-1)/2d_s} \tag{34}$$

The calculation of the internal concentration or filling fracton $\Im$ gives the following result [57]:

$$\Im = \frac{N^{d_s}}{d_s^2 R_{//}} \tag{35}$$

Hence, the main conclusion from the adduced above estimations is the fact that higher connectivity degree (d_s>1.0) of branched polymers imposes more strict restrictions on conformation and behavior of such polymers in restricted geometries [57].

KEYWORDS

- **Euclidean dimension**
- **Flory formula**
- **lattice animals**
- **Monte-Carlo method**
- **phantom fractal**
- **true self-avoiding walk**

REFERENCES

1. Mandelbrot, B. B.; The Fractals Geometry of Nature. San-Francisco, W. H.; Freeman and Company, **1982,** 459 p.
2. Stapleton, H. J.; Allen, J. P.; Flynn, C. P.; Stinson, D. G.; Kurtz, S. R.; Fractal form of proteins. *Phys. Rev. Lett.* **1980,** *45(17),* 1456–1459.
3. Peliti, L.; Random walks with memory. In book: Fractals in Physics. Ed. Pietronero, L.; Tosatti, E.; Amsterdam, Oxford, New York, Tokyo, North-Holland, **1986,** 106–116.
4. Havlin, S.; Ben-Avraham, D.; New approach to self-avoiding walks as a critical phenomenon. *J. Phys. A*, **1982,** *15(4),* L321-L328.
5. Havlin, S.; Ben-Avraham, D.; Theoretical and numerical study of fractal dimensionality in self-avoiding walks. *Phys. Rev. A*, **1982,** *26(3),* 1728–1734.
6. Family, F.; Fractal dimension and grand universality of critical phenomena. *J. Stat. Phys.* **1984,** *36(5/6)*, 881–896.
7. De Gennes, P. G.; Scaling Concepts in Polymer Physics. Ithaca, Cornell University Press, **1979,** 368 p.
8. Flory, P.; Principles of Polymer Chemistry. Ithaca, Cornell University Press, **1971,** 432 p.
9. Isaacson, J.; Lubensky, T. C.; Flory exponents for generalized polymer problems. *J. Phys. Lett.* (Paris), **1980,** *41(19),* L469–L471.
10. Kobayashi, M.; Yoshioka, T.; Imai, M.; Itoh, Y.; Structural ordering on physical gelation of syndiotactic polystirene dispersed in chloroform studied by time-resolved measurements of small angle neutron scattering (SANS) and infrared spectroscopy. Macromolecules, **1995,** *28(22),* 7376–7385.
11. Tsvetkov, V. N.; Eskin, V. E.; Frenkel, S. Ya. Structure of Macromolecules in Solution. Moscow, Science, **1964,** 719 p.
12. Baranov, V. G.; Frenkel, S. Ya.; Brestkin, Yu.V.; Dimensionality of different states of linear macromolecule. Reports of Academy of Sciences of SSSR, **1986,** *290(2),* 369–372.
13. Shiyan, A. A.; The stable distributions of macromolecular fractals mass in diluted polymer solutions. High-Molecular Compounds. B, **1995,** *37(9),* 1578–1580.
14. Colvin, J. T.; Stapleton, H. J.; Fractal and spectral dimensions of biopolymer chains: solvent studies of electron spin relaxation rates in myoglobin azide. *J. Chem. Phys.* **1985,** *82(10),* 4699–4706.
15. Muthukumar, M.; Winter, H. H.; Fractal dimension of a cross-linking polymer at the gel point. Macromolecules, **1986,** *19(4),* 1284–1285.
16. Chu, B.; Wu, C.; Wu, D.-Q.; Phillips, J. C.; Fractal geometry in branched epoxy polymer kinetics. Macromolecules, **1987,** *20(10),* 2642–2644.
17. Hess, W.; Vilgis, T. A.; Winter, H. H.; Dynamical critical behavior during chemical gelation and vulcanization. Macromolecules, **1988,** *21(8),* 2536–2542.
18. Witten, T. A.; Sander, L. M.; Diffusion-limited aggregation as kinetical critical phenomena. *Phys. Rev. Lett.* **1981,** *47(19),* 1400–1403.
19. Witten, T. A.; Sander, L. M.; Diffusion-limited aggregation. *Phys. Rev. B*, **1983,** *27(9),* 5686–5697.

20. Meakin, P.; Diffusion-controlled cluster formation in 2–6-dimensional space. *Phys. Rev. A*, **1983,** *27(3),* 1495–1507.
21. Smirnov, B. M.; Fractal clusters. Acievements of Physical Sciences, **1986,** 1*49(2),* 177–219.
22. Magomedov, G. M.; Kozlov, G. V.; Zaikov, G. E.; Structure and Properties of Cross-Linked Polymers. Shawbury, A Smithers Group Company, **2011,** 492 p.
23. Sutherland, D. N. A theoretical model of floc structure. J.; Colloid Interf. Sci.; **1967,** *25(3),* 373–380.
24. Meakin, P.; Formation of fractal clusters by irreversible diffusion-limited aggregation. *Phys. Rev. Lett.* **1983,** *51(13),* 1119–1122.
25. Kolb, M.; Botet, R.; Jullien, R.; Scaling of kinetically growing clusters. *Phys. Rev. Lett.* **1983,** *51(13),* 1123–1126.
26. Kokorevich, A. G.; Gravitis, Ya. A.; Ozol-Kalnin, V. G.; Development of scaling approach at lignin supramolecular structure study. Chemistry of Wood, **1989,** *1,* 3–24.
27. Jullien, R.; Kolb, M.; Hierarchical method for chemically limited cluster–cluster aggregation. *J. Phys. A*, **1984,** *17(12),* L639–L643.
28. Brown, W. D.; Ball, R. C.; Computer simulation of chemically limited cluster–cluster aggregation. *J. Phys. A*, **1985,** *18(9),* L517–L521.
29. Ozol-Kalnin, V. G.; Kokorevich, A. G.; Gravitis, Ya.A.; Estimation of fractal and chemical dimensions of lulk- and end-wise polymers. Chemistry of Woof, **1986,** *5,* 108–109.
30. Karmanov, A. P.; Monakov, Yu.B.; Fractal structure of bulk- and end-wise-dehydropolymers. High-Molecular Compounds. B, **1995,** *37(2),* 328–331.
31. Pietronero, L.; Peliti, L.; Probability of survival and factor of intensification in polymers statistics. In book: Fractals in Physics. Ed. Pietronero, L.; Tosatti, L.; Amsterdam, Oxford, New York, Tokyo, North-Holland, **1986,** 117–121.
32. Family, F.; Daoud, M.; Experimental realization of true self-avoiding walks. *Phys. Rev. B*, **1984,** *29(3),* 1506–1507.
33. Kaufman, J. H.; Baker, C. K.; Nazzal, A. I.; Flickner, M.; Melroy, O. R.; Kapitulnik, A.; Statics and dynamics of the diffusion-limited polymerization of the conducting polymer polypyrrole. *Phys. Rev. Lett.* **1986,** *56(18),* 1932–1935.
34. Kaufman, J. H.; Melroy, O. R.; Abraham, F. F.; Nazzal, A. I.; Growth instability in diffusion controlled polymerization. *Solid State Commun.* **1986,** *60(9),* 757–761.
35. Patel, S. S.; Takahashi, K. M.; Polymer dynamics in dilute and semidilute solutions. Macromolecules, **1992,** *25(17),* 4382–4391.
36. Muthukumar, M.; Dynamics of polymeric fractals. *J. Chem. Phys.* **1985,** *83(6),* 3161–3168.
37. Vilgis, T. A.; Swollen and condensed states of polymeric fractals. *Phys. Rev. A*, **1987,** *36(3),* 1506–1508.
38. Vilgis, T. A.; Polymeric fractals and the unique treatment of polymers. J.; Phys. France, **1988,** *49(9),* 1481–1483.
39. Vilgis, T. A.; Flory theory of polymeric fractals—intersection, saturation and condensation. *Physica A* **1988,** *153(2),* 341–354.
40. Cates, M. E.; Brownian dynamics of self-similar macromolecules. J.; Phys. France, **1985,** *46(7),* 1059–1077.

41. Alexander, S.; Orbach, R.; Density of states on fractals: "fractons." *J. Phys. Lett.* (Paris), **1982,** *43(17),* L625-L631.
42. Kozlov, G. V.; Zaikov, G. E.; Structure of the Polymer Amorphous State. Leiden, Boston, Brill Academic Publishers, **2004,** 465 p.
43. Alexandrowicz, Z.; Kinetics of formation and mean shape of branced polymers. *Phys. Rev. Lett.* **1985,** *54(13),* 1420–1423.
44. Alexandrowicz Z. In book: Fractals in Physics. Ed. Pietronero, L.; Tosatti, E.; Amsterdam, Oxford, New York, Tokyo, North-Holland, **1986,** 172–178.
45. Kozlov, G. V.; Yanovskii, Yu. G.; Zaikov, G. E.; Structure and Properties of Particulate-Filled Polymer Composites: The Fractal Analysis. New York, Nova Science Publishers, Inc.; **2010,** 282 p.
46. Mikitaev, A. K.; Kozlov, G. V.; Zaikov, G. E.; Polymer Nanocomposites: Variety of Structural Forms and Applications. New York, Nova Science Publishers, Inc.; **2008,** 319 p.
47. Alexander, S.; Laermans, C.; Orbach, R.; Rosenberg, H. M.; Fracton interpretation of vibrational properties of cross-linked polymers, glasses and irradiated quartz. *Phys. Rev. B*, **1983,** *28(8),* 4615–4619.
48. Chabra, A.; Herrmann, H. J.; Landau, D. P.; Fractal dimensions of skeletons and clusters in gelation kinetic model. In book: Fractals in Physics. Ed. Pietronero, L.; Tosatti, E.; Amsterdam, Oxford, New York, Tokyo, North-Holland, **1986,** 179–183.
49. Gefen Yu.; Aharony, A.; Alexander, S.; Anomalous diffusion on percolating clusters. *Phys. Rev. Lett.* **1983,** *50(1),* 77–80.
50. Rammal, R.; Toulouse, G.; Random walks on fractal structures and percolation clusters. *J. Phys. Lett.* (Paris), **1983,** *44(1),* L13-L22.
51. Rammal, R.; Toulouse, G.; Vannimenus, J.; Self-avoiding walks on fractal spaces: exact results and Flory approximation. J.; Phys. France, **1984,** *45(3),* 389–394.
52. Aharony, A.; Harris, A. B.; Flory approximant for self-avoiding walks on fractals. *J. Stat. Phys.* **1989,** *54,* (¾), 1091–1097.
53. Vannimenus, J.; Phase transitions for polymer on fractal lattices. *Physica D*, **1989,** *38(2),* 351–355.
54. Kozlov, G. V.; Yanovskii, Yu.G.; Lipatov, Yu.S.; Fractal model for description of polymer matrix structural changes in particulate-filled composites. Mechanics of Composite Materials and Structures, **2002,** *8(4),* 467–474.
55. Kozlov, G. V.; Lipatov, Yu. S.; The change of polymer matrix structure in particulate-filled composites: the fractal treatment. Mechanics of Composite Materials and Structures, **2004,** *40(6),* 827–834.
56. Lhuillier, D. A simple model for polymeric fractals in a good solvent and an improved version of Flory approximation. *J. Phys. France*, **1988,** *49(5),* 705–710.
57. Vilgis, T. A.; Haronska, P.; Benhamou, M.; Branched polymers in restricted geometry: Flory theory, scaling and blobs. *J. Phys. II France*, **1994,** *4(12),* 2187–2196.

CHAPTER 2

FRACTAL PHYSICS OF POLYMER SOLUTIONS

CONTENTS

2.1 Fractal Variant of the Mark-Kuhn-Houwink Equation 27
2.2 The Intercommunication of Macromolecular Coil in Solution Structure and Characteristics 55
2.3 The Temperature Dependence of Fractal Dimension of a Macromolecular Coils in Diluted Solutions 96
2.4 The Physical Significance and Evaluation Methods of Structural Parameters of Low-Molecular Solvents in Diluted Polymer Solutions 103
2.5 The Thermodynamics of Polymers Dissolution 113
2.6 The Intercommunication of Structures in Diluted Solution and Polymers Condensed State 117
2.7 The Influence of Macromolecular Coil Structure on the Processes of Polymers Synthesis in Solution 128
2.8 The Intercommunication of Macromolecule Branching and Macromolecular Coil Structure in Diluted Solutions 157
2.9 The Fractal Analysis Of Flocculating Ability of Poly(Dimethyl Diallyl Ammonium Chloride) 175
2.10 The Mechanisms of Films Formation From Polymer Solutions 185
2.11 The Theoretical Description of Polymers Molecular Weight Distribution 198

2.12 The Analysis of a Solution of Polymers with Complex Molecular Architecture 211
Keywords 238
References 238

2.1 FRACTAL VARIANT OF THE MARK-KUHN-HOUWINK EQUATION

Mark-Kuhn-Houwink equation, derived on the basis of large experimental material analysis, obtained wide spreading for polymers average viscosity molecular weight determination by their solutions intrinsic viscosity [η] measured values [1]. This equation has a look like:

$$[\eta]=K_{\eta}M_{\eta}^{a_{\eta}}, \tag{1}$$

where K_{η} and a_{η} are constants characterizing a polymer in the given solvent at the given temperature in certain molecular weight *MM*.

At the experimental data analysis it turns out that for many flexible-chain polymers in good solvents the correlation between constants K_{η} and a_{η} in the Eq. (1) is observed [2]. The following relationship was received:

$$K_{\eta}\approx\frac{21}{m_0}\left(\frac{4\times10^{-4}MM}{m_0}\right)^{a_{\eta}}, \tag{2}$$

where m_0 is mean weight of polymer "elementary link" (without substituents).

In some cases similar empirical relationships usage allows to estimate *MM* value without specimens detailed study performation. A more exact relationship for K_{η} was proposed by Budtov [1]:

$$K_{\eta}=\frac{0.46A^{2}l\times10^{24}}{m_0\left(7.3sm_0\right)^{a_{\eta}}}, \tag{3}$$

where *A* is Kuhn segment length, *l* is monomer link length, *s* is a monomer links number per segment.

For polymers with the same chain rigidity the Eq. (3) is very close to the empirical one—the Eq. (2). As it was noted in Ref. [1], the Eqs. (1)–(3) proved to be very useful at polymers individual samples study, which are not characterized in the required for the analysis solvent, but with the known structural parameters.

The fractal analysis application has demonstrated [3] that the constant a_η is not empirical, but serves as a macromolecular coil structure in solution characteristic and it is linked with the coil fractal dimension D_f by a simple relationship:

$$a_\eta = \frac{3 - D_f}{D_f} \tag{4}$$

The last circumstance allows obtaining the fractal (structural) variant of Mark-Kuhn-Houwink equation by using of well-known relations between $[\eta]$, *MM* and macromolecular coil gyration radius R_g. As the initial relationship for this purpose the following well-known formula was used [1].

$$[\eta] = 6^{1/2}\Phi(\alpha)\frac{\langle R_g^3 \rangle}{MM}, \tag{5}$$

where Φ is a form factor of a macromolecular coil in the conditions of rotatory friction, α is the coil swelling coefficient.

For convenience α is accepted equal to viscous coefficient of swelling α_η, which is determined as follows [1]:

$$\alpha_\eta^3 = \frac{[\eta]}{[\eta]_\theta}, \tag{6}$$

where $[\eta]$ and $[\eta]_\theta$ are the intrinsic viscosities of a polymer in arbitrary and θ-solvent, respectively.

The simplest relationship between Φ and α can be written as follows [1]:

$$\Phi(\alpha) = \Phi_\theta\left(0.753 + \frac{0.247}{\alpha^3}\right), \tag{7}$$

where $\Phi_\theta = \Phi(1)$.

The relation between R_g and *MM* can be expressed by the following approximation [4]:

$$R_g = BN^{1/D_f} = B\left(\frac{MM}{m_0}\right)^{1/D_f}, \tag{8}$$

where B is the proportionality coefficient, N is the degree of polymerization.

The value B can be determined by the following simple enough method. For polyarylate F-2 solution in 1,1,4,4-tetrachlorethane Mark-Kuhn-Houwink equation has a look like [5]:

$$[\eta] = 2.421\times10^{-4} MM^{0.696} \tag{9}$$

Accepting MM magnitude arbitrarily equal to 8×10^4, let us determine from the Eq. (9) [η] value for F-2, proving to be equal to 0.626 dl/g. Further from the Eq. (5) at the indicated values MM and [η] and accepting $\Phi(\alpha)=2\times10^{23}$ [1], let us find R_g=234 Å. And lastly, at m_0=440 from the equation (8) let us determine the value B, which proves to be equal to 12.4 Å. Coefficient B value can be obtained similarly for other polyarylates.

The Eqs. (5)–(8) combination allows to obtain the relationship between [η] and MM, similar to Mark-Kuhn-Houwink equation [6–9]:

$$[\eta] = c(\alpha)\frac{MM^{(3-D_f)/D_f}}{m_0^{3/D_f}}, \tag{10}$$

where c is some constant, dependent on α, since it includes the parameter $\Phi(\alpha)$ as a constituent part. This circumstance allows to estimate the coefficient $c(\alpha)$ at α change according to the Eq. (7). The value $\alpha=\alpha_\eta$ itself can be estimated according to the Eqs. (1), (2), and (6) of Chapter 1, if $[\eta]_\theta$ is determined at a_η=0.5 (or D_f=2.0) [10].

As it is known through Ref. [11], the ratio $[\eta]/[\eta]_\theta$ is D_f unequivocal function, that follows from the equation:

$$D_f = \frac{5([\eta]/[\eta]_\theta)-3}{3([\eta]/[\eta]_\theta)-2} \tag{11}$$

Thus, from the Eq. (10) it follows, that the proportionality coefficient K_η depends on two parameters only: D_f and m_0. Similar empirical dependence of [η] on *MM* was obtained in Ref. [12]. The equation for K_η estimation accepted finally the following form [6]:

$$K_\eta = \frac{8.1\left(0.753 + 0.247/\alpha^3\right)}{m_0^{3/D_f}} \tag{12}$$

Coefficient K_η in the Eq. (12) is correct for [η] dimension in dl/g and m_0—in g/mole. The value 8.1 in this equation is chosen from the considerations of the theory and experiment best correspondence, since *B* variation is possible for different polyarylates.

In Fig. 1, the comparison of the theoretical values K_η (K_η^T) calculated according to the Eq. (12) and corresponding experimental values K_η^e according to the data [5] for polyarylates is adduced. As one can see, a good correspondence is observed between K_η^T and K_η^e. In the given case the values D_f were calculated according to the Eq. (4) of Chapter 1.

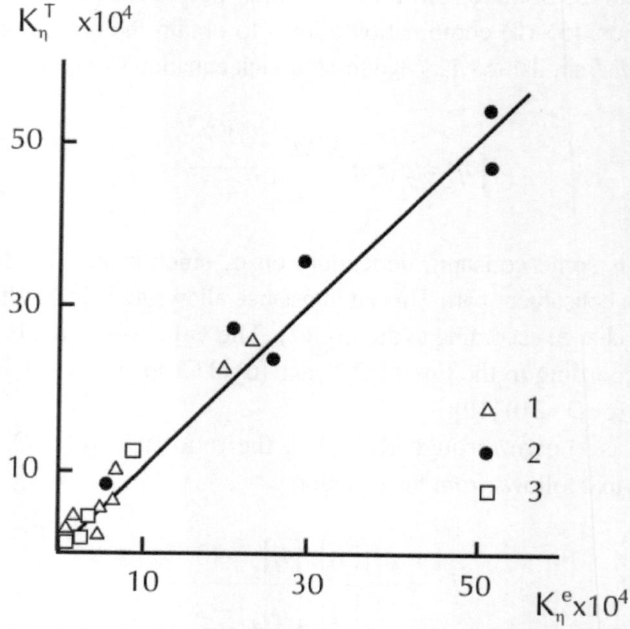

FIGURE 1 The relation between experimental K_η^e and calculated according to the Eq. (12) Mark-Kuhn-Houwink equation constants for solutions of polyarylates series [5] in 1,1,4,4-tetrachlorethane (1), tetrahydrofuran (2) and 1,4-dioxane (3).

It is known through Ref. [13], that the value D_f is defined by interactions of macromolecular coil elements between themselves and interactions polymer-solvent. The regular solutions theory absence complicates the exact definition of interactions of the second group and therefore the following approximate expression was used for D_f determination [15]:

$$D_f \approx 1.5 + 0.35\left(\Delta\delta_f\right)^{2/(1+\delta_c)}, \tag{13}$$

obtained within the framework of two-component solubility parameter [14] (the Eq. (13) conclusion in detail will be given lower). The component δ_f characterizes dispersive interactions energy and energy of dipole bonds interactions, δ_c—energy of interaction between atoms with deficiency (acceptor) and surplus (donor) of electrons. The values of δ_m and δ_c are accepted according to the data of Ref. [14] and together with the calculated according to the Eq. (13) D_f values are adduced in Table 1. According to the Ref. [14], the solubility parameter δ is expressed by the relationship:

$$\delta^2 = \delta_m^2 + \delta_c^2, \tag{14}$$

and the value $\Delta\delta_f$ is determined as follows:

$$\Delta\delta_m = \left|\delta_m^p - \delta_m\right|, \tag{15}$$

where δ_m^p is the value of component δ_f for polymer (for polyarylate F-1 δ_m^p =18 $(\text{Mj/m}^3)^{1/2}$ [14]).

In Fig. 2, the comparison of the experimental dependence $[\eta]$ (δ_s) according to the data of Ref. [16] (solid line) and determined according to the Eq. (10) $[\eta]$ values (points) for polyarylate F-1 is adduced. The polymer MM was calculated according to the Eq. (1) using the values $[\eta]$ [16], a_η and K_η [5] for F-1 solution in 1,1,4,4-tetrachlorethane. As one can see, the calculations according to the Eq. (10) data correspond well to the experimentally obtained curve.

Hence, the results adduced above have shown the possibility of Mark-Kuhn-Houwink theoretical deduction with fractal analysis representations application and also the intrinsic viscosity [η] prediction for diluted solutions. The constant K_η in the indicated equation depends on fractal dimension D_f, characterizing macromolecular coil structure, and monomer link molecular weight m_0.

TABLE 1 The literary and calculated characteristics of polyarylate F-2 solutions in various solvents.

Solvent	δ_m, $(cal/cm^3)^{1/2}$, [14]	δ_c, $(cal/cm^3)^{1/2}$, [14]	D_f
Cyclohexanone	8.50	6.03	1.770
Tetrahydrofuran	8.22	4.80	1.806
N, N-dimethylformamide	8.52	8.15	1.783
Chloroform	8.65	3.01	1.675
Methylene chloride	9.03	4.0	1.733
N, N-dimethylacetamide	8.50	6.66	1.776
Dichlorethane	8.90	4.74	1.656

Let us consider further simple technique of determination of fractal dimension D_f of macromolecular coil in diluted polymer solution, within the framework of which the Eq. (11) was obtained. The determination of D_f value is the first stage of macromolecular coils study within the framework of fractal analysis (see chapter 1) and the similar estimations are performed by measurement of the exponents in Mark-Kuhn-Houwink type equations, linking intrinsic viscosity [η] (the Eq. (1)), translational diffusivity D_0 or rate sedimentation coefficient S_0 with polymers molecular weight MM [3]:

$$D_0 \sim MM^{-a_D}, \tag{16}$$

$$S_0 \sim MM^{a_s} \tag{17}$$

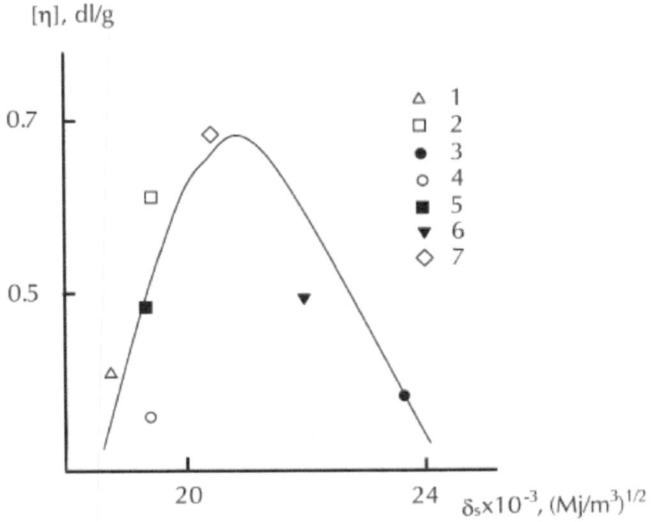

FIGURE 2 The experimental (solid curve) and theoretical (points) dependences of intrinsic viscosity [η] on solubility parameter δ_s of solvent for polyarylate F-1. The solvents: 1—cyclohexanone, 2—chloroform, 3—N, N-dimethylformamide, 4—tetrahydrofuran, 5—methylene chloride, 6—N, N-dimethylacetamide, 7—dichlorethane.

Then the value D_f can be calculated according to the Eq. (4) and according to the following relationships [3]:

$$D_f^D = \frac{1}{a_D}, \tag{18}$$

$$D_f^s = \frac{1}{1-a_s} \tag{19}$$

All considered methods required complex and laborious enough measurements [17, 18]. The simplest and not requiring complex equipment from the enumerated methods is [η] measurement, which can be performed practically in any laboratory. Proceeding from this, the authors [19–21] proposed the simple express-method of estimation of fractal dimension D_f of macromolecular coils in diluted solutions, based on the same principles are used at the Eqs. (4), (18) and (19) deduction.

As it is known through Ref. [22], macromolecular coil swelling coefficient α is determined as follows:

$$\alpha = \left(\frac{h^2}{h_\theta^2} \right)^{1/2}, \tag{20}$$

where h and h_θ are mean-quadratic distances between macromolecule ends in arbitrary and θ-solvent, respectively.

In its turn, the value α is linked with intrinsic viscosities of polymer in arbitrary [η] and θ-solvent $[\eta]_\theta$ according to the Eq. (6). The bulk interactions (which result in deviation of coil shape from ideal Gaussian one) parameter ε is determined as follows [22]:

$$\varepsilon = \frac{d \ln \alpha^2}{d \ln MM} = \frac{\alpha^2 - 1}{5\alpha^2 - 3} \tag{21}$$

In its turn, both ε and coil fractal dimension D_f depend on the exponent a_η value in Mark-Kuhn-Houwink Eq. (1) [18]:

$$\varepsilon = \frac{2a_\eta - 1}{3} \tag{22}$$

and the dependence $D_f(a_\eta)$ is given by the Eq. (4).

The Eqs. (4), (6) and (20)–(22) combination allows one to obtain the Eq. (11) for the value D_f determination from the intrinsic viscosities [η] and $[\eta]_\theta$ values only. The value $[\eta]_\theta$ can be estimated either directly from the experiment or according to the Eq. (1) at the condition a_η=0.5, which is valid for θ-point, if the constant K_η in this equation is known.

For the Eq. (11) verification the data of Ref. [17] for biopolymer of polysaccharides class (rodexman) were used, for which Mark-Kuhn-Houwink has a look like:

$$[\eta] = 2.33 \times 10^{-2} MM^{0.75} \tag{23}$$

In this case D_f calculation according to the Eq. (4) gives the value approx. 1.71. D_f estimation according to the Eq. (11) is adduced in Table

2 for eight rodexman fractions with molecular weight within the range of (27–103)×10^3 g/mole, where the values [η] [17] and $[\eta]_\theta$, calculated according to the Eq. (23) at the condition a_η=0.5, are also adduced. As one can see, from this table data, the value D_f varies within a very narrow range of 1.68–1.70, that corresponds well to the estimation according to the Eq. (4). Some D_f increase is observed at *MM* reduction, which also corresponds to the known data [18].

TABLE 2 The comparison of calculated by two methods values of fractal dimension D_f of rodexman macromolecular coil in water solutions.

MM×10^3 g/mole	[η]×10^3, m^3/kg	$[\eta]_\theta$×10^3 m^3/kg	D_f, the Eq. (4)	D_f, the Eq. (11)
103	135	7.48	—	1.68
93	115	7.11	—	1.69
90	104	6.99	—	1.69
72	98	6.25	1.71	1.69
67	82	6.03	—	1.69
45	62	4.94	—	1.69
32	51	4.17	—	1.69
27	37	3.83	—	1.70

If for polymer the constants in the Eqs. (16) and (17) are known, then the proposed express-method allows to perform values D_0 and S_0 values by the known D_f magnitudes according to the Eqs. (16), (18) and (17), (19), respectively. The comparison of the experimentally received [17] and estimated by the indicated method values D_0 and S_0 is adduced in Table 3. In the given case a satisfactory correspondence of both datasets is obtained, although experimental values exceed systematically estimated ones by the proposed method. This discrepancy is due to the strong dependence of D_0 and S_0 on a_D and a_s, respectively, in virtue of power character of Mark-Kuhn-Houwink type equations [20].

TABLE 3 The comparison of experimental values of translational diffusivity D_0^e and rate sedimentation coefficient S_0^e and these parameters calculated magnitudes D_0^T and S_0^T for rodexman in water solutions.

MM×10³, g/mole	$D_0^e \times 10^{11}$, m²/s	$D_0^T \times 10^{11}$, m²/s, the Eq. (16)	$S_0^e \times 10^{13}$, m²/s	$S_0^T \times 10^{13}$, m²/s, the Eq. (17)
103	2.35	1.78	3.4	2.49
93	2.20	1.89	2.9	2.38
90	—	—	3.0	2.35
72	2.50	2.20	2.6	2.15
67	2.70	2.29	2.6	2.09
45	3.50	2.90	2.2	1.77
32	4.25	3.55	1.9	1.54
27	4.80	3.93	1.8	1.44

It is natural, that the proposed technique, having no specific suppositions, is true not only for biopolymers, but also for other types of macromolecules in solution. In Ref. [5] the values a_η for polyarylates number in different solvents are adduced. This allows calculating D_f value for these polymers according to the Eq. (4). Besides, in the same chapter the values of coefficient K_η in Mark-Kuhn-Houwink equations (the Eq. (1)) for the same polymers and approximate equation are adduced [5]:

$$K_\eta = 0.268\left(6.03\times10^{-5}\right)^{a_\eta} \quad (24)$$

allowing to estimate the value K_η for θ-solvent at the condition a_η=0.5.

Using these data, the values [η] and $[\eta]_\theta$ can be calculated for the same molecular weight in case of the indicated polyarylates and different solvents and then to calculate D_f value according to the Eq. (11). The comparison of the calculated according to the Eqs. (4) and (11) D_f values for seven polyarylates (accepted for them in Ref. [5] conventional signs were maintained) are adduced in Table 4. As it follows from this table data, both methods gave satisfactory correspondence of D_f values and relative discrepancy between them does not exceed 10% [20].

TABLE 4 The comparison of macromolecular coil fractal dimensions D_f, calculated by different methods, for polyarylates.

Polymer conditional sign	Solvent	D_f		Discrepancy, %
		The Eq. (4)	The Eq. (11)	
PF-1 M	Tetrachloroethane	1.80	1.93	7.2
PF-2 M	1,4-dioxane	1.67	1.70	1.8
PF-2 M	Tetrahydrofuran	1.64	1.69	3.0
PF-2P	Tetrachloroethane	1.77	1.95	10.0
PF-7P	Tetrachloroethane	1.78	1.87	5.1
PD-10 M	Tetrachloroethane	1.83	1.95	6.6
PD-10P	Tetrachloroethane	2.00	1.94	3.0

The Eq. (24) is an individual case of a more general Eq. (2). Thus, knowing polymer chain chemical structure for m_0 determination and D_f value for the magnitude a_η estimation according to the Eq. (4), Mark-Kuhn-Houwink equation parameters (the Eq. (1)) can be determined and polymer molecular weight can be estimated by a simple method, using the values $[\eta]$ and $[\eta]_\theta$ only. In Table 5 the comparison of the indicated parameters (K_η and a_η), determined experimentally [5] and estimated by the indicated above method, is adduced for the polyarylates, enumerated in Table 4. As one can see, a satisfactory correspondence of theory and experiment is obtained again.

TABLE 5 Mark-Kuhn-Houwink equation parameters.

Polymer conditional sign	Solvent	$K_\eta \times 10^4$ [5]	a_η [5]	$K_\eta \times 10^4$, the Eq. (24)	a_η, the Eq. (4)
PF-1 M	Tetrachloroethane	3.350	0.670	12.30	0.554
PF-2 M	1,4-dioxane	0.883	0.800	1.59	0.765
PF-2 M	Tetrahydrofuran	0.724	0.828	1.44	0.775
PF-2P	Tetrachloroethane	2.421	0.696	1.44	0.538
PF-7P	Tetrachloroethane	4.095	0.684	7.58	0.604
PD-10 M	Tetrachloroethane	23.05	0.640	14.40	0.538
PD-10P	Tetrachloroethane	5.220	0.495	13.30	0.546

Further six arbitrary values [η] within the range of 0.1–1.2 dl/g were accepted and according to these values average viscous molecular weight M_η magnitudes were estimated using Mark-Kuhn-Houwink equation and both constants K_η and a_η sets (determined experimentally [5] and calculated according to the represented above technique, see Table 5) for solution of polyarylate PF-2 M in tetrahydrofuran. Comparison of the obtained by such method M_η values is adduced in Table 6, from which their satisfactory correspondence follows.

TABLE 6 The comparison of average viscous molecular weights of polyarylate PF-2 M in tetrahydrofuran.

[η], dl/g	M_η^e, g/mole	M_η^T, g/mole
0.1	6300	4600
0.3	14,600	19,100
0.5	44,200	36,900
0.7	66,400	57,000
1.0	102,300	90,300
1.2	127,500	114,200

M_η^e—K_η and a_η according to the data of Ref. [5]; M_η^T—K_η and a_η according to the Eq. (24) and Eq. (4) of Chapter 1, respectively. The values M_η^e and M_η^T were determined with the aid of Mark-Kuhn-Houwink equation.

Hence, the considered above express-method of macromolecular coils in diluted solution fractal dimension determination allows to estimate simply and exactly enough the value D_f and also a parameters number, depending on D_f. It is assumed, that this method will be the most useful at the comparative analysis of D_f value of a polymers number (or the same polymer in several solvents), since the comparison with the data, obtained by other methods, can give large enough discrepancy, inherent in a dependences of power kind. Let us note, that the indicated restriction is valid for all similar methods (see Table 3 in Ref. [3]). The knowledge of polymer chemical structure and values D_f allows to estimate parameters K_η and a_η in Mark-Kuhn-Houwink equation and to calculate polymer average viscous molecular weight values, without resorting to more laborious methods.

As it has been noted above, the value D_f is defined by two factors: interactions of coil elements between themselves and interactions polymer-solvent [13]. However, at present the prediction of the second from the indicated factors runs across certain difficulties, the absence of polymer solutions adequate theory is the main cause [22]. Therefore, for the indicated task solution semiempirical methods are often used, for example, prediction of polymer solubility degree by the difference of solubility parameters of polymer δ_p and solvent δ_s ($\Delta\delta=|\delta_{p-}\delta_s|$) [16]. As it has been shown in Ref. [22], such conception is very simplified. It can give the indication only to the fact, in which solvents polymer does not undoubtedly dissolved and does not have predicting power in regard to polymers [22]. This has resulted in the development of the improved schemes within the same conception, so-called two- and three-solubility parameters [14, 23]. These schemes essence consists of solubility parameter (δ_p or δ_s) representation as a vector, are components of which contributions in it interactions different groups (dispersive, polar, hydrogen and so on). Various authors group these contributions by different modes. In Ref. [14] it has been shown, that the proposed in it simple scheme of two-dimensional (two-component) solubility parameter allows to predict simply enough polymer solubility degree in different liquids (or liquids mixtures). The simplicity and successful scheme [14] application assumed the possibility of its usage for the value D_f prediction (see the Eq. (13)) [24].

At present it has been supposed (with the indicated above restrictions [22]) that the increase of polymer and solvent solubility parameters difference $\Delta\delta$ results in polymer solubility deterioration and a_η reduction [5, 16]. In its turn, D_f increase corresponds to a_η decrease according to the Eq. (4). So, $D_f=5/3$ corresponds to a good solvent and $D_f=2.0$ to θ-solvent [25]. Therefore, D_f increasing should be expected at $\Delta\delta$ growth. This assumption is confirmed by the plots of Fig. 3, where the dependences $D_f(\Delta\delta)$ are adduced for copolymers of stirene with acrylonitrile and methyl methacrylate (SAN-MMA) [24]. However, the data large scattering in Fig. 3 allows speaking only about tendency, but not about correlation. The relationships $D_f(\Delta\delta)$ approximation by the curves, shown in Fig. 3, has demonstrated one interesting feature of these plots, which consists of the fact, that at $\Delta\delta=0$ the values D_f (designated further as D_f^0) were obtained differently for various solvents groups and are located within the range of approx.

1.60–1.73. Before to determine this effect causes, let us consider the main principles of solubility two-dimensional parameter scheme [14].

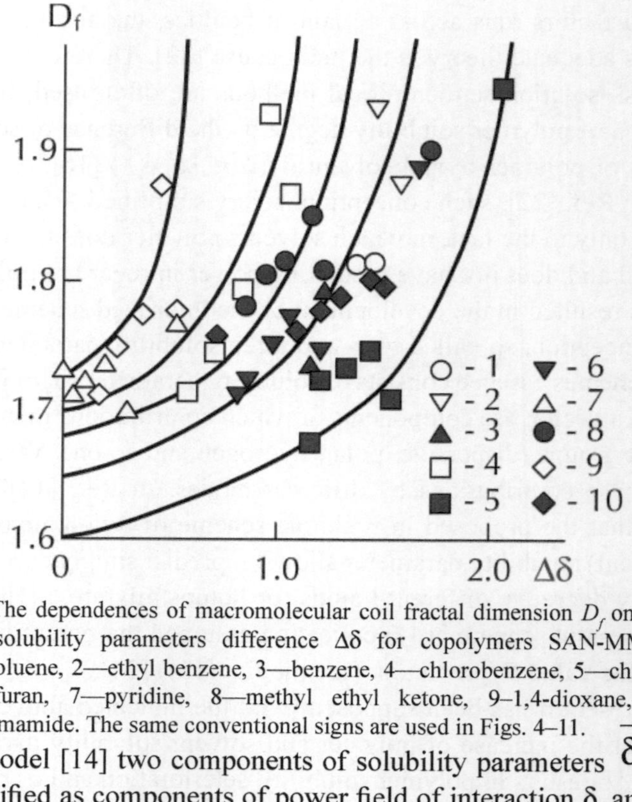

FIGURE 3 The dependences of macromolecular coil fractal dimension D_f on polymer and solvent solubility parameters difference $\Delta\delta$ for copolymers SAN-MMA. The solvents: 1—toluene, 2—ethyl benzene, 3—benzene, 4—chlorobenzene, 5—chloroform, 6—tetrahydrofuran, 7—pyridine, 8—methyl ethyl ketone, 9–1,4-dioxane, 10—N, N-dimethylformamide. The same conventional signs are used in Figs. 4–11.

In the model [14] two components of solubility parameters $\vec{\delta}$ (δ_p or δ_s) are identified as components of power field of interaction δ_f and component of complexing δ_c. The main postulates of model [14] in respect to the present chapter can be expressed as follows:

1. Solvent ability to dissolve any polymer is defined by two and only two conditions:

$$\delta_m \geq 0, \tag{25}$$

$$\delta_c \geq 0. \tag{26}$$

2. The two components δ_m and δ_c for each solvent are related to the value $\vec{\delta}$ by the Eq. (14).
3. The probability for the polymer A being dissolved in solvent B exceeds at reduction of difference absolute values their solubility parameters components:

$$\Delta\delta_m = \left|\delta_{mA} - \delta_{mB}\right|, \tag{27}$$

and

$$\Delta\delta_c = \left|\delta_{cA} - \delta_{cB}\right|. \tag{28}$$

The component δ_m of solubility parameter includes dispersive interactions energy and dipole bonds interaction energy and the component δ_c – hydrogen bonding interaction and the interaction energy between an electron-rich atom of one molecule (donor) and an electron-deficient atom of another molecule (acceptor), which requires a certain orientation of these two molecules. In such treatment there is no need to introduce a separate component for interaction between polar molecules description.

Returning to the data of Fig. 3, let us point out, that the certain intercommunication between D_f^0 and δ_c was found out. This intercommunication consists of D_f^0 systematic reduction at δ_c growth. The indicated observation is important itself. It means, that the intensification of interactions, defining the component δ_c, results in D_f reduction that accelerates essentially the polymerization process [27, 28]. This allows constructing the dependence of D_f on the difference $\Delta\delta_m = |\delta_p - \delta_m|$, eliminated δ_c influence, which is adduced in Fig. 4. From the data of this figure it follows that now the value D_f^0 has the same value for all solvents and is close to D_f value for a good solvent (D_f=5/3 [25]). Besides, the curves $D_f(\Delta\delta_m)$ shape itself assumes a possibility of both their superposition and scaling description. The last possibility will be considered lower and the superposition possibility allows predicting D_f and therefore it should be considered in more detail.

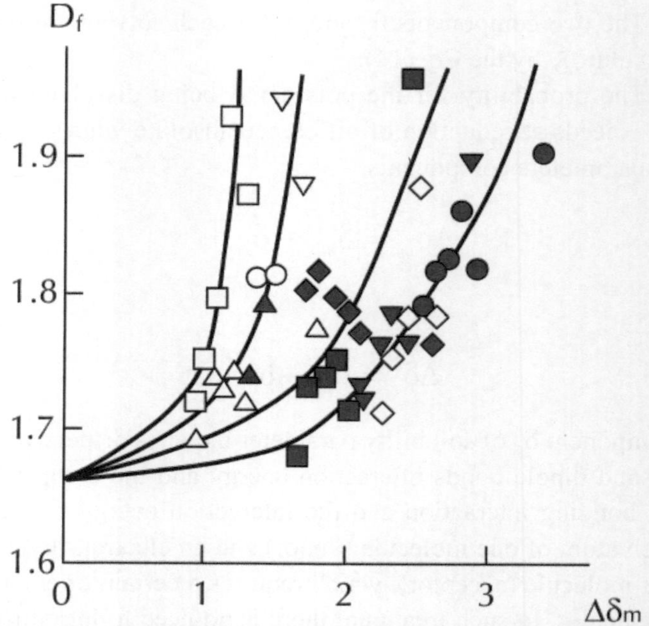

FIGURE 4 The dependence of macromolecular coil fractal dimension D_f on the difference $\Delta\delta_m = |\delta_{p-}\delta_m|$ for copolymers SAN-MMA.

The curves $D_f(\Delta\delta_m)$ can be matched by purely empirical mode, selecting a shift factor over $\Delta\delta_m$ axis ($\Delta\Phi_e$) according to the principle of the curves best matching. In this case the matching is achieved by the dependence $D_f(\Delta\Phi_e\Delta\delta_m)$ construction. However, it has been noticed, that $\Delta\Phi_e$ value (which varies within the range of 2.5–1.0) systematically reduces at δ_c growth. This observation is reflected in Fig. 5, where the relationship $\Delta\Phi_e(\delta_c)$ (points) is adduced and the solid curve gives this relationship smooth approximation. The lack of the shown in Fig. 5 relationship is large enough $\Delta\Phi_e$ scatter in respect of their approximate curve $\Delta\Phi_{ap}(\delta_c)$. Despite this, the plot of Fig. 5 is a useful one, since it points out the factor, influencing on $\Delta\Phi_e$ and gives its change tendency.

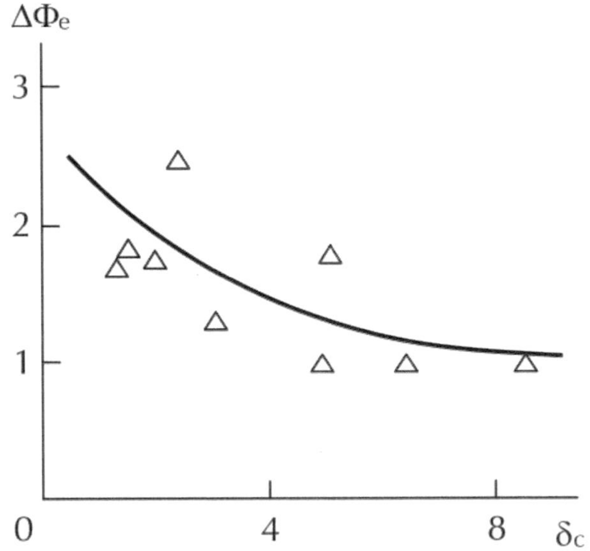

FIGURE 5 The dependence of stift factor $\Delta\Phi_e$ on solubility parameter component δ_c value for copolymers SAN-MMA.

The obtained according to the considered technique the matched curves $D_f(\Delta\Phi_{ap}\Delta\delta_m)$ and $D_f(\Delta\Phi_e\Delta\delta_m)$ are shown in Figs. 6 and 7, respectively. As it follows from Figs. 6 and 7 plots comparison, the better matching (small scatter) is achieved at empirical stift factor $\Delta\Phi_e$ using (Fig. 7), but $\Delta\Phi_{ap}$ using gives possibility of the value D_f prediction, that is the purpose of Ref. [24].

A common aspect of the adduced in Figs. 6 and 7 assumes approximately quadratic dependence of D_f on $(\Delta\Phi_{ap}\Delta\delta_m)$ or $(\Delta\Phi_e\Delta\delta_m)$. Therefore, there is possibility to obtain more clear and useful for practical application (again for D_f prediction) linear correlation. Such correlation can be realized by using square of the corresponding argument for D_f, that is, either $(\Delta\Phi_{ap}\Delta\delta_m)$ or $(\Delta\Phi_e\Delta\delta_m)$. The dependence of D_f on $(\Delta\Phi_e\Delta\delta_m)^2$, shown in Fig. 8, justifies completely this supposition. The good enough linear correlation is received, which is described analytically by the relationship [24]:

$$D_f = 1.5 + 0.042\left(\Delta\Phi_e\Delta\delta_m\right)^2 \tag{29}$$

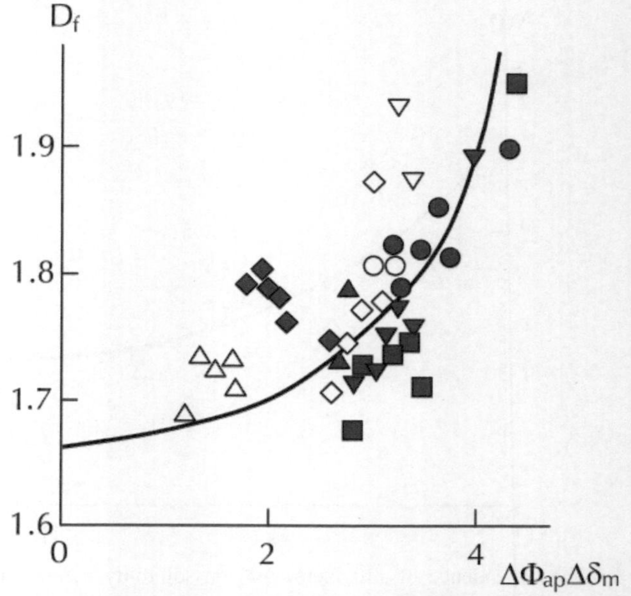

FIGURE 6 The dependence of macromolecular coil fractal dimension D_f on corrected difference ($\Delta\Phi_{ap}\Delta\delta_m$) for copolymers SAN-MMA.

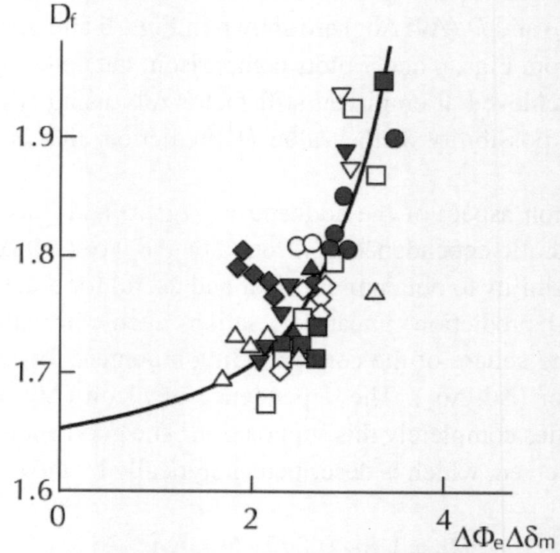

FIGURE 7 The dependence of macromolecular coil fractal dimension D_f on corrected difference ($\Delta\Phi_e\Delta\delta_m$) for copolymers SAN-MMA.

The similar dependence is obtained in case of the dependence of D_f on $(\Delta\Phi_{ap}\Delta\delta_m)^2$ as well, but with the data larger scatter in virtue of the indicated above causes.

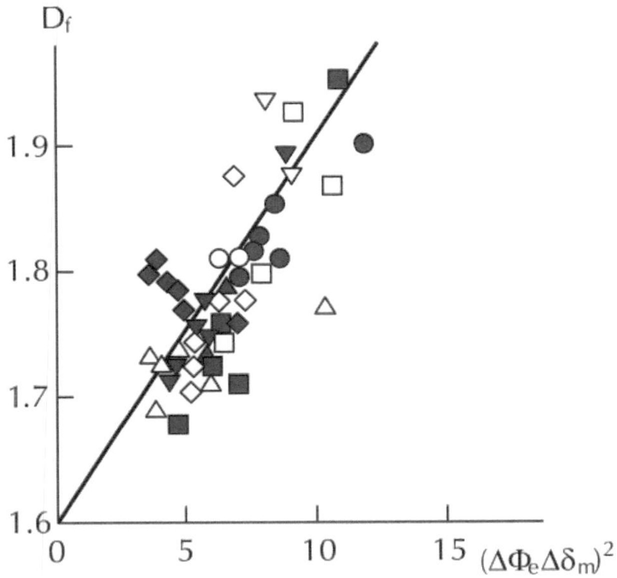

FIGURE 8 The dependence of macromolecular coil fractal dimension D_f on the parameter $(\Delta\Phi_e\Delta\delta_m)^2$ value for copolymers SAN-MMA.

Let us note, that the minimum (at $\Delta\delta_m$=0) D_f value in the Eq. (29) is equal to 1.5, which corresponds to the dimension of "leaking" coil (a_η=1.0) [10]. This D_f value can be realized at the condition $\Delta\delta_m$=0 or $\delta_p=\delta_m$ only (since $\Delta\Phi_e\geq1.0$, see Fig. 5) and therefore is unattainable one for the considered copolymers (for them δ_p=10.30–11.17 (cal/cm^3)$^{1/2}$ [26] and δ_m=9.39–7.0 (cal/cm^3)$^{1/2}$ [14]). As a matter of fact, from 40 adduced in Fig. 8 points only 8 points have essential deviation—4 points for N, N-dimethylformamide and one point apiece—for chlorobenzene, chloroform, pyridine and methyl ethyl ketone, that assumes the indicated deviations randomness [24].

The Eq. (29) allows one theoretical estimation (prediction) of D_f value, if the values δ_p, δ_m and δ_c (or $\Delta\Phi_e$) are known. In Fig. 9, the comparison of experimental (estimated according to the Eq. (4)) D_f^e and calculated according to the Eq. (29) D_f^T values of macromolecular coils fractal dimension for copolymers SAN-MMA in different solvents with $\Delta\Phi_e$ using is adduced. As one can see, the good correspondence of theory and experiment is obtained (the correlation coefficient is equal to approximately 0.80). As it was to be expected, the correspondence between D_f^e and D_f^T in case $\Delta\Phi_{ap}$ using proves to be worse, but it is also acceptable.

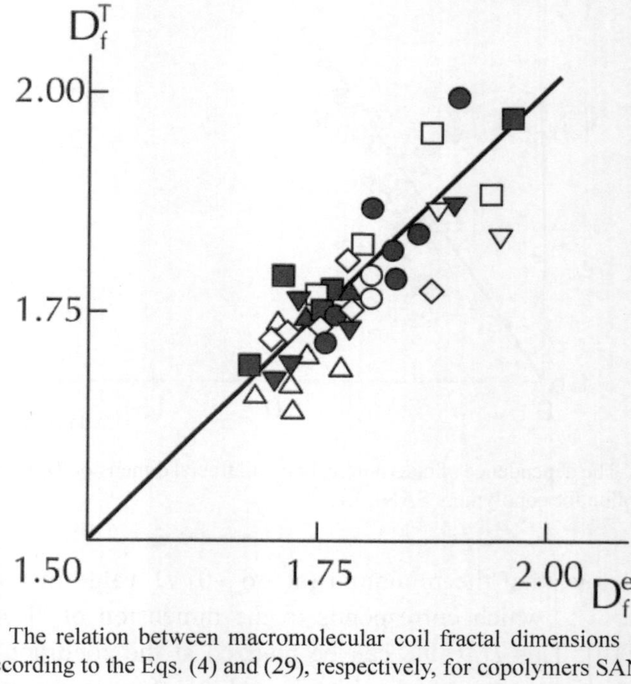

FIGURE 9 The relation between macromolecular coil fractal dimensions D_f^e and D_f^T, calculated according to the Eqs. (4) and (29), respectively, for copolymers SAN-MMA.

For the proposed technique community verification, namely, the equation (29), the authors [24] used the literary data for polymer-solvent 16 pairs in reference to polymers of different kinds: polycarbonate (PC), poly(methyl methacrylate) (PMMA), poly(vinyl chloride) (PVC), polydimethylsiloxane (PDMS), polyarylate (PAr) and polystirene (PS) [5, 17, 18, 20]. The solubility parameters δ_p of the indicated polymers is accepted according to the data [16, 32]. The comparison of calculated according to the Eqs. (4) and (29) D_f values for these polymer-solvent pairs with using in the latter from the equations shift factor $\Delta\Phi_{ap}$ (according to the data of Fig. 5) is adduced in Table 7. As one can see, and in this case the reasonable correspondence of both sets of D_f values is obtained despite the indicated above lacks of $\Delta\Phi_{ap}$ application. The average discrepancy between theoretical and experimental D_f values is equal to 5.5% and maximum—9.5 % [24].

TABLE 7 The comparison of calculated according to the Eqs. (4) and (29) values of macromolecular coil fractal dimension D_f.

Polymer	Solvent	D_f, the Eq. (4)	D_f, the Eq. (29)	Relative error, %
	Tetrahydrofuran	1.765	1.740	1.4
Polycarbonate	1,4-dioxane	1.796	1.726	3.9
	Methylene chloride	1.648	1.570	4.7
	Methyl ethyl ketone	1.744	1.725	1.1
Poly(methyl methacrylate)	Chloroform			
	Toluene	1.667	1.590	4.6
	Benzene	1.734	1.583	8.3
	Acetone	1.705	1.546	9.3
		1.765	1.790	1.4
Poly(vinyl chloride)	Cyclohexanone	1.695	1.826	7.7
Polydimethylsiloxane	Benzene	1.676	1.707	1.8
	Toluene	1.563	1.712	9.5
Polyarylate F-1	Tetrahydrofuran	2.158	2.010	6.9
Polyarylate F-2	Tetrahydrofuran	1.641	1.740	6.0
	1,4-dioxane	1.667	1.744	4.6
Polystirene	Toluene	1.676	1.530	8.7
	Tetrahydrofuran	1.714	1.564	8.8

Let us return now to the possibility of the dependences $D_f(\Delta\delta_m)$, shown in Fig. 4, scaling description. The general expression for this scaling has a look like [24]:

$$D_f \sim (\Delta\delta_m)^{\beta} \tag{30}$$

where β is an exponent, characteristic for each solvent.

The value β estimation according to the slope of double logariphmic dependences ln D_f[ln ($\Delta\delta_m$)], which has proved to be linear, has shown that for 10 studied solvents it varies within the limits of approx. 0.06–0.42. It has also been noted, that the value β increases systematically at δ_c reduction. This observation allows constructing the dependence β($\delta_{\tilde{n}}$), shown in Fig. 10, which proves to be linear and passing through coordinates origin. According to this dependence the value β can be expressed according to the following empirical relationship [24]:

$$\beta = \frac{0.60}{\delta_c} \tag{31}$$

and the Eq. (30) now can be written as follows [24]:

$$D_f \sim (\Delta\delta_m)^{0.60/\delta_c} \tag{32}$$

$D_f \sim (\Delta\delta_m)^{0.60/\delta_c}$.

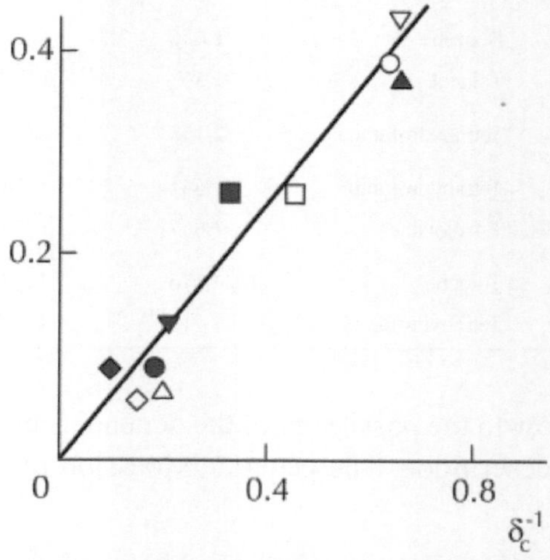

FIGURE 10 The dependence of parameter β on reciprocal value of solubility parameter component δ_c for copolymers SAN-MMA.

The Eq. (32) displays an interesting fact, namely, antibatness of the dependences of fractal dimension D_f on components of solvent solubility parameter δ_m and δ_c. As it follows from the indicated relationship, $\Delta\delta_m$ increase (i.e., δ_m enhancement for fixed polymer) results to D_f growth and δ_c increase—in its reduction. In addition the Eq. (32) form assumes stronger (power one) dependence D_f on δ_c than on δ_m. At the same time polymer solubility range, that is, the range $\Delta\delta_m$, in which D_f changes within the limits of 1.50–2.0, becomes wider at δ_c growth [24].

In Fig. 11 the comparison of the scaling (calculated according to the Eq. (32)) and experimental dependences $D_f(\Delta\delta_m)$ for copolymers SAN-MMA in chlorobenzene, chloroform and tetrahydrofuran. As one can see, the good correspondence of D_f calculations is obtained according to the Eqs. (4) and (32), that indicates scaling approach perspectivity for the value D_f prediction.

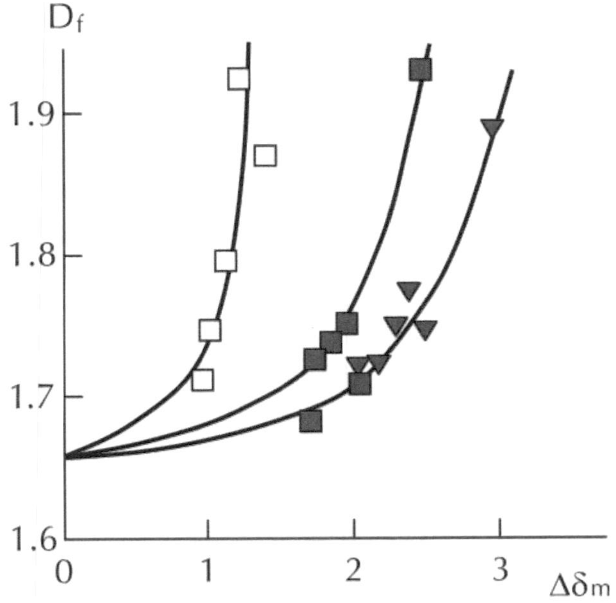

FIGURE 11 The dependences of macromolecular coil fractal dimension D_f on difference $\Delta\delta_m$, calculated according to the Eqs. (4) (the solid curves) and (32) (points) for copolymers SAN-MMA.

However, the scaling expression cannot be used directly in the form, given by the Eq. (32), for D_f prediction is impossible by two reasons, as minimum. Firstly, this relationship assumes D_f=0 at $\Delta\delta_m$=0, that does not have physical significance. Secondly, at δ_c=0 and $\Delta\delta_m$>0 the Eq. (32) gives $D_f\to\infty$, that also does not have physical significance. Therefore, the indicated relationship was modified for two indicated disparities elimination that allows one to obtain the Eq. (13).

The prediction of D_f according to the Eq. (13) results is presented in Table 8 for the indicated 16 polymer-solvent pairs. As it follows from the Table 8 data, a good enough correspondence of theory and experiment is obtained. The average discrepancy between theoretical and experimental D_f values is equal to 3.9% and maximum—10.1%. In other words, such error is comparable with calculation according to the Eq. (29) error (see Table 7).

Let us note, that the Eq. (13) supposes the possibility of "leaking" coil realization (D_f=1.50 [10]) for all polymers. However, in the general case this does not occur. Therefore, if to assume, that for PAr F-1 and PS the state "coil in good solvent" with D_f=5/3 [10] is only attainable, then the first member in the equation (13) right-hand part should be replaced on 1.667 and in this case the correspondence will be improved noticeably.

Thus, the main conclusion from the results adduced above is perspectivity of both considered models on the basis of two-dimensional solubility parameter for macromolecular coils in diluted solutions fractal dimension prediction as the first stage of a polymers synthesis computer simulation. The important conclusion from the Eq. (32) is the value D_f prediction hopelessness on the basis of solvent integral solubility parameter δ_s (or $\vec{\delta}$) only, since the dependences of D_f on its components are antibate. It is also obvious, that the further chapter for a tasks number solution is necessary for the proposed methods practical application. For example, it is obvious, that the more precise theoretical estimation of shift factor $\Delta\Phi_{ap}$ and theoretical prediction of proportionality coefficients in the Eqs. (13) and (32) are such tasks [24].

TABLE 8 The comparison of calculated according to the Eqs. (4) and (13) macromolecular coil fractal dimensions D_f.

Polymer	Solvent	D_f, the Eq. (4)	D_f, the Eq. (13)	Relative error, %
	Tetrahydrofuran	1.765	1.734	1.8
Polycarbonate	1,4-dioxane	1.796	1.736	3.3
	Methylene chloride	1.648	1.680	1.9
	Methyl ethyl ketone	1.744	1.739	0.3
Poly(methyl methacrylate)	Chloroform			
	Toluene	1.667	1.684	1.0
	Benzene	1.734	1.641	5.4
	Acetone	1.705	1.663	2.5
		1.765	1.749	0.9
Poly(vinyl chloride)	Cyclohexanone	1.695	1.788	5.5
Polydimethylsiloxane	Benzene	1.676	1.745	4.0
	Toluene	1.563	1.739	10.1
Polyarylate F-1	Tetrahydrofuran	2.054	1.943	5.4
Polyarylate F-2	Tetrahydrofuran	1.641	1.772	7.4
	1,4-dioxane	1.667	1.758	5.2
Polystirene	Toluene	1.676	1.701	1.5
	Tetrahydrofuran	1.714	1.823	6.0

Footnote: for PAr F-1 and PS in the Eq. (13) right-hand part as the first member of sum the value 1.667 was used.

For the value D_f and, hence, the exponent a_η in Mark-Kuhn-Houwink equation the Eq. (18) of Chapter 1 can be used, from which it follows, that the value D_f=5/3 for a good solvent is reached at δ_f=0 and D_f=2.0 for a poor (θ–) solvent—at δ_f=1.0. At the same time a large number of experimental observations has shown the common tendency—the closer the solubility parameters of polymer δ_p and solvent δ_s, are the better polymer dissolves in the given solvent. This allows to suppose the following correlation between δ_f, δ_p and δ_s form [33]:

$$\delta_f \sim \left|\delta_p - \delta_s\right| \tag{33}$$

The proportionality coefficient in the Eq. (33) can be received from the following considerations. As it is known through Ref. [22], a polymer ceases to dissolve in solvent at $|\delta_{p}-\delta_{s}| \geq 2.5$ (cal/cm^3)$^{1/2}$. This is the criterion for poor solvent with δ_f=1.0, that allows one to accept the indicated coefficient as equal to approx. 0.4. The comparison of theoretical and experimental data of a_η value for 18 polymer-solvent pairs has shown their good correspondence [33].

Cationic water-dissolved polyelectrolytes find wide application in industry different fields, namely, for the ecological problems solution [34]. For understanding these polymers action mechanism and synthesis processes it is necessary to define their molecular characteristics and water solutions properties too [35]. Therefore, the authors [36] performed description of cationic polyelectrolytes in solution behavior within the framework of fractal analysis on the example of copolymer of acrylamide with trimethylammonium methacrylate chloride (PAA-TMAC). The data [35] for four copolymer PAA-TMAC with TMAC contents of 8, 25, 50 and 100 mol. % were used. In Ref. [35] the equations Mark-Kuhn-Houwink type were obtained, which linked intrinsic viscosity [η] (the Eq. (1)) and macromolecular coil gyration radius R_g with average weight molecular weight of polymer [35]:

$$\left\langle R_g^2 \right\rangle^{1/2} = K_R \overline{M}_w^{a_R}, \tag{34}$$

where a_η, a_R, K_η and K_R are constants for each copolymer.

The value of macromolecular coil fractal dimension D_f in solution can be determined according to the Eq. (4). In Fig. 12 the dependence of D_f value on TMAC contents c_{TMAC} in copolymers PAA-TMAC is adduced, allowing one to judge about interactions of macromolecular coil elements between themselves. As it is known through Ref. [13], D_f value of macromolecular coil in solution is controlled by factors two groups: interactions polymer-solvent and interactions of macromolecular coil elements between themselves. Since coils behavior in one solvent (NaCl water solution [35]) is considered, then it should be supposed, that the shown in Fig. 12 D_f variation is due to the factors, belonging to the second group. Quite small content of component with charged groups causes D_f sharp increase, that is, it results in coil essential compactness enhancement. This means strong attraction interactions availability between PAA and TMAC fragments. However, at TMAC contents increase D_f reduction is observed, in addition within contents TMAC (c_{TMAC}) range of 8–50 mol. % this reduction is close to a linear one. Such D_f reduction supposes strong repulsion interactions availability between TMAC fragments, compensating attraction interactions between PAA and TMAC fragments [36].

In Fig. 12, the dependence of macromolecular coil gyration radius R_g on c_{TMAC} for the considered copolymers is also adduced. The similarity of the dependences $D_f(c_{TMAC})$ and $R_g(c_{TMAC})$, which is obvious at their comparison, draws its attention. Let us note, that R_g values were calculated according to the Eq. (34) in supposition of the same value $\overline{M}_w=10^5$ for all copolymers. Proceeding from coils compactness degree reduction at D_f decrease at c_{TMAC} enhancement one should expect the increase, but not reduction R_g. For this apparent discrepancy explanation let us consider the fractal relationship between R_g and D_f (the Eq. (8)), in which a polymerization degree N is determined as follows [36]:

$$N = \frac{\overline{M}_w}{M_0}, \tag{35}$$

where M_0 is molecular weight of polymer chain "elementary link."

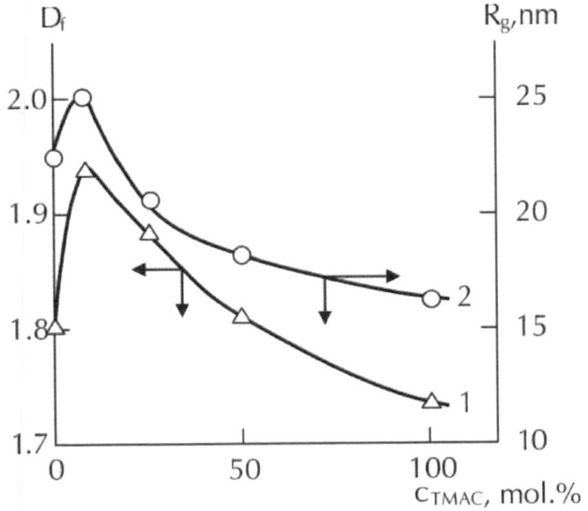

FIGURE 12 The dependence of macromolecular coil fractal dimension D_f (1) and gyration radius R_g (2) on TMAC contents c_{TMAC} for copolymers PAA-TMAC.

From the Eqs. (8) and (35) it follows, that for the dependences of D_f and R_g on c_{TMAC}, shown in Fig. 12, simultaneous fulfillment it is necessary N decrease at c_{TMAC} growth or, that is equivalent, M_0 increase, which in its turn means copolymer chain rigidity enhancement. The Eqs. (8) and (35) combination allows to calculate M_0 value and its dependence on c_{TMAC} for copolymers PAA-TMAC, which is shown in Fig. 13. As it was to be expected, the essential increase M_0 (approximately in 4 times) is observed at TMAC contents enhancement from 8 up to 100 mol. %.

The same conclusion follows from the coefficient K_η in the Eq. (1) change. As the data of Ref. [35] have shown, in the indicated variation c_{TMAC} range K_η reduction in more than 20 times is observed. The parameter m_0 in the Eq. (2), according to which K_η is determined, can be calculated as follows [37]:

$$m_0 = \frac{M_0}{\rho N_A}, \tag{36}$$

where ρ is polymer density, N_A is Avogadro number.

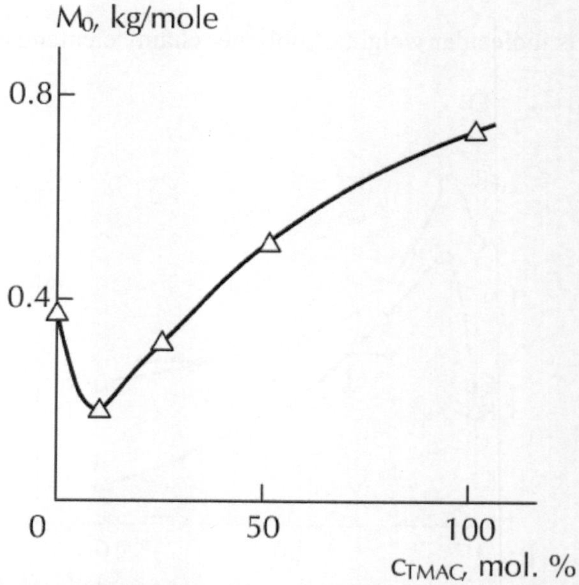

FIGURE 13 The dependence of "elementary link" molecular weight M_0 on TMAC contents c_{TMAC} for copolymers PAA-TMAC.

One more method of the coefficient K_η calculation, namely, the Eq. (12), gives Mark-Kuhn-Houwink equation fractal variant [6]. In Table 9 the comparison of experimental and calculated according to the Eqs. (2), (12) K_η values are adduced. As one can see, the experimental K_η values correspond better to calculation according to the Eq. (12), than according to the Eq. (2).

Hence, the increase of TMAC contents in copolymers PAA-TMAC has two consequences for macromolecular coil structure: the chain rigidity increase and macromolecular coil fractal dimension D_f reduction is observed. These changes influence essentially on both synthesis processes and flocculation of low-molecular admixtures processes of the considered copolymers [36].

TABLE 9 Comparison of the coefficient K_η values in Mark-Kuhn-Houwink equation, obtained by different methods.

TMAC contents, mol. %	$K_h \times 10^4$		
	Experiment [35]	The Eq. (2)	The Eq. (12)
0	2.57	2.09	3.20
8	12.80	27.60	20.0
25	6.30	7.10	6.21
50	1.85	1.45	1.86
100	0.60	0.38	0.66

2.2 THE INTERCOMMUNICATION OF MACROMOLECULAR COIL IN SOLUTION STRUCTURE AND CHARACTERISTICS

As it is known through Ref. [10], macromolecular coil gyration radius R_g scales with its molecular weight *MM* as:

$$R_g \sim MM^{\nu_F} \quad (37)$$

where ν_F is Flory exponent.

For statistical polymer chain, subordinating to distribution Gaussian function, ν_F=0.5 [38]. However, it was found out experimentally [10], that the value ν_F varies within the limits of 1/3–1 and this was explained by bulk interactions between randomly approached chain links and also between chain links and solvent molecules. These interactions degree can be characterized by the parameter ε, which is defined as follows [38]:

$$\left\langle \overline{h}^2 \right\rangle \sim MM^{1+\varepsilon} \tag{38}$$

where $\left\langle \overline{h}^2 \right\rangle$ is mean-square distance between macromolecule ends.

Within the framework of fractal analysis the intercommunication between R_g and MM is defined by the Eq. (8). As well as the exponent ν_F, the dimension D_f in the Eq. (8) characterizes bulk distribution of macromolecular coil elements [25]. For polymers the value D_f is usually determined according to the Eq. (4). The parameter ε is also linked with the exponent a_η in Mark-Kuhn-Houwink equation according to the Eq. (22). The combination of Eqs. (4) and (22) allows to obtain the following relationship between D_f and ε [39]:

$$D_f = \frac{2}{\varepsilon + 1} \tag{39}$$

Thus, the dimension D_f is also characteristic of volume effects for macromolecular coil in solution in virtue of its unequivocal dependence on ε.

Using the adduced in Ref. [10] values D_f, the authors [39] constructed the dependence $\varepsilon(D_f)$ according to the Eq. (39). This dependence, adduced in Fig. 14, allows making several conclusions. Firstly, the parameter ε variation range is somewhat wider, that it was supposed earlier [38]. So, according to the data of Ref. [38] a_η value changes from approx. 0.50 up to 0.80, that gives the variation range of ε=0–0.20. Using the permissible values D_f range of 1–3, the authors [39] obtained the variation of ε=-1/3–1.0. Secondly, as it follows from the plot of Fig. 14, the dependence $\varepsilon(D_f)$ has in fact one characteristic point: ε=0 at D_f=2.0. From the point of view of the dependence $\varepsilon(D_f)$ a stretched chain, leaking coil and coil in good solvent qualitatively are not different from anything else, excluding the quantitative characteristic—the interaction parameter ε. At the point D_f=2.0 parameter ε changes its sign, that corresponds to interactions type change from repulsive forces (positive ε) to attractive forces (negative ε). As it has been noted above, the value ε=0 corresponds to D_f=2.0 or ν_F=0.5, that is, to coil state in θ-solvent. In this state the interactions between macromolecule separate elements and also between macromolecule and solvent regain balance completely [25].

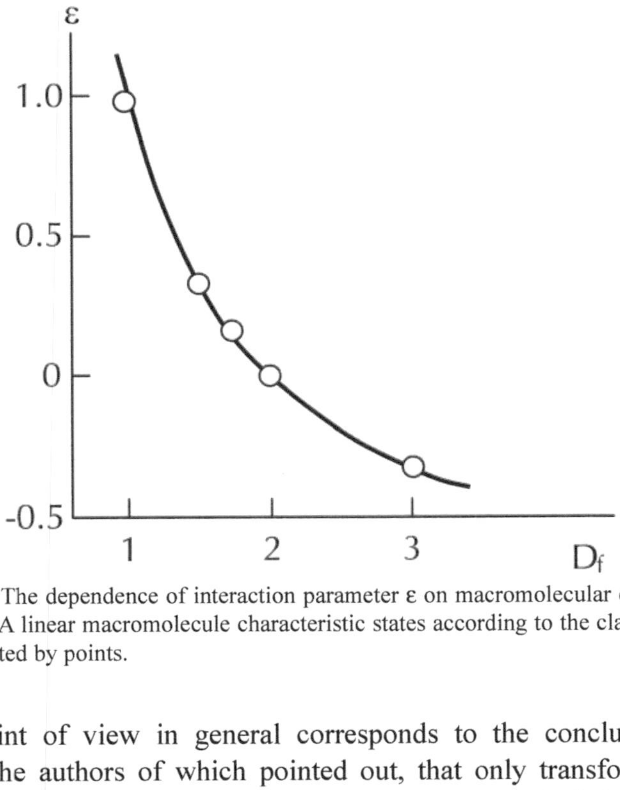

FIGURE 14 The dependence of interaction parameter ε on macromolecular coil fractal dimension D_f. A linear macromolecule characteristic states according to the classification [10] are indicated by points.

This point of view in general corresponds to the conclusions of Ref. [10], the authors of which pointed out, that only transformations coil↔stretched chain (D_f=1.0) and coil↔globule (D_f=3.0) had character of transformation "disorder"–"order" of gas-liquid type and other transitions corresponded to "disorder"–"disorder" transformations and came rather insignificant changes of fractal dimension. They have also noted, that in case of transformations coil in θ-solvent↔coil in good solvent↔leaking coil fractal transitions occur. With the positions of Fig. 14 plot the region D_f=2–1 can be considered as extended transition coil↔stretched chain and the region D_f=2–3—as extended transition coil↔globule. These transitions are characterized by ε different signs, that is, by different types of volume interactions [39].

Thirdly, ε decay rate change at D_f growth at D_f=1.5 can be attributed to qualitative changes of the dependence $\varepsilon(D_f)$ with certain reservations. This bending point is not accidental and can be characterized within the framework of fractal analysis [25], where intersections number of two arbitrary fractals with dimensions D_{f_1} and D_{f_2} can be estimated according to the Eq. (13) of Chapter 1. The indicated equation predicts the possibility of two fractals interpenetration (their transparency and opaqueness. Fractals penetrate freely one into another (are transparent), if N_{int} decreases at R_g growth, that is, if $D_{f_1}+D_{f_2}\le d$. Assuming $D_{f_1}=D_{f_2}$ and d=3, let us obtain the transparence (leakiness) condition for macromolecular coils: $D_f\le 1.5$. Thus, the repulsive interactions intensification of either macromolecule elements or macromolecule and solvent results in the end in fractals (macromolecular coils) transparence one for another.

The Eqs. (38) and (39) combination allows to obtain the following expressions [39]:

$$\left\langle \overline{h}^2 \right\rangle \sim MM^{2/D_f} \tag{40}$$

or

$$\left\langle \overline{h} \right\rangle \sim MM^{1/D_f} \tag{41}$$

These relationships in combination with the Eq. (8) follow from the macromolecular coil self-similarity property, which is obligatory one for any fractal [40], namely, two sizes of the same coil ($\left\langle \overline{h} \right\rangle$ and R_g) are scaled with *MM* change equally, that is, they have the same scaling exponent $1/D_f$.

Let us consider now the examples, demonstrating, how the fractal dimension D_f can characterize volume effects for biopolymers (and polymers at all) [39, 41]. With this purpose the available in literature values a_η for two polysaccharides (polymaltotrioze (pullulan), for which $a_\eta\approx 0.66$ [18], and rodexman, for which $a_\eta\approx 0.75$ [17]) and also dimethyldiallylammonium chloride (DMDAAC), for which $a_\eta\approx 0.82$ in water solutions [42] were used. According to the Eq. (4) estimation the values D_f are equal to

1.81, 1.71 and 1.65 for pullulan, rodexman and DMDAAC, respectively. This means, that for them the macromolecular coil repulsive interactions intensification in the rank of pullulan→rodexman→DMDAAC is observed, since solvent for the indicated polymers is the same one. The data of Ref. [43] allow one to determine the value a_η for pullulan solution in dimethyl sulfoxide (DMSO), which is equal to approx. 0.73, that corresponds to $D_f \approx 1.73$. Since the same biopolymer (pullulan) is considered, then D_f reduction from 1.81 up to 1.73 in water and DMSO solutions, respectively, should be attributed at the expense of repulsive interactions intensification between pullulan macromolecule and DMSO molecules in comparison with water molecules.

As it is known through Ref. [44], for the fractal objects characterization as distinct from Euclidean (compact) ones the usage of three dimension, as minimum, is required—d, D_f and spectral (fraction) dimension d_s, which characterizes an object connectivity degree. The value d_s can be estimated with the aid of the Eq. (12) of Chapter 1, according to which the values d_s are equal to 1.13, 1.04 and approx. 1.0 for pullulan, rodexman and DMDAAC in water solutions, respectively. This means that the macromolecular coil connectivity degree is reduced in the rank of pullulan→rodexman→DMDAAC, in addition for the latter the value d_s is equal to the corresponding value for linear polymer isolated chain, that is, to the minimum possible value $d_s=1.0$ [45]. Such d_s variation supposes a macromolecule contacts number decrease at macromolecular coil sizes increase (see Eqs. (8) and (41)), that is, at D_f reduction and repulsive interactions level enhancement.

Hence, the results stated above have shown that the macromolecular coil fractal dimension D_f can serve as volume effects measure. The value D_f for coil in θ-solvent characterizes such effects absence ($D_f=2.0$, $\varepsilon=0$). $D_f<2.0$ means repulsive interactions availability, $D_f>2.0$—attractive interactions between randomly drawing closer to one another chain links and also between chain links and solvent molecules. The repulsive interactions weakening and, respectively, attractive interactions intensification means macromolecular coil connectivity degree enhancement, characterized by the spectral dimension d_s. Thus, the dimensions D_f and d_s variation (at fixed d) characterizes completely enough biopolymers (and polymers at all) macromolecular coil behavior in diluted solutions [39].

The authors [46] performed the hydro dynamical and molecular characteristics of aromatic copolyethersulfoneformals (PESF) [47] in diluted solutions (the solvents—tetrachloroethane and 1,4-dioxane) determination and their intercommunication with fractal characteristics. Kuhn segment size A for PESF can be determined with the aid of Flory-Fox equation [48]:

$$[\eta]=\Phi(h,\varepsilon)\left[\left(NA^2\right)^{3/2}/MM\right]\alpha^3, \tag{42}$$

where $\Phi(h, \varepsilon)$ is Flory constant, N is Kuhn segments number per one macromolecule.

Let us consider the estimation methods of parameters, including in the Eq. (42). The value $\Phi(h, \varepsilon)$ for arbitrary conditions is determined according to Ptitsyn-Eizner approximation [38]:

$$\Phi(h,\varepsilon)\approx\Phi(\varepsilon)=\Phi_0\left(1-2.63\varepsilon+2.86\varepsilon^2\right), \tag{43}$$

where ε is the volume interactions parameter, which is determined according to the Eq. (39). In its turn, the value D_f was determined according to the formula (11), where the value of PESF intrinsic viscosity in tetrachloroethane was accepted as $[\eta]$ and the value $[\eta]$ in 1,4-dioxane – as $[\eta]_\theta$. The swelling coefficient value was calculated according to the Eq. (6) at the similar conditions. The values N are linked with α and ε as follows [48]:

$$\alpha \approx N^{\varepsilon/2} \tag{44}$$

And at last, the molecular weight for PESF is estimated according to Mark-Kuhn-Houwink equation (the Eq. (1)), in which the coefficient K_η is determined according to the Eq. (2).

The dependence of calculated by the indicated method A values on sulfone fragments contents in PESF is shown in Fig. 15. As one can see, A absolute values, characterizing PESF chain rigidity, are within the expected range of 11.3–16.0 Å [49]. The fact, that the values A for copolymers PESF are located higher than the additive dependence, draws attention. The authors [46] supposed that was due to comparatively rigid and very polar diphenylsulfone fragments contents increase in polymer chain that contributed to both chain rigidity growth and intermolecular interaction intensification.

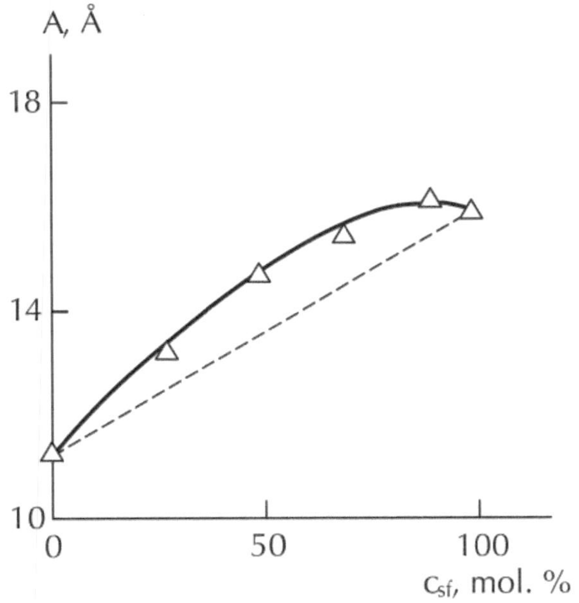

FIGURE 15 The dependence of Kuhn segment length A on sulfone fragments contents c_{sf} for PESF.

The chain rigidity is one of the most important properties of macromolecule, to a large extent defining a solid-phase polymer structure and properties [50]. Therefore, the influence of this parameter characterized by the value A, on macromolecular coil structure characterized by the dimension D_f, should be expected. Actually, the adduced in Fig. 16 dependence $D_f(A)$ confirms this supposition. Thus, the Fig. 16 data pointed out that polymer chain rigidity enhancement results in macromolecular coil compactness degree reduction. In its turn, this circumstance should be reflected on solid-phase block polymer properties, since it is well-known through Ref. [50], which chain rigidity increase results in polymer glass transition temperature T_g enhancement. In Fig. 17 the dependence $T_g(A)$ for PESF is adduced, from which the indicated above rule confirmation follows. Thus, the data of Figs. 15–17 allow to outline one from variants of block polymer glass transition temperature regulation. T_c enhancement can be achieved by the polymer chain rigidity and/or the dimension D_f reduction, for example, at the expense of solvent with larger thermo-dynamical affinity to polymer choice.

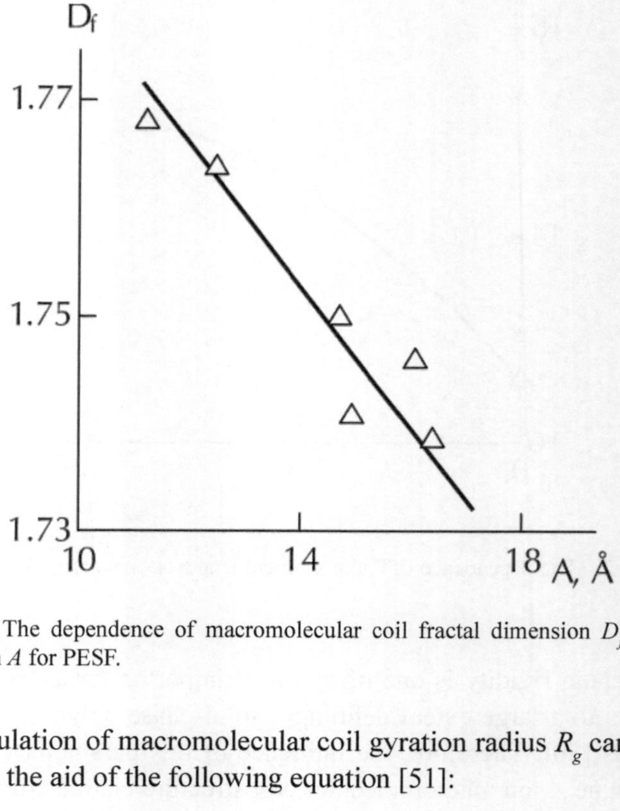

FIGURE 16 The dependence of macromolecular coil fractal dimension D_f on Kuhn segment length A for PESF.

The calculation of macromolecular coil gyration radius R_g can be performed with the aid of the following equation [51]:

$$[\eta] = \Phi \sigma^{1/2} \frac{R_g^3}{MM} \quad (45)$$

where the parameters Φ and MM were determined as earlier.

Within the framework of fractal analysis the intercommunication between R_g and polymerization degree N gives the Eq. (8). In Fig. 18 the dependence $R_g(N^{1/D_f})$ is adduced, from which the gyration radius R_g

classical and fractal definitions correspondence follows. The correlation $R_g(N^{1/D_f})$ linearity and the fact, that it passes through coordinates origin, allows one to obtain the following fractal relationship for R_g calculation in PESF case [46]:

$$R_g^{fr} = 37.5N^{1/D_f}\text{, Å.} \qquad (46)$$

$$R_g^{fr} \quad 37.5N^{1/D_f}\text{, Å.}$$

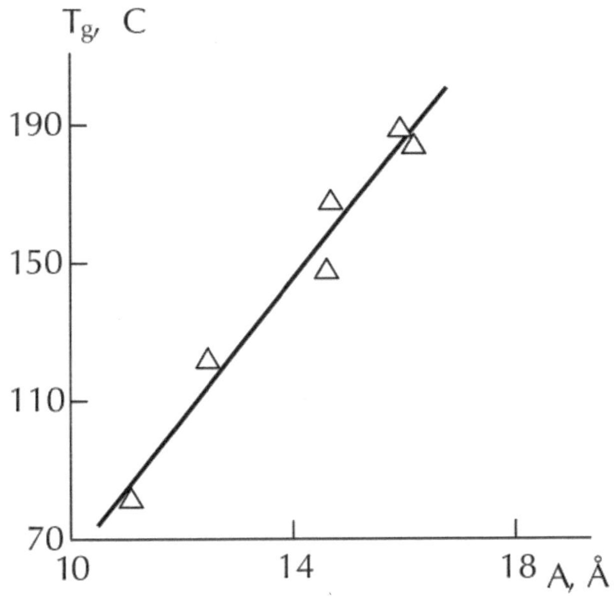

FIGURE 17 The dependence of glass transition temperature T_g on Kuhn segment length A for PESF.

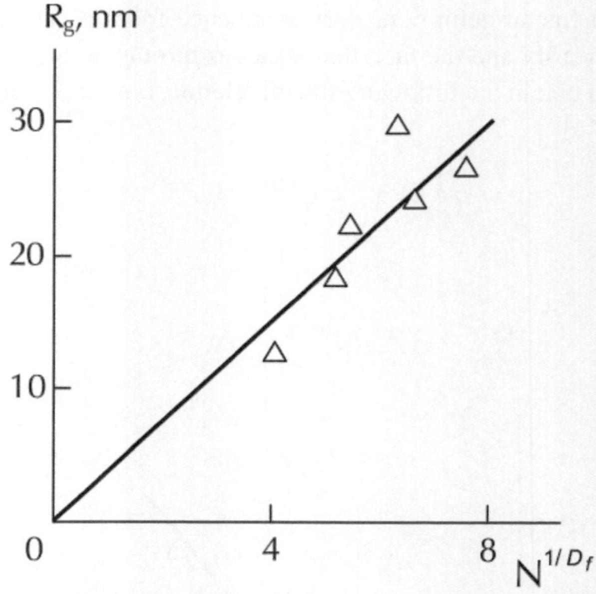

FIGURE 18 The relation between gyration radius R_g and fractal parameter N^{1/D_f} for PESF.

In Fig. 19 the comparison of gyration radius R_g and R_g^{fr} values, obtained with the aid the classical and fractal approaches (the Eqs. (45) and (46)), respectively, is adduced. As one can see, both approaches given a good correspondence of gyration radius values for PESF macromolecular coil. The range R_g of 150–280 Å is obtained, in addition PESF molecular weight reduction in them at sulfone fragments contents in them decrease is the main cause of the indicated variation.

Hence, the values of Kuhn segment length A and macromolecular coil gyration radius R_g within the framework of both classical and fractal approaches are estimated. All calculations performed on the basis of very simple experimental measurements of PESF two intrinsic viscosities: in good solvent and θ-solvent. These parameters variation ranges make up: A=11.3–16.0 Å, R_g=150–280 Å. The value A, characterizing PESF chain rigidity, is located above the additive dependence, that results in corresponding reduction of dimension D_f and enhancement of glass transition temperature T_g of solid-phase copolymer. The calculation R_g within the framework of classical and fractal approaches gave the magnitudes, agreed well by absolute value.

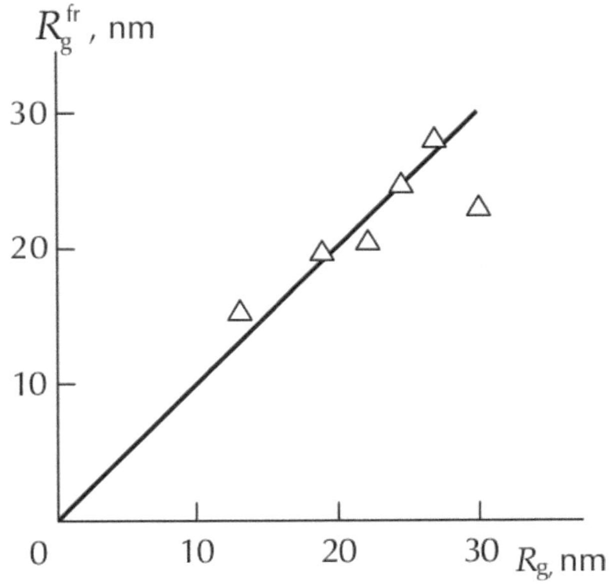

FIGURE 19 The relation between macromolecular coil gyration radius values R_g and R_g^{fr}, calculated according to the Eqs. (45) and (46), respectively.

The authors [52] studied the dependence of macromolecular coil fractal dimension D_f on Huggins constant k_x, characterizing interactions polymer-solvent [1]. These studies were performed on the example PAr of different chemical structure [5], PVC and PMMA [53], polysulfone (PSF) [54] and PC [29] solutions. All indicated polymers belong to flexible—and semirigidchain polymers classes and for them the values of parameters K_η and a_η of Mark-Kuhn-Houwink equation were accepted according to the literary sources. The dependence of reduced viscosity η_{red} on concentration c for diluted polymer solutions is described usually by Huggins equation [1]:

$$\eta_{red} = [\eta] + k_x [\eta]^2 c + ... \tag{47}$$

Besides, the relation between specific viscosity η_{sp} and c can be obtained within the framework of Shultz-Blashke equation [55]:

$$[\eta]=\frac{\eta_{sp}/c}{1+K\eta_{sp}} \tag{48}$$

where K is constant, accepted in the first approximation equal to 0.28.

The Eqs. (47) and (48) at the condition c=constant (the value c is accepted further equal to 0.5 mass %) allows to obtain a simple expression for k_x estimation [52]:

$$k_x=\frac{0.14}{1-0.14[\eta]} \tag{49}$$

At k_x calculation it has been assumed, that all the indicated above polymers have the value $MM=5\times10^5$. Such MM value was chosen for the reason that smaller MM values gave close k_x values that increases estimations error. The values [η], corresponding to the indicated MM magnitude, were calculated according to Mark-Kuhn-Houwink equation.

In Fig. 20 the dependence of dimension D_f, calculated according to the Eq. (4), on parameter k_x^{-2} (such dependence form was chosen with the purpose of linear correlation obtaining) is adduced. As one can see, the good linear correlation for 30 different experimental points is obtained (the correlation coefficient is equal to 0.930), allowing one to predict simply enough the value D_f. It has been expected, that for other MM values the correlation $D_f(k_x^{-2})$ will have a similar look, but a different slope. The dependence $D_f(k_x^{-2})$, adduced in Fig. 20, allows to make several important conclusions [52, 56].

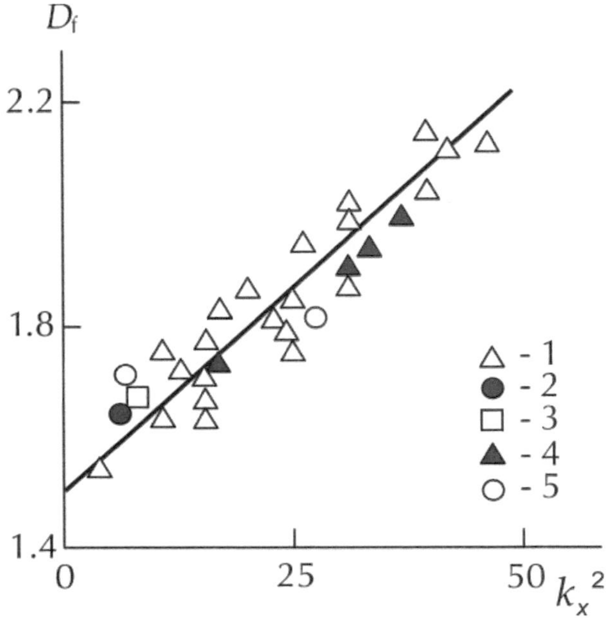

FIGURE 20 The relation between macromolecular coil fractal dimension D_f and parameter k_x^{-2} for PAr (1), PMMA (2), PVC (3), PSF (4) and PC (5).

At first, there the impression is made that this correlation gives the dependence of D_f on interactions polymer-solvent, characterized by Huggins parameter k_x, and does not take into consideration interactions of coil elements between themselves. However, now this correlation linearity itself supposes, that it takes into consideration the second group factors, indicated above. For example, it is well-known through Ref. [5], that chain rigidity enhancement results in the exponent a_η in Mark-Kuhn-Houwink equation increase and, hence, to D_f reduction (the Eq. (4)). Simultaneously chain rigidity enhancement results in [η] growth at other equal conditions. Thus, both chain rigidity increase and thermo-dynamical quality of solvent in respect of polymer give the same effect—[η] enhancement and corresponding k_x growth according to the Eq. (49). In its turn, this results in k_x^{-2} reduction and D_f decrease, which was to be expected. This supposition is confirmed experimentally as well in Ref. [57] Kuhn segment length A, characterizing polymer chain thermo-dynamical rigidity, increase at solvent thermo-dynamical quality improvement for two polyarylates is shown.

Secondly, as it follows from the Eq. (49), the minimum value k_x=0.14 (or maximum value $k_x^{-2}\approx$51) is achieved at [η]=0. From the plot of Fig. 20 it follows, that D_f=2.25 corresponds to this value k_x. As it is known through Ref. [13], the excluded volume interactions screening results in D_f increase and at complete screening (the indicated effects compensation) the value D_f^c correspond to the so-called compensated state of macromolecular coil. Within the framework of Flory theory, when compensation is realized at the expense of interactions with other coils, D_f^c=2.5 (for three-dimensional Euclidean space). The other mode of repulsive interactions weakening is attractive interactions introduction. For this case, corresponding to the isolated coil (to diluted solution) the value D_f^c is given by the Eq. (4) of Chapter 1 and then for case d=3 the value $D_f^c\approx$2.286, that corresponds practically exactly to the limiting D_f value, obtained from the plot of Fig. 20.

The other limiting case, corresponding only to repulsive interactions, can be obtained from the Eq. (49) at [η]=7.14 dl/g, that corresponds to the condition k_x^{-2}=0. In this case the plot of Fig. 20 extrapolation to k_x^{-2}=0 gives D_f=1.5, that corresponds to the leaking macromolecular coil dimension [10]. Thus, the limiting values D, obtained from the plot of Fig. 20, correspond completely to theoretical conclusions.

Thirdly, at present the exponent a_η in Mark-Kuhn-Houwink equation reduction at polymer molecular weight MM growth is well-known through Ref. [17, 48]. The Eqs. (1), (4) and (49) combination allows to obtain the following relationship [52]:

$$D_f = \frac{3\ln MM}{\ln MM + \ln\left(7.14k_x - 1\right) - \ln K_\eta - \ln k_x} \tag{50}$$

Thus, the Eq. (50) gives the analytical intercommunication of D_f and MM. The authors [52] estimated the value D_f for polyamide acid in case of values MM=4×10^4 and 15×10^4 g/mole, corresponding to two ranges of a_η values according to the data of Ref. [48] and obtained the exact correspondence of the values D_f (or a_η), obtained experimentally and calculated according to the Eq. (50). Hence, a_η reduction (D_f increase) at MM growth has law-governed character.

Thus, the results stated above have shown the possibility of the linear dependence of D_f on parameter k_x^{-2}, characterizing polymer-solvent interactions level, plotting. The indicated correlation limits correspond to the theoretical limiting values D_f. These circumstances allow using the correlation $D_f(k_x^{-2})$ for prediction structure of macromolecular coil in diluted solutions [56].

Let us consider further in more detail the dependence of macromolecular coil structure on polymer chain rigidity, which is the most important characteristic of polymer in virtue of its structure specific features. In Ref. [58] the estimation of important molecular parameters including macromolecular coil gyration radius R_g and Kuhn segment length, characterizing polymer chain rigidity, for three rigid-chain polymers was performed. Let us note, that similar estimations have, as a rule, very large scatter. So, for the considered in Ref. [58] three rigid-chain polymers the values A, determined by different methods, have approximately quintuple variation for each from the indicated polymers. Therefore, the authors [59] used the fractal analysis concepts for closer definition of Kuhn segment length A estimations, obtained in Ref. [58]. For this purpose the data adduced in Ref. [58] for three rigid-chain aromatic polyamides-poly n-benzamide (PBA), poly n-phenyleneterefthalamide (PPTA) and aromatic polyamide with benzimideazol fragment in chain (PA) were used. The polymerization degree was determined as follows [59]:

$$N = \frac{L}{A}, \tag{51}$$

where L is polymer chain length, accepted according to the data of Ref. [58].

The choice of Kuhn segment length A for using in the Eq. (51) is defined by the fact, that this parameter characterizes the fragment of the polymer chain, whose mobility is independent practically on other segments [60]. In Kuhn segment polymer chain several monomer links are included, which do not maintain their individuality, since their mobility is possible only within the framework of Kuhn segment displacements. Kuhn segment length A is linked with a polymer chain rigidity characteristic—the persistent length a by a simple relationship [60]:

$$A = 2a, \tag{52}$$

which is fulfilled approximately for different models of polymer chain.

In Fig. 21, the correlation $N(R_g^{D_f})$, corresponding to the Eq. (8), for the three indicated above polyamides is adduced. As it follows from the data of this figure, although supposed by the Eq. (8) proportionality is performed, but for each polymer the individual correlation is obtained and the straight lines slopes, describing this correlation, differ essentially. This observation supposes the corrections necessity for the values A, determined in Ref. [28], which were obtained by an approximate enough method. Let us note, that using in the Eq. (51) of mean values A, obtained in different chapters and cited in Table 2 of Ref. [58], results in essential approaching of the straight lines, shown in Fig. 21. Therefore, the authors [59] corrected the values A in order to obtain the sole linear correlation $R_g^{D_f}(L/A)$, which is shown in Fig. 22. The corrected with accounting for this correlation A values are adduced in Table 10, where the corresponding magnitudes, obtained in Ref. [58] and estimated by other methods (Table 2 of Ref. [58]) are cited, too. The comparison of A values, obtained in different chapters, has shown, that the corrected according to the plot of Fig. 22 A values much better correspond to the range of the obtained earlier values A, than the estimations performed in Ref. [58]. So, for PBA the corrected value A is located approximately in the middle of the values A range, adduced in column 4, whereas the estimation A of Ref. [58] does not get into the indicated range. For PPTA the estimations of chapters [58] and [59] coincide approximately and correspond to lower boundary of the literary data A range. For PA the corrected A value also corresponds better to the literary data than to the estimation of Ref. [58].

There is one more indirect method of values A correctness checking within the framework of fractal analysis. As it is known through Ref. [5], enhancement of the polymer chain rigidity, characterized by A increase, is accompanied by the exponent a_η growth in Mark-Kuhn-Houwink equation. As it follows from the Eq. (4), D_f reduction at a_η increase should be expected and, hence, A growth.

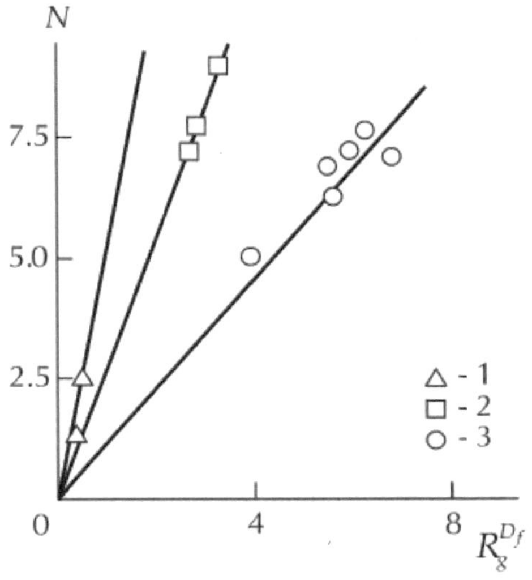

FIGURE 21 The relation between Kuhn segments number per macromolecule N and parameter $R_g^{D_f}$, corresponding to the Eq. (8), for PBA (1), PPTA (2) and PA (3). The values A are accepted according to the Ref. [58].

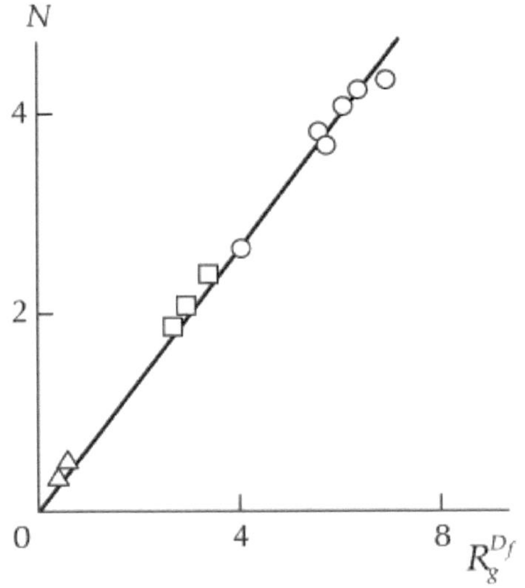

FIGURE 22 The relation between Kuhn segments number per macromolecule N on parameter $R_g^{D_f}$, corresponding to the Eq. (8), for PBA (1), PPTA (2) and PA (3). The corrected values A were used.

TABLE 10 The comparison of Kuhn segment length values, obtained by different methods.

Polymer	*A*, Å		
	Ref. [59], after first correction	Ref. [58]	The range of literary values
PBA	930	360–480	500–2100
PPTA	290	300–360	300–1300
PA	500	260–270	500–1200

In Fig. 23 the dependences of dimension D_f on the corrected and estimated in Ref. [58] values *A* are adduced. As it follows from the data of Fig. 23, in the first case the definite perfectly D_f reduction at *A* growth is observed, whereas the dependence of D_f on the obtained in Ref. [58] values *A* is practically absent. This comparison assumes, that the corrected in Ref. [59] *A* values reflect aromatic polyamides chain rigidity more exactly, than the estimated ones in Ref. [59].

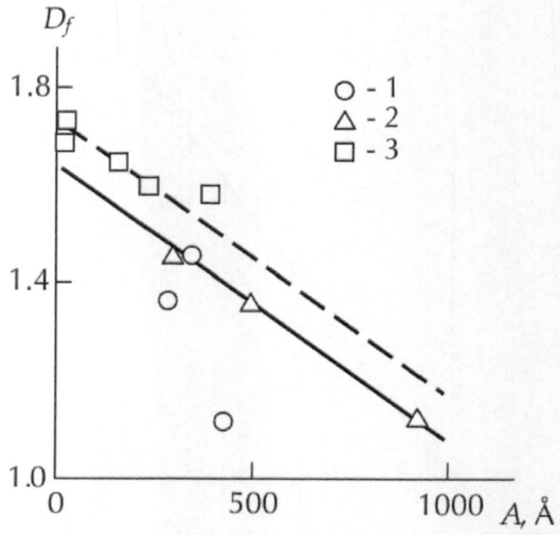

FIGURE 23 The dependences of macromolecular coil fractal dimension D_f on Kuhn segment length *A* for aromatic polyamides (1, 2) and other polymers according to the data of chapters [5, 57, 61] (3). The values *A* for aromatic polyamides were accepted according to the data of chapters [58] (1) and [59] (2).

It should be noted, that the considered above correction can be obtained and without fractal analysis application as well, using well-known Flory concept, that follows from the Eq. (1) of Chapter 1. However, obtaining of the analytical correlation between Flory exponent and structural characteristics of polymers in solid phase is very difficult, if possible at all. At the same time this can be made within the framework of fractal analysis, since both macromolecular coil [25] and solid-phase polymer structure [62] are fractal objects. Hence, the possibility of solid-phase polymers properties quantitative prediction appears in such sequence: molecular characteristics (for example, A)→structure of macromolecular coil in solution→polymer condensed state structure→polymer properties. In Section 2.6, this problem will be considered in detail. These considerations predetermined the choice of fractal analysis in Ref. [59] as a mathematical calculus.

Let us also note, that the choice of the mean straight line in Fig. 21 as the initial data for the values A correction gives guarantee of these values ratio correctness for the studied aromatic polyamides, but does not guarantee the absolute values A correctness. Therefore, one more empirical correction of the Ref. [58] data is possible. In Fig. 23 the five data points for different polymers were traced, D_f values of which were calculated according to the Eq. (4) and the values a_η and A were accepted according to the literary sources [5, 57, 61]. As one can see, the straight line is obtained, which is parallel to the received one for aromatic polyamides, but displaced in larger values A side on about 300 Å. It is interesting to compare the values A after two such corrections with mean values of the values A range, cited in literature (the fourth column of Table 10). Such comparison performed in Table 11, from which a good correspondence of values, estimated by both indicated above methods. This correspondence allows to assume the adduced in Table 11 values A as the most probable.

Thus, the corrections performed above allowed making more precisely the carried out earlier estimations of Kuhn segment length for aromatic polyamides. The comparison of the corrected and estimated in Ref. [58] A values has shown, that the proposed in the last chapter the estimation A mode results in understated about in 2–3 times magnitudes. The correspondence of mean values of A range, cited in literature, and corrected by the indicated mode allows to assume the latter as the most probable values of Kuhn segment length for aromatic polyamides [59].

TABLE 11 The comparison of Kuhn segment length values with mean magnitude of a literary values range.

Polymer	*A*, Å	
	The Ref. [59] after the second correction	**The mean value of literary values range**
PBA	1230	1300
PPTA	590	800
PA	800	850

The model of freely connected chain, introduced by Kuhn and Mark, has a great significance for polymers conformational properties description [1]. In this model a polymer chain is replaced by an equivalent one, consisting of rod-like segments, spatial orientation of which is mutually independent. The freely connected chain statistics application possibility to real macromolecules is based on the fact, that any real chain at sufficient length acquires freely connected chain properties.

For Gaussian chain in ideal θ-solvent the condition is performed [1]:

$$\bar{h}^2 = NA \tag{53}$$

where $\bar{h}^2$ is mean-square end-to-end distance of the chain, N is Kuhn segments of length A number per chain.

For nonideal solvents the condition (53) is violated and macromolecular coil in solution conformation differs from Gaussian one. This entails A variation as a function of solvent thermo-dynamical quality in respect to polymer.

As it has been noted above, the value A is often used as a polymer chain rigidity measure. A macromolecular coil in diluted solution is the fractal and its structure can be characterized by the fractal dimension D_f. The larger D_f, is the more compact coil structure is. It is obvious, that the more rigid a polymer chain is, the more difficult it is to make compact of a corresponding macromolecular coil and the smaller the value D_f should be. Actually, such correlations between A and D_f were obtained in Refs. [46, 59, 63] (see Figs. 16 and 23). The indicated correlations the authors [64] used for the parameter A estimation in case of both flexible and rigid-chain polymers.

The necessary for calculations parameters (a_η and A) for flexible-chain polymers accepted according to the data of Ref. [65], for rigid-chain ones—according to the data of Ref. [66]. As it has been noted above, in Ref. [63] the empirical relationships between D_f and A were obtained. For flexible-chain polymers the indicated relationship has a look like [63]:

$$D_f = 2.0 - 1.32\times10^{-2} A \tag{54}$$

and for rigid-chain ones [63]:

$$D_f = 1.667 - 4.45\times10^{-4} A \tag{55}$$

where the value A is given in Angströms.

The Eqs. (54) and (55) have the expected from the most general considerations form: A increase, that is, polymer chain rigidity enhancement, results in D_f reduction, that is, in the decrease of compactness degree of a macromolecular coil in solution. Let us pay attention to the principal distinction of the Eq. (53) on the one hand and Eqs. (54), and (55) on the other hand. The Eq. (53) does not allow the condition of infinitely flexible chain, that is, A=0, whereas the Eqs. (54) and (55) give such condition at D_f=2.0 (i.e., in ideal θ-solvent) and D_f=1.667, respectively. This supposes, that the Eqs. (54) and (55) give the dependence D_f on two interactions groups: polymer-solvent and coil elements between themselves, that is, on those interactions groups, which define D_f value [13]. Without excepting the more exact dependences $D_f(A)$ will extrapolate for flexible-chain polymers at D_f=2.0 to A value in θ-solvent, for which macromolecular coil has Gaussian conformation. Nevertheless, the Eqs. (54) and (55) allow to make the value A estimation even as the obtained in Ref. [63] approximations.

In Figs. 24 and 25 the comparison of the dependences $A(D_f)$, calculated according to the Eqs. (54) and (55) for flexible- and rigid-chain polymers, respectively, and adduced in Refs. [65, 66], is shown. As one can see, in both indicated cases the expected tendency of the function $A(D_f)$ is obtained and for larger part of polymer-solvent pairs a good quantitative correspondence is received. The significant scatter of A values for two used estimation methods in case of 5 from 19 adduced in Figs. 24 and 25 polymer-solvent pairs is due to corresponding scatter of literary A magnitudes. So, even for well-studied flexible-chain polymers, for example, polyethylene, very differing values 20 Å [49] and 12 Å [65] are cited in literature. For rigid-chain polymers this distinction can reach fourfold one (see Tables 10 and 11).

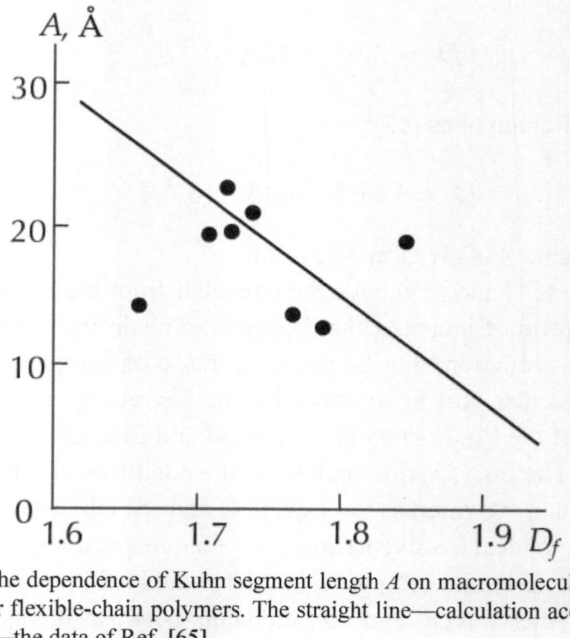

FIGURE 24 The dependence of Kuhn segment length A on macromolecular coil fractal dimension D_f for flexible-chain polymers. The straight line—calculation according to the Eq. (54), points—the data of Ref. [65].

Let us note one more important aspect. The Eqs. (54) and (55) comparison allows to give one more definition of flexible- and rigid-chain polymers. The indicated equations suppose, that the greatest values of macromolecular coil fractal dimension are equal to 2.0 (i.e., a coil in ideal θ-solvent [10]) for flexible-chain polymers and 1.667 (i.e., a coil in good solvent [10])—for rigid-chain ones. Hence, such polymers, chain rigidity of which (intramolecular interactions) allows to compensate it by polymer-solvent interactions (intermolecular ones) only to a certain degree, that is, up to D_f=1.667. Another criterion follows from the equation (54): at the minimum value D_f=1.0 let us obtain the value $A\approx75$ Å. In other words, polymers with $A\leq75$ Å are flexible-chain ones (for them D_f=2.0 can be reached by intramolecular interactions screening) and with $A>75$ Å—rigid-chain ones [64].

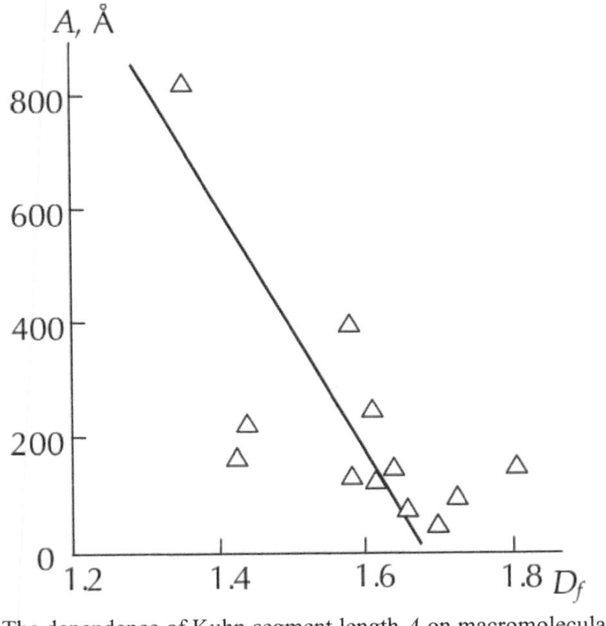

FIGURE 25 The dependence of Kuhn segment length A on macromolecular coil fractal dimension D_f for rigid-chain polymers. The straight line—calculation according to the Eq. (55), points—the data of Ref. [66].

Thus, the results obtained above confirmed the intercommunication of polymer chain rigidity and macromolecular coil in solution structure. The well-known classification of polymers (flexible- and rigid-chain ones) is confirmed also and new criteria of such division are received.

The authors [67] considered the possibility of application for dimension D_f prediction of Flory-Huggins semiempirical interaction parameter on the example of aromatic copolyethersulfoneformals solutions in chloroform and tetrachloroethane. The solubility parameter of PESF δ_p was calculated by a group contributions method [16] and corresponding values for solvents δ_s were accepted according to the data of Ref. [14]. D_f value for 8 polymer-solvent pairs was determined according to the Eq. (4) and the values a_η and Flory-Huggins interaction parameter χ_1 for the same polymer-solvent pairs are adduced in chapters [29] and [1], respectively.

As it has been noted above, the fractal dimension D_f of macromolecular coil in solution is determined by two interactions groups: interactions polymer-solvent and interactions of coil elements between themselves. Such definition allows to link between themselves the dimension D_f and Flory-Huggins interaction parameter χ_1, which was determined as follows [1]:

$$\chi_1 = \frac{E_{vap}}{RT}\left(1-\frac{\delta_p}{\delta_s}\right)^2 + \chi_s \tag{56}$$

where E_{vap} is solvent vaporization molar energy, R is universal gas constant, T is temperature, and χ_s is empirical parameter.

Sometimes the value χ_1 is determined as follows [1]:

$$\chi_1 = \chi_n + \chi_s, \tag{57}$$

where χ_n and χ_s are enthalpic and entropic components of Flory-Huggins parameter.

The Eq. (56) allows to suppose, that interactions of coil elements between themselves can be determined with the aid of parameter δ_p and interactions polymer-solvent—with the aid of the ratio δ_p/δ_s. Besides, the entropic component χ_s takes into account the effect of the system interacting elements ordering.

In Fig. 26, the relationship $D_f(\chi_1)$ is presented for 8 polymer-solvent pairs. As one can see, it is approximated well enough by a straight line and has expected limits. For χ_1=0 (polymer-solvent interactions absence, δ_s=δ_p) D_f=1.50, that is, the leaking coil is formed in solution [10]. It is obvious, that this condition is reached only at χ_s=0. For large enough $\chi_1 \geq 1$, that is, for transition solvent-nonsolvent [1], the value D_f=2.0, that is, it corresponds to θ-solvents [10]. The relationship $D_f(\chi_1)$ can be expressed analytically with the aid of the empirical equation [67]:

$$D_f = 1.50 + 0.45\chi_1 \tag{58}$$

Hence, the made above supposition about D_f and χ_1 intercommunication corresponds to the experimental data. Proceeding from this, let us consider the similar relationship for PESF.

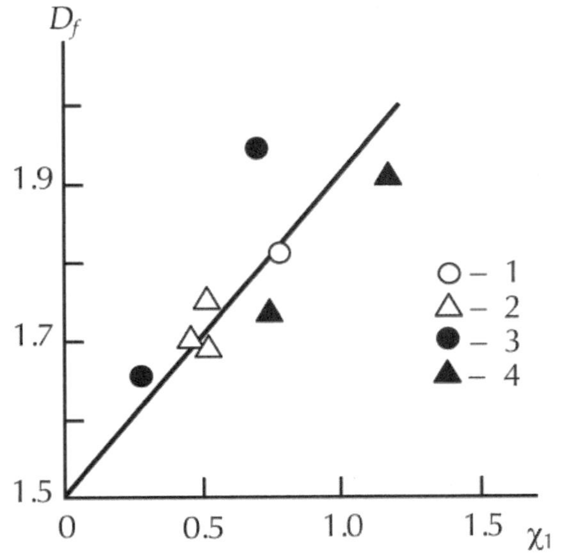

FIGURE 26 The dependence of macromolecular coil fractal dimension D_f on Flory-Huggins interaction parameter χ_1 for solutions of polystyrene (1), poly(methyl methacrylate) (2), polysulfone (3) and poly(vinyl acetate) (4).

In Fig. 27, the dependence of δ_p, calculated by a group contributions methods, on formal blocks contents c_{form} in PESF is adduced, which has nonadditive character. Such dependence reason consists of neighboring links intramolecular interaction role intensification owing to copolymer links difference.

The enthalpic component χ_n of the interaction parameter χ_1 calculation according to the Eqs. (56) and (57) for PESF in the indicated solvents (Fig. 28) shows opposite tendencies of χ_n change for PESF solutions in chloroform and tetrachloroethane. Besides, the dependence $\chi_n(c_{form})$ for solutions in chloroform has additive character and for solutions in tetrachloroethane—nonadditive one (Fig. 28). And at last, the values χ_n and χ_1 calculated according to the Eq. (58) comparison has shown that the condition $\chi_n < \chi_1$ is performed. This means the necessity of the entropic component χ_s accounting in the Eqs. (56) and (57).

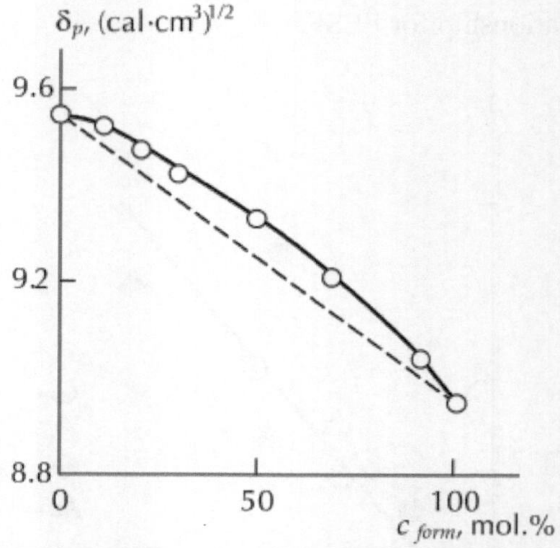

FIGURE 27 The dependence of solubility parameter δ_p of PESF on formal blocks contents in it c_{form}.

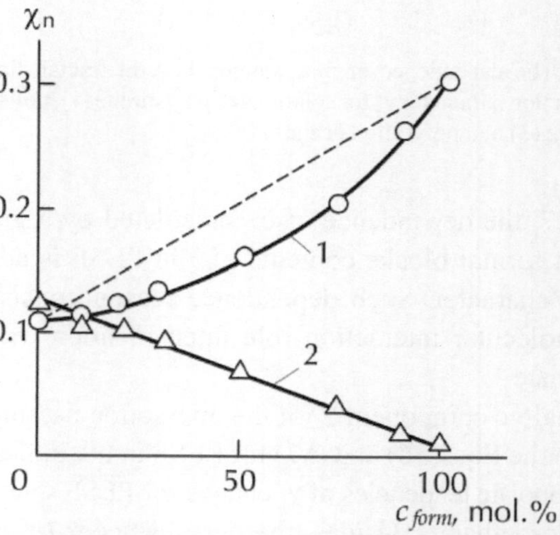

FIGURE 28 The dependences of enthalpic component χ_n of Flory-Huggins interaction parameter on formal blocks contents c_{form} for PESF solutions in tetrachloroethane (1) and chloroform (2).

In Fig. 29 the dependence $\chi_s(c_{form})$ is adduced, where χ_s was determined as difference (χ_1-χ_n). Different tendencies of χ_s with copolymers composition for their solutions in chloroform and tetrachloroethane are observed again. Let us note, that these tendencies are opposite to the dependences $\chi_n(c_{form})$ tendencies, adduced in Fig. 28.

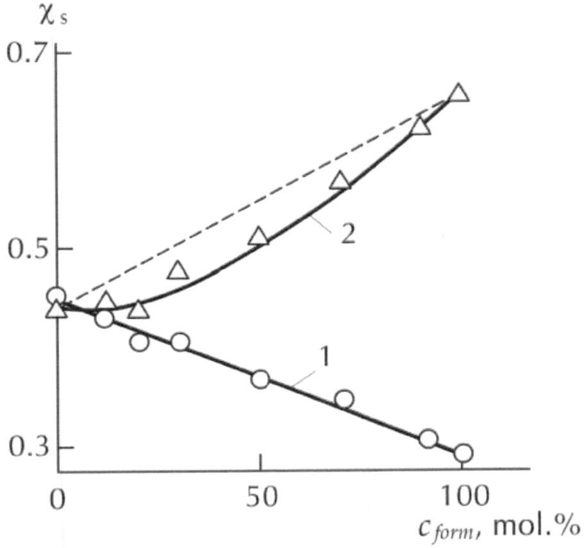

FIGURE 29 The dependence of entropic component of Flory-Huggins interaction parameter on formal blocks contents c_{form} for PESF solutions in tetrachloroethane (1) and chloroform (2).

The dependence $\chi_s(c_{form})$ for PESF solutions in tetrachloroethane is approximated well by a straight line that gives the possibility of χ_s estimation in this case as an additive value. In Fig. 30 the theoretical (calculation according to the Eq. (58) with the indicated above χ_s estimation) and experimental (calculation according to the Eq. (11)) dependences $D_f(c_{form})$ for PESF solutions in tetrachloroethane, which showed excellent correspondence (the discrepancy makes up no more than 0.3%). The dependence $D_f(c_{form})$ calculation for PESF solutions in chloroform at the same conditions, that is, in supposition of the additive dependence $\chi_s(c_{form})$, gives much larger discrepancy with the similar experimental dependence (Fig. 30).

Thus, the results stated above demonstrated that Flory-Huggins interaction parameters described exactly enough interactions system for macromolecular coil in solution, controlling its fractal dimension value. The main problem at the Eq. (58) using with the purpose of D_f prediction is the absence of the empirical parameter χ_s calculation technique [67].

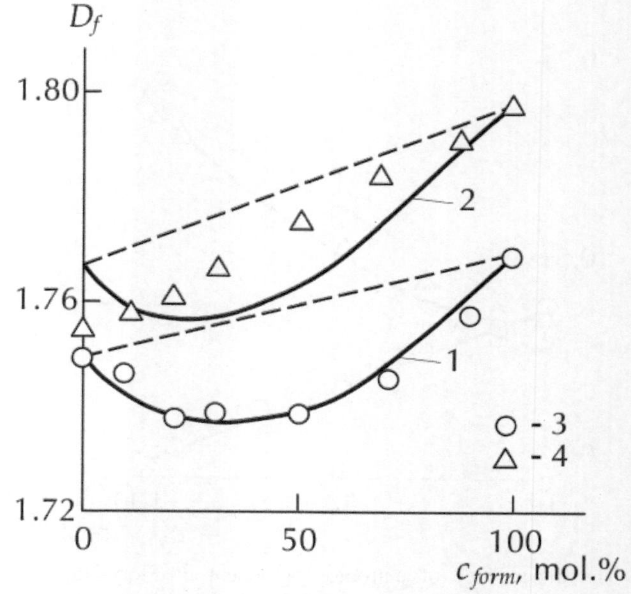

FIGURE 30 The dependences of macromolecular coil fractal dimension D_f on formal blocks contents c_{form} for PESF solutions in tetrachloroethane (1, 3) and chloroform (2, 4). Calculations were performed according to the Eqs. (11) (1, 2) and (58) (3, 4).

As it has been shown above, a macromolecular coil in solution fractal dimension is defined by two groups of factors: polymer-solvent interactions and macromolecular coil elements between themselves. As an approximation, the first from the indicated factors can be characterized by the difference of polymer δ_p and solvent δ_s solubility parameters: $\Delta\delta=|\delta_p-\delta_s|$ [16]. As to the second group of factors, then a parameters number exists, influencing on D_f value in some way: chain rigidity, bulk side groups availability, hydrogen bond and so on. Since at present the strict theory of polymer regular solutions has not still developed, then analytical relationships obtaining between D_f value and the indicated above parameters is impossible. Therefore, the authors [68] considered qualitatively the indicated parameters influence on D_f value and pointed out the last variation tendencies at the change of either from these parameters.

The authors [68] used the literary data for polymers number—PC and PMMA [29], PS [31], PDMS [30], butadiene-nitril rubber (SKN-18) [32], and polyarylates (PAr) di-(4-oxiphenyl)-methane, 2,2-di-(4-oxiphenyl)-propane, 9,9-di-(4′-oxiphenyl)-fluorene (the conventional sign PD), phenolphthaleine and phenolphthaleine anilide (the conventional sign PF) [5]. According to the indicated chapters the exponent a_η values in Mark-Kuhn-Houwink are accepted and according to them D_f values were calculated according to the Eq. (4). The characteristic ratio C_∞ values, which are a polymer chain statistical flexibility indicator [1], are accepted according to the data of chapters [69, 70] and polymers δ_p and solvent δ_s solubility parameters magnitudes according to the data of Ref. [16].

In Fig. 31, the dependences of D_f on the value $\Delta\delta=|\delta_p-\delta_s|$ are adduced, which characterizes as the first approximation the solvent thermo-dynamical affinity in respect to polymer [16]. As it follows from this figure data, certain laws are observed for these parameters relation. At first, $\Delta\delta$ increase or solvent thermo-dynamical affinity in respect to polymer change for the worse in all cases results in D_f increase or macromolecular coil compactness enhancement, that is defined by the Eq. (8) within the framework of fractal analysis. Thus, the larger $\Delta\delta$, is the smaller macromolecular coil gyration radius is at the same polymer molecular weight *MM* (or polymerization degree *N*).

Secondly, as it is known through Ref. [10], θ-conditions correspond to the value D_f=2.0. From the data of Fig. 31 it follows, that the larger polymer chain statistical flexibility (it changes for the adduced in Fig. 31 polymers PAr→PC→SKN-18→PDMS as C_∞=2.3→2.4→5.2→7.6) is the weaker the dependence $D_f(\Delta\delta)$ is or θ-conditions are reached within the larger range of $\Delta\delta$. It can be said accounting for the cited above reservations, that chain rigidity increase narrows the choice of solvents, in which a polymer can be dissolved. Let us also note, that at $\Delta\delta$=0 all adduced in Fig. 31 dependences are extrapolated to D_f=1.50, that is, for all indicated polymers the leaking coil realization is possible under certain circumstances [10].

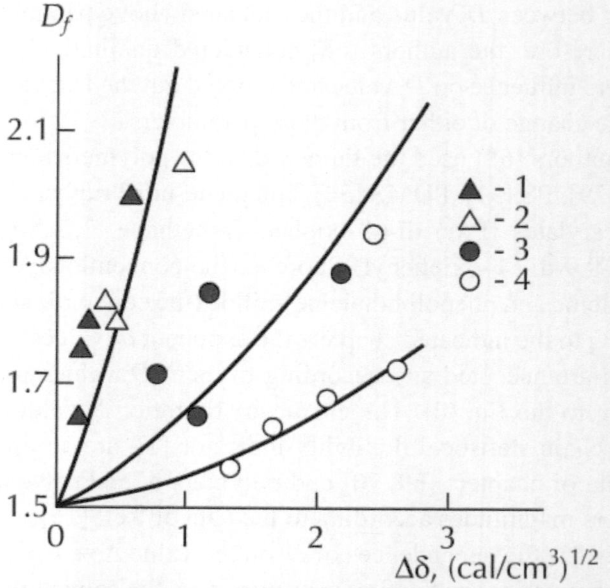

FIGURE 31 The dependences of macromolecular coil fractal dimension D_f on difference of polymer and solvent solubility parameters $\Delta\delta$ for PC (1), PAr (2), SKN-18 (3) and PDMS (4).

However, besides the indicated factors, that is, $\Delta\delta$ and C_∞, the other parameters can influence on D_f value. In Fig. 32, the dependence $D_f(\Delta\delta)$ for polyarylates PD and PF is adduced. As one can see, this dependence breaks down into two curves, in addition one from them includes PAr with rigid para-connections in the main chain (the polyarylates PF-2 and PF-7, Table 5, p. 120 in Ref. [5]) and the other—polyarylates with less rigid metha-connections. The indicated plot demonstrates again the importance of such characteristic as polymer chain rigidity for determination of D_f value of macromolecular coil in solution.

In Fig. 33 the dependence $D_f(\Delta\delta)$ is adduced for two polymers (PMMA and PS), the peculiarity of which is bulk side groups availability (and, as consequence, large cross-sectional area of macromolecule [26]) and high statistical flexibility of the chain ($C_\infty \approx 8$–10 [1, 69, 70]). Although in this very case the general regularity is observed—D_f growth at the change for the worse of solvent thermo-dynamical affinity in respect to polymer, that is, at $\Delta\delta$ increase, but at $\Delta\delta=0$ the values D_f are extrapolated not to

D_f=1.50 (as in Fig. 31), but to D_f≈1.67, that corresponds to the dimension of macromolecular coil in good solvent [10]. In other words, for these polymers (PMMA and PS) the leaking coil state reaching is impossible.

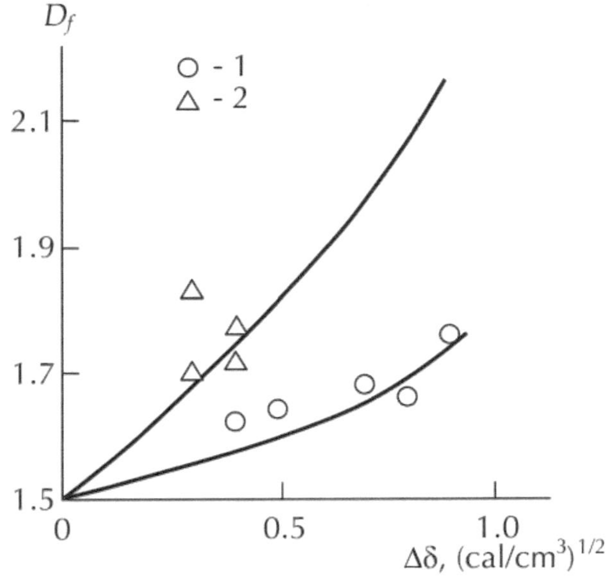

FIGURE 32 The dependences of macromolecular coil fractal dimension D_f on difference of polymer and solvent solubility parameters Δδ for polyarylates with metha-(1) and para-connections in the main chain.

The common shape of the curves D_f(Δδ), shown in Figs. 31–33, supposes their normalization possibility for the general dependence receiving, reflecting the influence all considered above factors on the value D_f. Such normalization requires displacement over both Δδ axis (Figs. 31 and 32) and D_f axis (Fig. 33). The analysis of Fig. 31 supposes that C_∞ can be a normalizing parameter and in this case we will obtain the dependence D_f(Δδ/C_∞) and Fig. 33 supposes the introduction of displacement over D_f axis for polymers, similar to PMMA and PS, by the difference value of fractal dimensions of a coil in good solvent (D_f≈1.67) and in leaking coil state (D_f=1.50), that is, by the value of ΔD_f≈0.17. The example of such normalization for the adduced in Figs. 31–33 polymers is shown in Fig. 34.

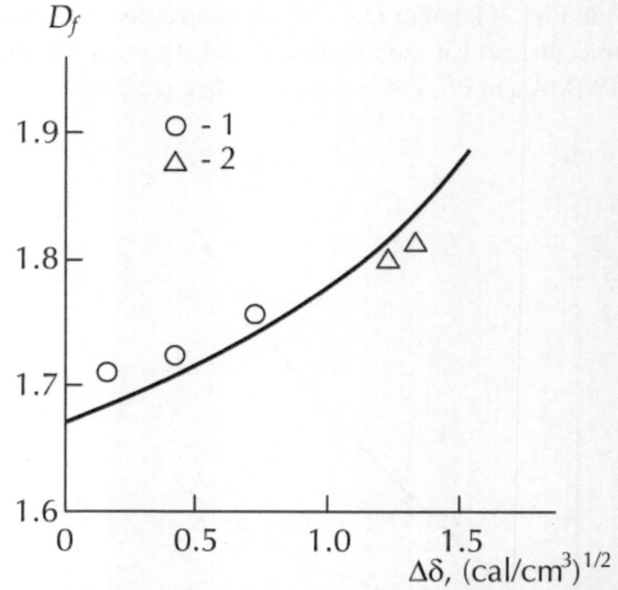

FIGURE 33 The dependences of macromolecular coil fractal dimension D_f on difference of polymer and solvent solubility parameters Δδ for PMMA (1) and PS (2).

One can see, that almost all polymers, excluding polyarylates, having para-connections in the main chain (PF-2 and PF-7) are placed on one general dependence $D_f(\Delta\delta/C_\infty)$, although with considerable scatter. For similar scatter both objective and subjective reasons exist. The noted above polymers regular solutions strict theory absence should be classified under the first ones, in virtue of this the adduced in Fig. 34 relation is crude enough approximation and reflects only change D_f tendencies. The large scatter of the cited in literature parameters values, used for the plot of Fig. 34 construction, that is, δ_p, δ_s and C_∞ should be classified as subjective reasons. The authors [68] could not find the dependences $D_f(\Delta\delta/C_\infty)$ normalization parameter for polyarylates, having para-connections in the main chain (having metha-connections that correspond well to the general correlation) and therefore the separate calibrating curve will be required for them. Nevertheless, the plots adduced in Fig. 34 reflect clear tendencies of the fractal dimension of macromolecular coil in solution as a function of factors, characterizing interactions of both polymer-solvent and coil itself.

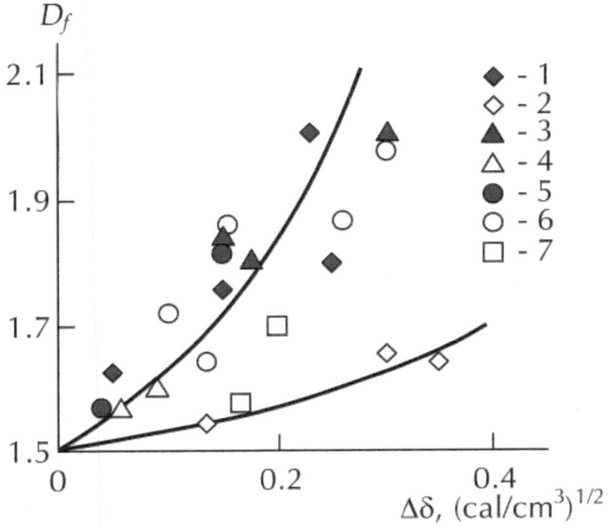

FIGURE 34 The dependences of macromolecular coil fractal dimension D_f on combined parameter ($\Delta\delta/C_\infty$) for PC (1), PAr with para-connections (2), PAr with metha-connections (3), PMMA (4), PS (5), PDMS (6) and SKN-18 (7).

However, one more specific factors number exists, which are capable to influence on the value D_f in the considered conditions, that confirms again macromolecular coils in solutions behavior complexity. In Ref. [26] an experimental data large set is adduced about the behavior of copolymer stirene-acrylonitril-methacrylate (AN-MMA) in different solvents (see Fig. 3). In Fig. 35, the dependences D_f ($\Delta\delta$) are adduced for AN-MMA in two solvents only: chloroform and 1,4-dioxane, which form two different dependences for the same copolymers. The authors [68] supposed that two dependences formation was due to 1,4-dioxane (unlike chloroform) ability to form hydrogen bonds polymer-solvent, which results in macromolecular coil compactness enhancement owing to the indicated interaction intensification. Besides, similar kind analysis for copolymers is more complex in general, since solubility of different blocks, making up copolymer, in the same solvent can turn out to be different [26, 57].

Thus, summing up the stated above results, the following tendencies of the fractal dimension of macromolecular coil in solution can be pointed out as a function of factors number:

(1) a solvent thermo-dynamical affinity in respect to polymer change for the worse always results in D_f enhancement;
(2) the larger polymer main chain flexibility is, the weaker tendency of D_f growth is at $\Delta\delta$ increase;
(3) in addition to item 2, it should be pointed out, that the introduction in the main chain of rigid para-connections instead of more flexible metha-connections results in D_f sharp reduction at other equal conditions;
(4) polymers with bulk side groups and flexible main chain will have the values D_f larger by about constant value $\Delta D_f \approx 0.17$ (presenting itself difference of dimensions of coil in good solvent and leaking coil state) in comparison with polymers, which do not have the indicated features;
(5) hydrogen bonds polymer-solvent availability results in much faster D_f growth at $\Delta\delta$ increase (about twice) in comparison with solvent, un capable to form the indicated bonds.

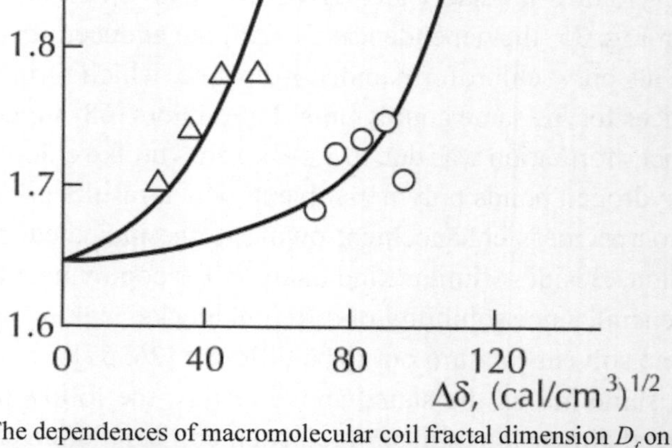

FIGURE 35 The dependences of macromolecular coil fractal dimension D_f on difference of polymer and solvent solubility parameters $\Delta\delta$ for copolymers AN-MMA in chloroform (1) and 1,4-dioxane (2).

These general rules allow having a goal-directed choice of solvent for either polymer synthesis. As it has been shown in Ref. [71], D_f reduction results in reaction essential acceleration at other equal conditions.

The fact, that macromolecular coil in diluted solution is a fractal object, allows to use the mathematical calculus of fractional differentiation and integration for its parameters description [72–74]. Within the framework of this formalism there is the possibility for exact accounting of such nonlinear phenomena as, for example, spatial correlations [74]. In the last years the methods of fractional differentiation and integration are applied successfully for polymer properties description as well [75–77]. The authors [78–81] used this approach for average distance between polymer chain ends calculation of polycarbonate (PC) in two different solvents.

For the calculation of the average end-to-end distance of polymer macromolecule $\langle \bar{h}^2 \rangle^{1/2}$, which is an important parameter in the theory of polymer solutions [1], there are a number of methods. So, the following empirical relationships for PC in solutions of methylene chloride (MCh) and tetrahydrofuran (THF) were obtained [82]:

$$\langle \bar{h}^2 \rangle^{1/2} = 0.66 M_\eta^{0.58} \quad \text{for MCh,} \tag{59}$$

and

$$\langle \bar{h}^2 \rangle^{1/2} = 1.04 M_\eta^{0.53} \quad \text{for THF,} \tag{60}$$

where M_η is the viscosity-average molecular weight.

Another method uses the following equation [1]:

$$[\eta] = \Phi(\alpha) \frac{\langle \bar{h}^2 \rangle^{1/2}}{M_\eta} \tag{61}$$

where $[\eta]$ is the intrinsic viscosity of polymer solution and $\Phi(\alpha)$ is a parameter dependent on the swelling coefficient α. At $\alpha^3 > 5$, $\Phi(\alpha)$ can be accepted equal to approximately 2×10^{23} [1].

Besides, the equation for the value $\left\langle \bar{h}^2 \right\rangle^{1/2}$ estimation can be obtained within the framework of fractional derivation and integration as follows [83]. Let $x=x(t)$ be the law of the change of some physical property at time t. The rate of change $x(t)$ has the following form [78]:

$$\frac{dx}{dt} = \vartheta(t)\,\tau_0 \tag{62}$$

where t is dimension less time and τ_0 is the characteristic time of the given process.

The Eq. (62) can be represented as follows [79]:

$$D_{0_t}^{\nu} D_{0_t}^{1-\nu} x(t) = \vartheta(t)\,\tau_0, \tag{63}$$

where

$$D_{0_t}^{\nu} f(t) = \frac{1}{\Gamma(1-\nu)} \frac{d}{dt} \int_0^t \frac{f(\tau)\,d\tau}{(t-\tau)^{\nu}} \tag{64}$$

is Rimman-Liouville fractional derivative of the order ν $(0<\nu<1)$.

If we introduce the notation $D_{0_t}^{1-\nu} x(t) = \bar{h}(t)$, the Eq. (63) becomes [80]:

$$D_{0_t}^{\nu} \bar{h}(t) = \vartheta(t)\,\tau_0 \tag{65}$$

The Eq. (65) solution looks like [80]:

$$\bar{h}(t) = \frac{\tau_0}{\Gamma(\nu)} \int_0^t \frac{\vartheta(t)\,d\tau}{(t-\tau)^{\nu}} \tag{66}$$

The integral in the Eq. (66) right part is easily calculated at $\vartheta(t)=\vartheta=\text{const}$, and returning to dimensional t, let us obtain [1]:

$$\bar{h}(t) = \frac{\vartheta \tau_0^{\nu}}{(1-\nu)\Gamma(\nu)}, \tag{67}$$

where $\Gamma(\nu)$ is Euler's gamma function.

Comparing the Eq. (67) with the known Einstein formula for the mean-square displacement of a particle making Brownian motion, the authors [83] have drawn the conclusion that the value of $\bar{h}(t)$ can be considered as $\bar{h}=\langle x^2\rangle^{1/2}$.

As it is known through Ref. [1], a polymer macromolecule can be divided into statistical segments of length l_{st}, and one can simulate it (in an elementary case) as Brownian motion of a segment. Then, the Eq. (67) can be used for the description of a polymer chain. In this case, rate ϑ can be treated as the rate of segment jumping, and $_0$ can be considered as the time of one jump. If $l_{st}=\vartheta\tau_0$ and t is equal to $\tau_0 N_{st}$ (where N_{st} is a number of statistical segments per a chain), let us obtain [83]:

$$\left\langle \bar{h}^2 \right\rangle^{1/2} = \frac{l_{st} N_{st}^{1-\nu}}{(1-\nu)\Gamma(\nu)}. \tag{68}$$

The purpose of further study is the calculation of $\left\langle \bar{h}^2 \right\rangle^{1/2}$ according to the Eqs. (59)–(61) and (68) and the subsequent comparison of the results on the example of PC solutions in two solvents (MCh and THF). For this purpose, five molecular weights (*MW*) for PC have been arbitrarily chosen: 2.5, 5.0, 7.5, 10.0 and 12.5×10^4, then according to the Eqs. (59) and (60) the empirical values $\left\langle \bar{h}^2 \right\rangle^{1/2}$ were calculated for the indicated MW_s. When the Eq. (61) is used for the determination of [η] values corresponding to the indicated MW_s, the following Mark-Kuhn-Houwink equations are used [82]:

$$[\eta] = 1.11\times10^{-4} MW^{0.82}\text{, for MCh,} \tag{69}$$

and

$$[\eta] = 3.99\times10^{-4} MW^{0.70}\text{, for THF,} \tag{70}$$

where [η] values are received in dl/g.

This allows to calculate $\left\langle \bar{h}^2 \right\rangle^{1/2}$ according to the Eq. (61). For the Eq. (68) application the parameters, included in it, that is, l_{st}, N_{st}, ν and Γ(ν) should be determined. The length of statistical segment l_{st} is determined according to the equation [70]:

$$l_{st} = C_\infty l_0 , \tag{71}$$

where C_∞ is the characteristic ratio, which is a chain statistical flexibility indicator [1], l_0 is the length of the skeletal bond of the main chain equal to 1.25 Å for PC [69].

Further two variants of C_∞ choice will be used. The first from them (static) is based on the application of a literary value C_∞ for PC, equal to 2.4 [69, 70]. The second (dynamical) variant supposes that C_∞ value depends on the structure of the macromolecular coil in solution, that is, D_f and in this very case C_∞ value is determined as follows [84]:

$$D_f = 2 - 4\left(\frac{2}{C_\infty^2 S}\right)^{1/2} , \tag{72}$$

where S is the cross-sectional area of a macromolecule, equal to 30.7 Å^2 for PC [85].

The volume of a macromolecule V_{mm} can be determined according to the known MW values as follows [37]:

$$V_{mm} = \frac{MW}{\rho N_A} , \tag{73}$$

where ρ is the density of a polymer (ρ=1200 kg/m^3 for PC [37]) and N_A is Avogadro number.

Further more, it is possible to calculate the length of a macromolecule L_m as V_{mm}/S and N_{st} as L_m/l_{st}.

As it has been shown in Ref. [72], the fractional exponent ν coincides with the fractal dimension of Cantor's set and indicates a fraction of the system states, being preserved during the entire evolution time t. Let us remind that Cantor's set is considered in one-dimensional Euclidean space (d=1) and therefore its fractal dimension d_f<1 by virtue of the fractal definition [86]. For fractal objects in Euclidean spaces with higher dimensions (d>1) as ν one should accept d_f fractional part or [76, 77]:

$$\nu = d_f - (d-1), \tag{74}$$

where d is the dimension of Euclidean space in which the fractal is considered.

Let us consider the physical significance of the value of a fractional exponent ν in the given context. As it is known through Ref. [87], the transition to the condensed polymer state occurs at D_f=2.5. This means that the limiting value of ν is reached at this value D_f, equivalent to d_f, and then let us obtain [77]:

$$\nu = D_f - (2.5-1), \tag{75}$$

or

$$\nu = D_f - 1.5. \tag{76}$$

And at last, Euler's gamma function Γ(ν) was calculated as follows [88]:

$$\Gamma(\nu) = \left(\frac{\pi}{2}\right)^{1/2} \nu^{-\nu} e^{-\nu} \tag{77}$$

In Fig. 36, the comparison of the dependences of $\left\langle \overline{h}^2 \right\rangle^{1/2}$ on *MW*, calculated by the three indicated methods, for solutions of PC in MCh is adduced. As one can see, a good correspondence (within the limits of 8%) for the values of $\left\langle \overline{h}^2 \right\rangle^{1/2}$ calculated according to the Eqs. (59), (61) and the Eq. (68) with the use of a dynamical variant of an C_∞ estimation, that is, the Eq. (72) has been obtained. The application of the statical variant, when C_∞ is constant, increases the error of the calculated values up to approximately 12%. As for the empirical Eq. (59), it gives values $\left\langle \overline{h}^2 \right\rangle^{1/2}$ much lower than the estimations according to the Eq. (61) [81].

As a matter of fact, a similar picture is obtained for solutions of PC in tetrahydrofuran (Fig. 37), although the discrepancy of the results received according to the Eqs. (59), (61) and (68) is somewhat greater and makes up approx. 17%. The error in the estimation of D_f may be its reason. So, reducing D_f from 1.765 to 1.70 (i.e., 3%) results in better correspondence of the indicated results (Fig. 37): the average disagreement is reduced up to 11.5%.

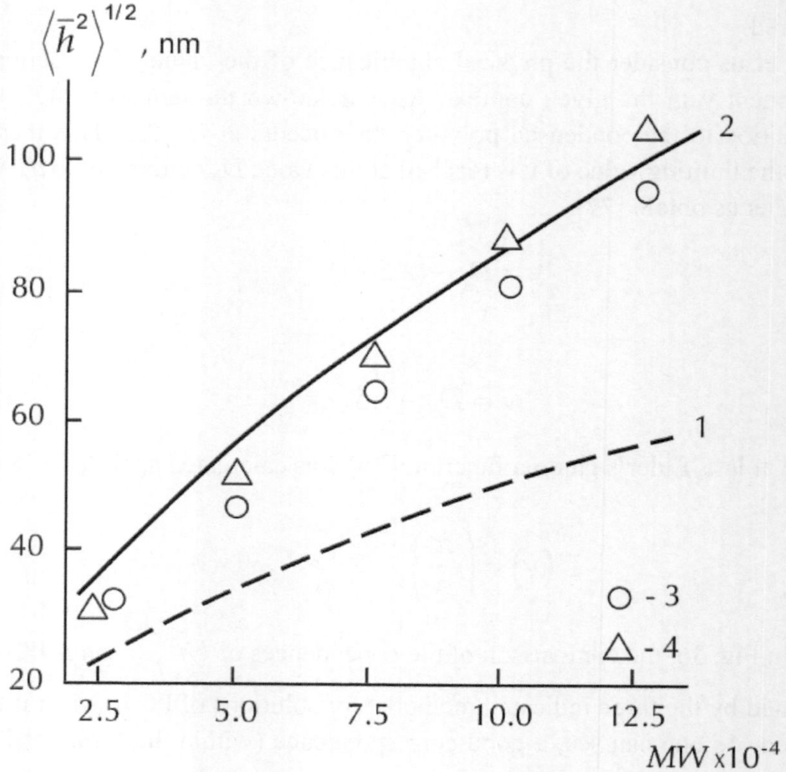

FIGURE 36 The dependences of polymer chain average end-to-end distance $\langle \bar{h}^2 \rangle^{1/2}$ on molecular weight MW for a solution of PC in methylene chloride. Calculation according to the Eqs. (59) (1), (61) (2), (68) at C_∞=const (3) and C_∞ estimated according to the Eq. (72) (4).

It is necessary to mark one more important feature of the dependences of $\langle \bar{h}^2 \rangle^{1/2}$ on MW, calculated according to the Eq. (61)—they grow faster than the similar dependences calculated according to the Eq. (60) at large

MW_s. As it is known through Ref. [1], at large MW_s the macromolecular coil becomes more compact, and this results in the decrease of the exponent a_η in Mark-Kuhn-Houwink equation and, according to the Eq. (4), in the increase of D_f. The estimations have shown that for solutions of PC in methylene chloride D_f enhancement from 1.648 up to 1.681, that is, 2% (or a_η decrease from 0.820 up to 0.785) results in $\langle \bar{h}^2 \rangle^{1/2}$ reduction from 107.4 up to 96.6 nm, that is, the indicated difference is eliminated.

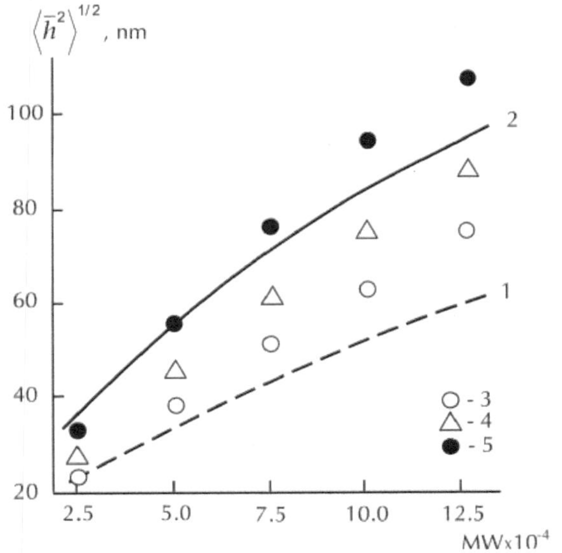

FIGURE 37 The dependences of polymer chain average end-to-end distance $\langle \bar{h}^2 \rangle^{1/2}$ on molecular weight *MW* for a solution of PC in tetrahydrofuran. Calculation according to the Eqs. (59) (1), (61) (2), (68) at C_∞=const (3), C_∞ estimated according to the Eq. (72) (4) and D_f=1.70 (5).

Let us return in conclusion to the correctness of $\Phi(\alpha)=2\times10^{23}$ used in the Eq. (61). As it is known through Ref. [6], α value is connected with D_f by the following relationship:

$$D_f = \frac{5\alpha^2 - 3}{3\alpha^2 - 2}. \tag{78}$$

For values D_f, obtained according to the Eq. (4) for solutions of PC in MCh and THF (D_f=1.648 and 1.764, respectively), the values α^3 are within the range of 2.5–25, that enables the usage of the indicated coefficient (α).

Thus, the stated above results allow one to make two main conclusions. Firstly, fractional differentiation and integration methods (the Eq. (68)) allow to calculate polymer chain molecular characteristics as precisely as other existing at present computational techniques. Secondly, accounting for the change of macromolecular coil structure, that is, its dynamics, at the external conditions variation, is needed for the correct calculation of the indicated characteristics.

2.3 THE TEMPERATURE DEPENDENCE OF FRACTAL DIMENSION OF A MACROMOLECULAR COILS IN DILUTED SOLUTIONS

The change of the fractal dimension D_f with temperature reflects the corresponding changes in sizes, degree of compactness and asymmetry of shape of a macromolecular coil in solution [89]. The importance of the temperature dependence of D_f study is determined by strong influence of this parameter on the processes of synthesis [28], catalysis [90], flocculation [91], so forth. At present as far as we know experimental evaluations of the temperature dependence of D_f of a macromolecular coil are absent. Theoretical estimations [13] suppose that the temperature enhancement makes the fractal less compact, that is, leads to D_f value reduction. Therefore, the authors [92] performed the experimental study of D_f dependence on temperature for the macromolecular coils of polyarylate F-1 [5] in diluted solutions and evaluation of D_f change influence on synthesis processes.

The polyarylate (PAr) on the basis of phenolphthaleine and dichloranhydride of isophthalic acid (F-1) [5] were used. Tetrachloroethane, N, N-dimethylformamide, tetrahydrofuran and 1,4-dioxane were used as solvents. As it is known through Ref. [89], the reason of changes in the structure of a macromolecular coil in diluted solution with temperature variation can be both the effects of long-range interaction (excluded volume effects) and the effects of short-range interaction leading to the specific influence of a solvent on unperturbed sizes of a macromolecule. For semirigid-chain polymers, to which polyarylates can be referred [5], macromolecules in solution are characterized by significant permeability and therefore the long-range effect can be neglected. For these polymers it is supposed that the temperature coefficient of viscosity dln []/*dT* is a negative one by the sign but by the absolute value it excesses the one for flexible-chain polymers of the same *MM*. In Fig. 38 the temperature dependences of [η] for F-1 solutions in three solvents: tetrachloroethane, tetrahydrofuran and N, N-dimethylformamide are adduced. These dependences have the expected character, that is, the decrease of [] with the increase of *T* is observed. As it has already been noted, this supposes, that the macromolecular coil structure is controlled mainly by the short-range interaction effect, which is the polymer-solvent interactions. The value of dln [η]/*dT* is equal to about—0.0053, that corresponds well to the corresponding parameter for more rigid-chain aromatic polyamides [89].

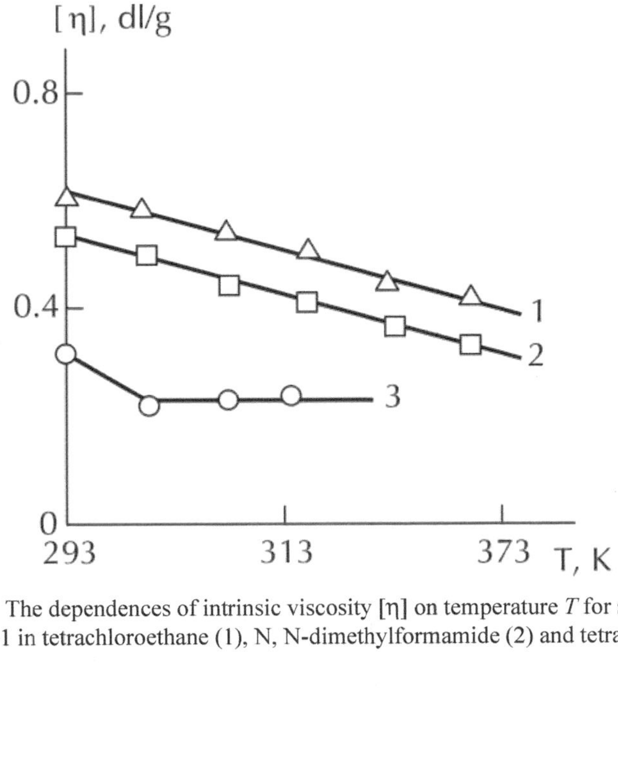

FIGURE 38 The dependences of intrinsic viscosity [η] on temperature *T* for solutions of polyarylate F-1 in tetrachloroethane (1), N, N-dimethylformamide (2) and tetrahydrofuran (3).

In Fig. 39, the dependences of η_{sp}/c on c for F-1 solutions in tetrachloroethane are adduced at three temperatures (293, 323 and 353 K), which correspond to the well-known Huggins Eq. (47). For F-1 the values k_n=0.397, 0.314 and 0.294 for T=293, 323 and 353 K were obtained, respectively. This means that the interaction polymer-solvent becomes weaker at the temperature growth [92].

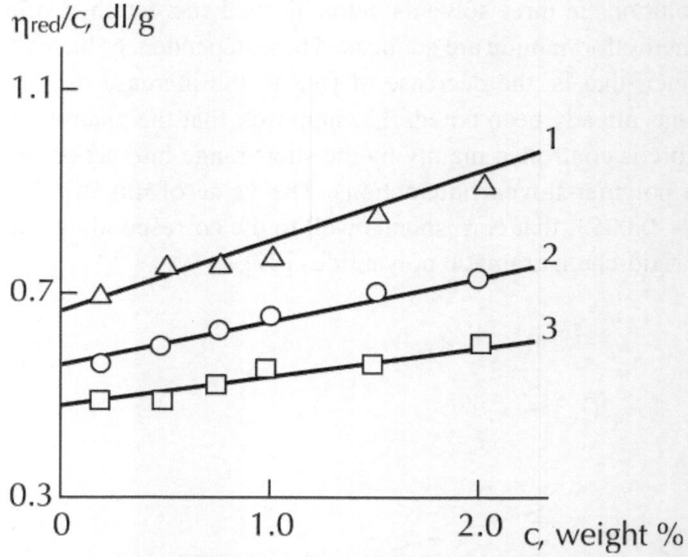

FIGURE 39 The dependences, corresponding to the Eq. (47), for polyarylate F-1 solution in tetrachloroethane at temperatures 293 (1), 323 (2) and 353 K (3).

As it is known, the plots of Fig. 39 allow one to determine the values of intrinsic viscosity [η] by the extrapolation to the polymer zero concentration c. Another method of [η] evaluation is Shultz-Blashke Eq. (48). In Table 12, the comparison of [η] values, evaluated by two indicated methods, is adduced for polyarylate F-1 at three testing temperatures and polyamidobenzymidazole [94] at two concentrations of solution in sulfuric acid. As it follows from the data of Table 12, the values [η], calculated according to the Eqs. (47) and (48) showed a good correspondence, that allows one to use Shultz-Blashke equation for [η] values of semirigid- and rigid-chain polymers estimation.

TABLE 12 The comparison of [η] values, evaluated according to Huggins and Shultz-Blashke equations.

Polymer	Solvent	*c*, mass %	*T*, K	[η], dl/g	
				The Eq. (47)	The Eq. (48)
Polyarylate F-1	Tetrachloroethane	0.5	293	0.63	0.61
	Tetrachloroethane	0.5	323	0.56	0.54
	Tetrachloroethane	0.5	353	0.47	0.41
Polyamidobenzymidazole	Sulphuric acid	0.5	293	2.60	2.60
	Sulphuric acid	0.2	293	2.60	2.40

In Fig. 40, the dependences of D_f on testing temperature T have been shown for F-1 solutions in tetrachloroethane and N, N-dimethylformamide (c=0.5 mass %). As one can see, D_f monotonous increasing at T growth is observed, that is, macromolecular coil compactness degree enhancement is realized. The plots of Fig. 40 extrapolation to D_f=2.0 [10] allows one to estimate θ-temperature values for the indicated solvents. Let us note, that the achievement of θ-conditions does not mean any critical state of the macromolecular coil. Such state for the coil in diluted solution (practically isolated macromolecule) can be reached at the fractal dimension critical value D_f^c, determined by the Eq. (4) of Chapter 1. As it follows from the indicated relationship, $D_f^c \approx 2.285$. Let us note, that approximately at this D_f value the dependence [η](T) for polyarylate F-1 in tetrahydrofuran (Fig. 38) becomes parallel to abscissa axis, that is, this D_f value is a critical one.

Let us also note, that in general case the dependence $D_f(T)$ is supposed to be rather complicated. So, for flexible-chain polymers one can observe at first the increase of [η] at T enhancement and then its decrease [1]. This supposes at first D_f decrease at T enhancement and then its increase.

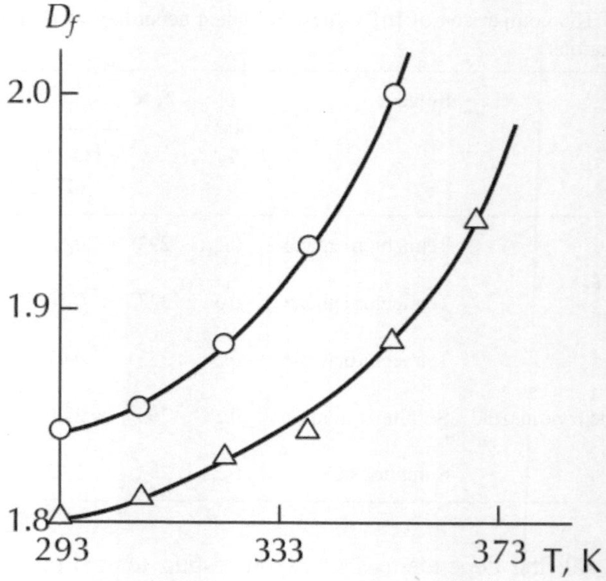

FIGURE 40 The dependences of macromolecular coil fractal dimension D_f on temperature T for polyarylate F-1 solutions in tetrachloroethane (1) and N, N-dimethylformamide (2).

The authors [92] calculated the dependence $D_f(T)$ with the aid of the Eq. (4) for the solution of triacetate cellulose in benzyle spirite according to the data of Ref. [95], which was adduced in Fig. 41. As it follows from the plot of Fig. 41, at first the D_f decrease is observed and the within the range of T=363–423 K fractal dimension increase takes place. Such course of the dependence $D_f(T)$ for triacetate cellulose can be explained by complex interactions polymer-solvent: D_f reduction at $T \leq 363$ K is explained by disolvation process at the expense of solvent "squeezing" out a macromolecular coil at the pairwise interactions intensification and D_f enhancement at $T > 363$ K (Fig. 41) is linked with the solvent structure failure [95]. For diacetate cellulose in different solvents [η] decrease at T enhancement is observed [61], that assumes D_f corresponding increasing according to the Eq. (11).

As it is known through Ref. [16], the polymer solubility parameter δ_p is determined according to the dependence of [η] or k_x on the solvent solubility parameter δ_s. In Fig. 42 the dependences [η](δ_s) for polyarylate F-1 in the four indicated above solvents at temperatures 293 and 338 K are adduced. As one can see, the qualitative changes in the dependences $[\eta](\delta_s)$ at temperature variation do not occur. This means that δ_p value will change with temperature as well as parameter δ_s.

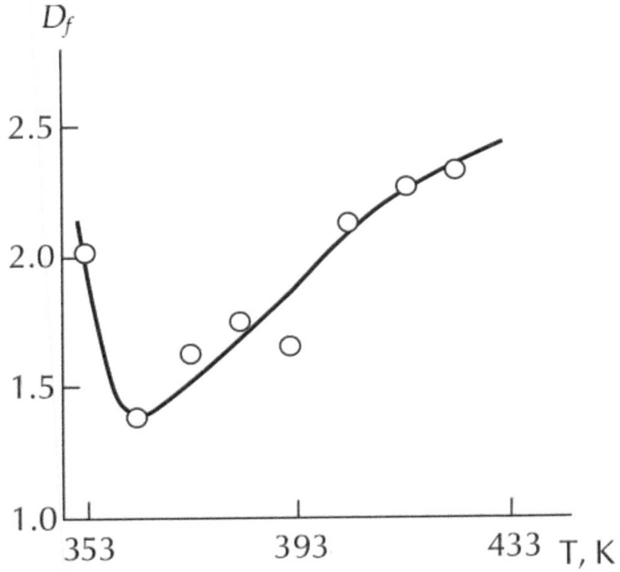

FIGURE 41 The dependence of macromolecular coil fractal dimension D_f on temperature T for triacetate cellulose solution in benzyle spirit according to the Ref. [95] data.

The obtained dependences $D_f(T)$ have great practical significance. As it has been shown earlier [28], the conversion degree Q at polymers synthesis can be described according to the following fractal relationship:

$$Q \sim t^{(3-D_f)/2}, \quad (79)$$

where t is synthesis reaction duration.

The Ref. [96] contains the data showing that the increase of synthesis temperature from 350 up to 368 K while obtaining polyhexamethylene-1,3-benzenesulfonamide leads to Q decrease from 0.72 up to 0.54, that is, approximately by 33%. If we suppose that for this polymer D_f change at T growth is similar to the one shown in Fig. 40, then we obtain

Q decrease according to the relationship (79) approximately by 30% at typical synthesis process duration *t*=60 min. Hence, the main reason of *Q* decrease in the adduced above example can be the dependence of macromolecular coil D_f on temperature.

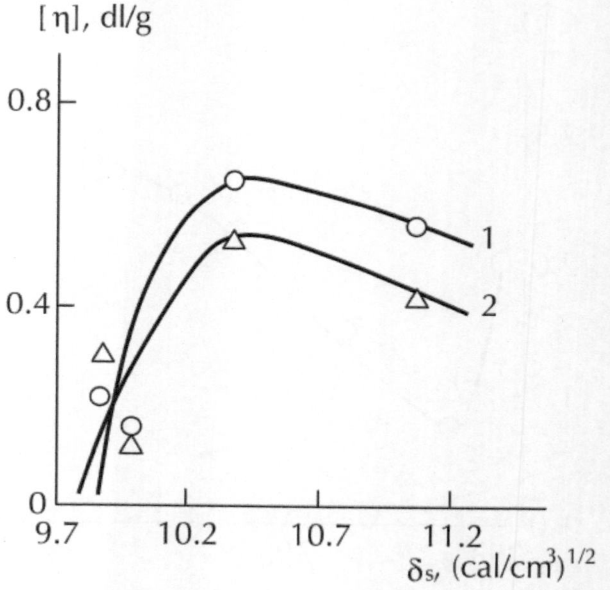

FIGURE 42 The dependence of intrinsic viscosity [η] on solvent solubility parameter δ_s for polyarylate F-1 solutions in different solvents at temperatures 293 (1) and 338 K (2).

Hence, the adduced above results suppose, that the dependence of the fractal dimension of a macromolecular coil in diluted solutions on temperature has complex enough character and for semirigid- and rigid-chain polymers D_f increasing with *T* enhancement is observed in general. However, the availability of specific effects of the interactions polymer-solvent can strongly affect on the dependence $D_f(T)$ character (as it has been shown in Ref. [95]). D_f change at *T* variation may have significant influence on the main characteristics of the polymer behavior in solutions, for example, at their synthesis.

2.4 THE PHYSICAL SIGNIFICANCE AND EVALUATION METHODS OF STRUCTURAL PARAMETERS OF LOW-MOLECULAR SOLVENTS IN DILUTED POLYMER SOLUTIONS

As it has been shown above, the solvent change in diluted polymer solution leads to a macromolecular coil fractal dimension D_f change. This effect is due to variation of the interactions polymer-solvent and within the framework of fractal analysis the indicated effect is described by the following equation [25]:

$$D_f = \frac{d_s(d+2)}{(1-\alpha)d_s+2}, \qquad (80)$$

where d_s is spectral (fraction) dimension of a macromolecular coil, characterizing its connectivity degree [45], d is dimension of Euclidean space, in which a fractal is considered (it is obvious, that in our case d=3) and the parameter α is expressed as follows [25]:

$$\alpha = \frac{\delta_f}{D_f^{ph}} \qquad (81)$$

where δ_f is the fractal dimension of solvent molecule, D_f^{ph} is the phantom (which does not take into account excluded volume effects) fractal dimension of a macromolecular coil, determined according to the Eq. (11).

It is easy to see, that in case of point zero-dimensional solvent molecules (δ_f=0) α=0 and the Eq. (80) gives the fractal dimension of swollen (i.e., accounting for the excluded volume effects) macromolecular coil, determined according to the Eq. (12) of Chapter 1.

Hence, at any rate for linear polymers (for which d_s=1.0 [45]) the problem of D_f value prediction in different solvents consists of prediction of α value or, since in this case D_f^{ph} =2.0, δ_f value. Therefore, the authors [97] developed the techniques of the dimension δ_f evaluation on the example of polycarbonate on the basis of bisphenol A (PC) in solutions of 14 different solvents.

For the dimension δ_f theoretical evaluation the following technique was used. As it has been shown in Ref. [24], D_f value within the framework of two-dimensional (two-component) solubility parameter [14] can be determined according to the Eq. (13). The solvents enumeration and corresponding to them δ_s and D_f values are adduced in Table 13. δ_f values were calculated according to the Eq. (80) and Eq. (11) of Chapter 1 at the condition d_s=1.0 and D_f^{ph} =2.0, since PC macromolecule is a linear one.

TABLE 13 The used in calculations solvents, fractal dimension of their molecules δ_f and PC macromolecular coils in these solvents fractal dimension D_f

№	Solvent	δ_f, (cal/cm^3)$^{1/2}$	D_f
1	Carbon tetrachloride	8.55	1.75
2	1,1,1-trichloroethane	8.57	1.74
3	Toluene	8.93	1.75
4	Benzene	9.16	1.80
5	Chloroform	9.16	1.73
6	Trichloroethylene	9.16	1.80
7	Tetrahydrofuran	9.52	1.86
8	Chlorobenzene	9.67	1.82
9	1,4-dioxane	10.13	1.80
10	Nitrobenzene	10.62	1.88
11	N, N-dimethylacetamide	10.80	1.86
12	Nitromethane	11.09	1.95
13	N, N-dimethylformamide	11.79	1.93
14	Nitroethane	12.90	1.97

The simplest from the possible correlations between dimension δ_f and solvent characteristics the dependence $\delta_f(\delta_s)$ proves to be [97], which has shown $_f$ growth at δ_s increasing approximately in quadratic form. For this relationship linearization the correlation $\delta_f(\delta_s^{1/2})$ was chosen, shown in

Fig. 43. As one can see, such dependence of δ_f on δ_s is a linear one actually and can be expressed as follows [97]:

$$\delta_f = 1.57\left(\delta_s^{1/2} - 2.83\right), \tag{82}$$

where δ_s is given in units $(cal/cm^3)^{1/2}$.

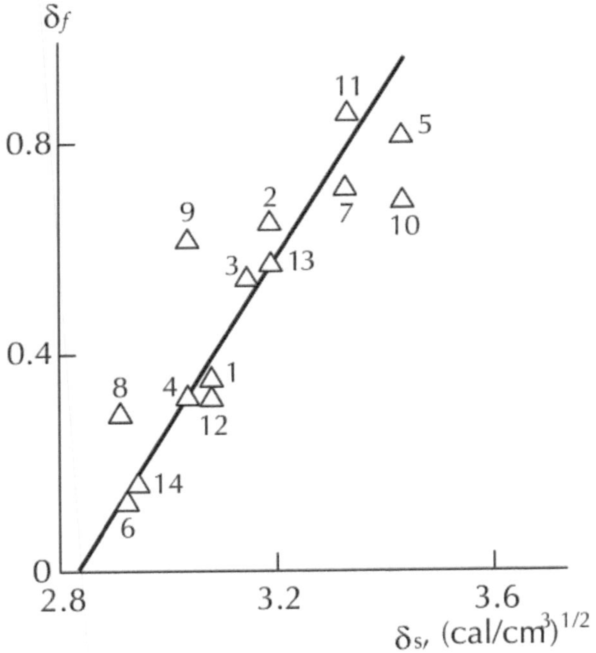

FIGURE 43 The dependence of organic solvents molecules fractal dimension δ_f on their solubility parameter δ_s. Figures at points correspond to solvent numbers in Table 13.

Such shape of the dependence $\delta_f(\delta_s)$ has certain physical grounds as it is known through Ref. [14], the parameter δ_s is determined as follows:

$$\delta_s^2 = \frac{E_{vap}}{V_{mol}}, \tag{83}$$

where E_{vap} is vaporization energy, V_{mol} is molar volume and the ratio E_{vap}/V_{mol} represents the solvent cohesion energy density.

From the Eqs. (82) and (83) it follows, that E_{vap} increasing leads to δ_s growth, that is, molecule (molecules totality) of solvent compactness enhancement. Since E_{vap} increasing means intensification of attractive interactions between solvent molecule elements (or between totality molecules), then δ_f growth tendency corresponds to the general tendency of such kind, obtained in Ref. [98] for fractal objects. V_{mol} enhancement means the increase of space fraction, occupied by solvent molecule, its structure compactness degree reduction and, hence, δ_f decreases [86].

Let us note an important consequences number, following from the Eqs. (80) and (82) and also the plot of Fig. 43 analysis. Firstly, the solvents with $\delta_s^{1/2} \leq 2.83$ (cal/cm^3)$^{1/4}$ or $\delta_s < 8$ (cal/cm^3)$^{1/2}$ (that corresponds to $\delta_s = 16.33$ (j/cm^3)$^{1/2}$ [16]) consist of molecules, which can be simulated by zero-dimensional point object. The solvent molecules (molecules totality) structure complication, characterized by δ_f increasing within the range of 0–1 results in δ_s growth from 8.0 up to approx. 12 (cal/cm^3)$^{1/2}$. Secondly, for D_f value obtaining according to the Eq. (12) of Chapter 1, that is, D_f value of a macromolecular coil in good solvent ($D_f \approx 1.667$), realization of the condition $\alpha = 0$ or $\delta_f = 0$ is required. Thirdly, for realization of a macromolecular coil θ-conformation, that is, $D_f = 2.0$ [10], the condition $\alpha = 0.5$ or $\delta_f = 1.0$ follows from the Eq. (80) (for linear polymers). Let us note, that -conditions reaching for a macromolecular coil of a very branched polymer ($d_s = 1.33$ [45], $D_f = 2.29$ [13]) requires already the condition $\delta_f \approx 1.36$ according to the Eq. (80). Fourthly, according to the known empirical rule [22], for polymer dissolution in low-molecular solvent the difference of these two substances solubility parameters should not exceed ±(1.5–2.0) (cal/cm^3)$^{1/2}$. From the data of Fig. 43 and the Eq. (82) it follows, that according to this rule PC ($\delta_p = 9.67$ (cal/cm^3)$^{1/2}$ [14]) is dissolved in the solvents, having the range of $\delta_s \approx 7.67$–11.67 (cal/cm^3)$^{1/2}$ or, according to the Eq. (82), the range of $\delta_f \approx 0$–0.92. In other words, PC is not dissolved in polymer solvents ($_f > 1.0$) [97].

The Eq. (82) applicability with the purpose of prediction the fractal dimension of a macromolecular coil in different solvents can be verified by the comparison of D_f values, calculated according to the Eq. (13) (Table 13) and the Eqs. (80), (82). Such comparison of D_f values is adduced in Fig. 44. As it follows from the data of this figure, between the received by the two indicated methods D_f values a good correspondence is observed (the average discrepancy makes up 2.5%, maximum—6.5%, correlation coefficient is equal to 0.8).

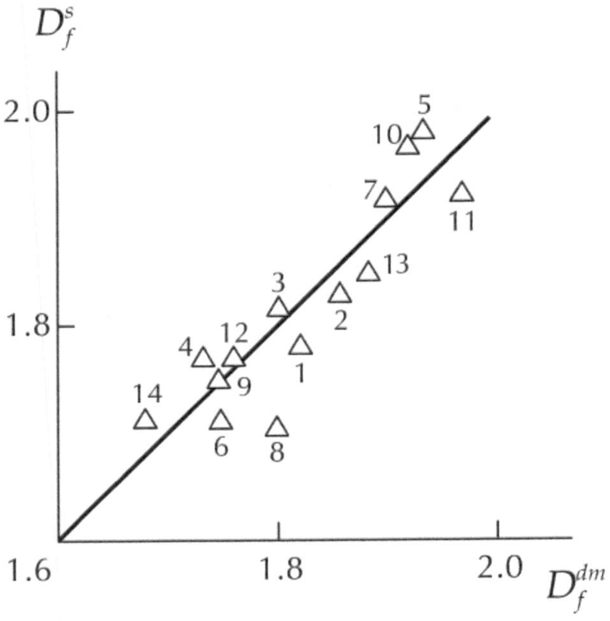

FIGURE 44 The comparison of fractal dimensions of PC macromolecular coil in different solvents, calculated according to the Eqs. (13) D_f^{dm} and (80), (82) D_f^s. Figures at points correspond to solvent numbers in Table 13.

Hence, the stated above results demonstrated the possibility of the prediction of molecule (molecules totality) structure of organic solvents, characterized by its fractal dimension, on the basis of one characteristic only, that is, the solvent solubility parameter δ_s. In its turn, this circumstance gives the possibility of prediction of a macromolecular coil structure, characterized by its fractal dimension D_f in diluted solutions with different solvents participation. A number of specific features of systems polymer-solvent was obtained and considered, is due to their structural characteristics.

It is known through Ref. [99, 100], that a polymer products obtaining from different solvents leads to the essential variation of their structure and properties. So, polyarylatedimethylsiloxane film samples production by different modes and from different solvents results in fracture stress variation within the limits of 1.5–25 MPa and fracture strain—within the limits of 115–395% [99]. PC films production from different solvents changes the thermo-degradation beginning temperature on 120 K [101]. It is supposed, that this effect depends essentially on the solvent nature, since polymers concentrated solutions already contain structures, similar to the simple types of polymers condensed state structures [100]. This intercommunication was described analytically within the framework of fractal analysis [102], where a simple relationship between fractal dimensions of the structure of a macromolecular coil in solution and polymers condensed state structure was obtained, which will be considered in detail in Section 2.6.

Thus, the dimension D_f prediction, characterizing the structure of a macromolecular coil in solution, should be considered as the first stage of polymers, produced from different solvents [25]. One from possible methods of D_f evaluation is the Eq. (18) of Chapter 1 application, from which it follows, that for D_f evaluation the ability of δ_f values prediction is required, for that the physical significance of this dimension should be understood. These two problems solution was performed in Refs. [103, 104].

The necessary for calculations experimental data (solubility parameters of polymers δ_p and solvents $_s$, Flory-Huggins interaction parameters χ_1, the exponents a_η in Mark-Kuhn-Houwink equation) were accepted according to the data of chapters [1, 14, 16, 29, 32]. The experimental values of fractal dimension of a macromolecular coil in diluted solution were determined according to the equation (4).

At the Eq. (18) of Chapter 1 derivation the supposition has been made, that the sizes of a polymer and solvent clusters are comparable. Since in case of low-molecular solvents a macromolecular coil size exceeds deliberately a solvent molecule size, then the conclusion should be made, that the solvent molecules totality structure, interacting with a macromolecular coil ("swarm" of solvent molecules) should be considered, but not a separate molecule. In favor of such conclusion the fact reveals, that δ_f value for the same solvent, but different polymers, has different values [24]. Therefore, it should be supposed, that the dimension δ_f characterizes interactions polymer-solvent. In Fig. 45 the dependence of δ_f on Flory-Huggins interaction parameter χ_1 for 4 polymers in different solvents has been shown, which confirms this supposition. The indicated correlation proves to be linear and is approximated by the following empirical correlation [104]:

$$\delta_f \approx 1.15\chi_1 - 0.30. \tag{84}$$

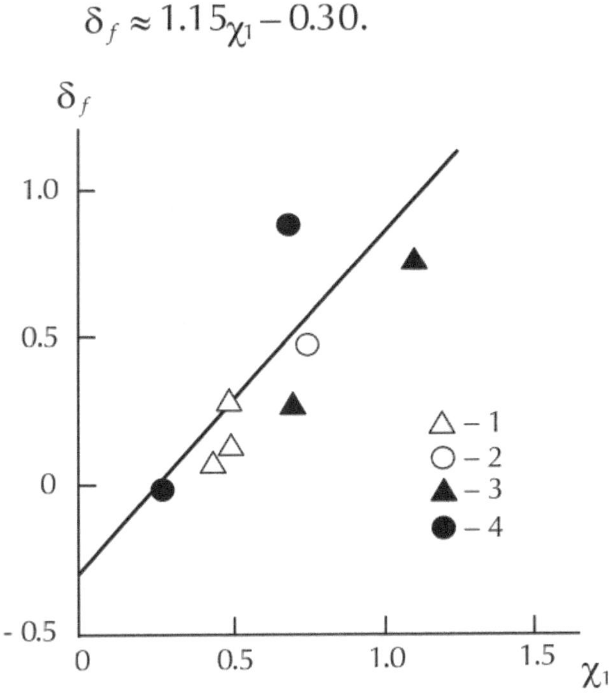

FIGURE 45 The relationship between fractal dimension of solvent molecules "swarm" δ_f and Flory-Huggins interaction parameter χ_1 for poly(methyl methacrylate) (1), polystirene (2), poly(vinyl acetate) (3) and polysulfone (4).

The most interesting feature of the Eq. (84) is the possibility of the dimension δ_f negative values receiving. As it is known through Ref. [86], the classical fractal analysis supposes only positive fractal dimensions, describing either structure. Therefore, the natural question arises in respect to the dimension $_f$ physical significance. The Eq. (13) of Chapter 1, describing intersections number N_{int} of two fractals with arbitrary dimensions D_{f_1} and D_{f_1}, can give a certain analogy. At $D_{f_1}+D_{f_2}>d$ fractals are opaque for one another, at $D_{f_1}+D_{f_1}\leq d$—transparent ones. Let us note, that the Eq. (13) of Chapter 1 gives the condition of permeability (transparence) of macromolecular coils with $D_{f_1}=D_{f_1}$ at d=3: $D_{f_1}=D_{f_2}$=1.50. In case of permeable (transparent) coils the exponent in the Eq. (13) of Chapter 1 right-hand part becomes a negative one. This allows to suppose, that the dimension δ_f characterizes the structure (distribution in space) of low-molecular solvent molecules, forming "swarm", that is, interacting with a macromolecular coil and that each contact macromolecular coil and solvent molecules should be considered as two fractals intersection point: coil and "swarm." It is quite obvious, that solvent molecules should occupy the same space region, as in a macromolecular coil with gyration radius R_g and therefore they do not form Euclidean object, that is, $\delta_f<d$. As it follows from the Eq. (18) of Chapter 1, the greatest value $_f$, at which the compact globule is formed ($D_f=d$), is equal to approx. 2.667. Hence, the dimension δ_f characterizes interactions polymer-solvent number or low-molecular solvent molecules number n per "swarm" of the radius R_s according to the general fractal relationship [86]:

$$n \sim R_s^{\delta_f} . \tag{85}$$

It is easy to see, that a macromolecular coil will be opaque for solvent molecules (accounting for the Eq. (18) of Chapter 1) at the condition $D_f\geq 2$, $\delta_f\geq 1$.

It is obvious, that the sizes of a macromolecular coil and low-molecular solvent molecules "swarm" should be approximately equal to ($R_g\approx R_s$) and taking into account the known fractal Eq. (8), let us obtain the following relationship for n [104]:

$$n \sim N^{\delta_f / D_f} , \tag{86}$$

where N is polymerization degree of macromolecule.

From the Eq. (86) it follows, that δ_f reduction, leading to weaker D_f decrease according to the Eq. (18) of Chapter 1, means n reduction at the condition N=const, that is, the indicated relationship describes a well-known effect of solvent "squeezing out" from a macromolecular coil.

Using the Eqs. (18) of Chapter 1, (84) and the literary values χ_1 [1], the limiting values δ_f can be estimated. From the Eq. (18) of Chapter 1 it follows, that for the macromolecular coil in good solvent (D_f≈1.667 [10]) the value δ_f=0. This means that the solvent with such dimension screens completely attractive interactions of macromolecular coil elements between themselves. For the transparent coil (D_f=1.50 [10]) realization the value δ_f= –0.645 is already required, that is, in this case the solvent should not only screen the indicated above coil elements attractive interactions, but also introduce repulsive interactions polymer-solvent. Thus, the value δ_f<0 means repulsive interactions polymer-solvent availability, δ_f>0—attractive interactions existence and δ_f=0—any interactions polymer-solvent absence. The cited in the monograph [1] χ_1 values for 74 polymer-solvent pairs are varied within the limits of −0.30–1.55. According to the Eq. (84) this means D_f variation within the limits of 1.50–2.21. Such variation for flexible-chain polymers as a matter of fact includes a complete range of D_f possible change for real macromolecular coils: from transparent coil (D_f=1.50 [16]) up to the so-called coil compensated state (D_f≈2.28 [13]).

Since it is supposed, that the dimension δ_f is a measure of interactions polymer-solvent, then it is natural to expect a certain intercommunication $_f$ with polymer $_p$ and solvent δ_s solubility parameters, characterizing these substances cohesion energy [1, 14]. The simple empirical Eq. (82) was received above for PC, which was verified for 14 solvents and showed a good correspondence (within the limits of 2.5%) to the experimental data (see Fig. 44).

The Eq. (82) analog for different polymers can be written as follows [104]:

$$\delta_f = \left(\delta_s^{1/2} - \frac{27}{\delta_p} \right), \tag{87}$$

where δ_s and δ_p are given in (cal/cm^3)$^{1/2}$.

In Fig. 46 the relationship between calculated according to the equation (87) δ_f^T and determined according to the Eq. (18) of Chapter 1 δ_f^T values of low-molecular solvent molecules "swarm" dimension for different polymer solutions in six solvents is assumed. As one can see, between the values δ_f^T and δ_f^e a good correspondence (within the limits of 10%) was obtained.

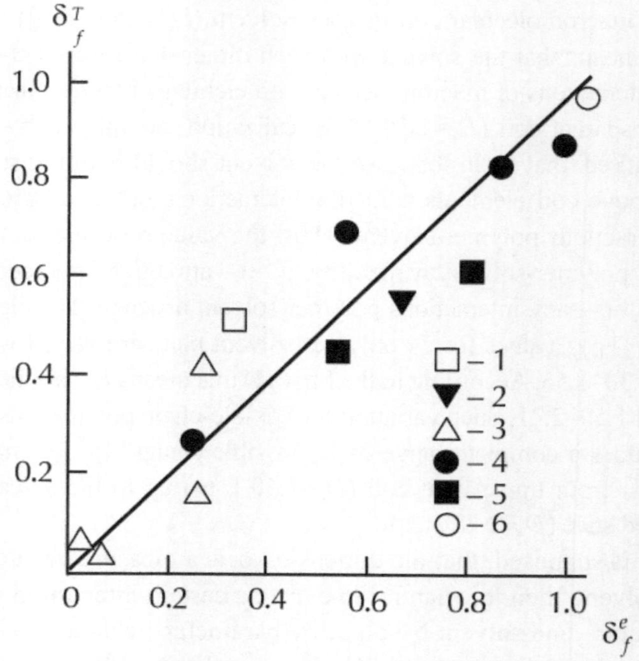

FIGURE 46 The relationship between calculated according to the Eq. (87) δ_f^e and Eq. (18) of Chapter 1 δ_f^e fractal dimension values of solvent molecules "swarm" for polymers solution in tetrachloroethane (1), 1,4-dioxane (2), chloroform (3), N, N-dimethylformamide (4), tetrahydrofuran (5) and dimethylsulfoxide (6).

The Eq. (87) differs somewhat from classical view on polymers dissolution problem. At present it is supposed, that a good solvent corresponds to the condition $\delta_p=\delta_s$, that answers to $D_f\approx 1.667$ or, according to the Eq. (18) of Chapter 1, $\delta_f=0$. However, according to the Eq. (87) this criterion will have another form [104]:

$$\delta_s^{1/2} = \frac{27}{\delta_p}. \tag{88}$$

In Ref. [14] a large number of polymers and solvents for them are adduced, in addition the latter are divided conditionally in four groups: good, indifferent, poor (where polymer dissolves partly or swells) and nonsolvent. In Table 14 the comparison of solvent type estimations according to the criterions $\delta_s = \delta_p$ and (88) in PC case, for which the deviation quantitative measures $|\delta_p - \delta_s|$ and $|\delta_s^{1/2} - 27/\delta_p|$, respectively were used. As it follows from the data of Table 14, solvent quality is described much better by the criterion (88), whereas the transition solvent-nonsolvent is determined completely by the value $|\delta_p - \delta_s|$. Let us note, that the constant 27 in the Eq. (87) variation is possible for different polymers as their characteristics function.

TABLE 14 Characteristics of solvent quality according to two criterions.

Solvent quality	$\|\delta_p - \delta_s\|$, (cal/cm³)$^{1/2}$	$\left\|\delta_s^{1/2} - \dfrac{27}{\delta_p}\right\|$
Good	0.46–1.82	0.02–0.22
Indifferent	0–3.30	0.28–0.41
Poor	0.10–0.78	0.11–0.79
Nonsolvent	1.43–2.35	0.05–0.11

Thus, the stated above results have shown, that the dimension δ_f characterizes the structure (distribution in space) of low-molecular solvent molecules "swarm" and can accept negative values. The empirical techniques of this dimension determination and evaluation of solvent quality in respect to polymer were proposed.

2.5 THE THERMODYNAMICS OF POLYMERS DISSOLUTION

Amorphous polymers heats of dissolution studies were performed often enough. It is supposed, that the heat effect of system transition from metastable (nonequilibrium) glassy state in quasiequilibrium solution give the dominant contribution in the heat of dissolution ΔH_d value. In addition, the component, determined by polymer structure, namely, by the ordered regions availability, can give the essential contribution. The authors of communication [105] performed the value ΔH_d quantitative estimation within the framework of the cluster model of polymers amorphous state structure [106, 107].

The cluster model assumes availability in the amorphous polymers structure of local order domains (clusters), consisting of several densely packed collinear segments of different macromolecules (amorphous analog of crystallite with stretched chains). These clusters are connected between themselves by "tie" chains, forming by virtue of this physical entanglements network, and are surrounded by loosely packed matrix, in which all fluctuation free volume is concentrated [107].

The authors [105] used amorphous glassy polymers: PMMA, polycarbonate (PC) and polyarylate (PAr). The necessary for further evaluations shear modulus values G were obtained according to Young's modulus E experimental values with the formula using [88]:

$$G = \frac{E}{d_f} \tag{89}$$

where d_f is a fractal dimension of polymer structure.

Several common features of ΔH_d draw ones attention [108, 109]. Firstly, above glass transition temperature T_g the value ΔH_d is independent on disolvation temperature T_d. Within the framework of a cluster model the clusters thermo-fluctuation decay at T_g is supposed [110], that is, ΔH_d behavior change at T_g controlled by just this effect should be expected. Secondly, T_g enhancement results in ΔH_d growth at the fixed T_d and within the framework of cluster model polymers with higher T_g have higher values of macromolecular entanglements cluster network density ν_{cl} at the fixed testing temperature [106]. The value ν_{cl} can be calculated with the aid of the following formula [107]:

$$\nu = 0.5 - 2.75 \times 10^{-15} \left(\nu_{cl} \right)^{1/2}, \tag{90}$$

where ν is Poisson's ratio.

Thirdly, the disolvation temperature enhancement leads to ΔH_d reduction and the cluster model supposes ν_{cl} decrease at temperature growth. But the greatest interest the linear dependence of ΔH_d on temperatures difference $(T_g - T_d)$ is presented. Such dependence for different polymers

[108] supposes, that ΔH_d is independent on amorphous polymers structure features, but is determined by approaching to T_g degree only. The dependences of ν_{cl} on (T_g-T) value (where T is the testing temperature) for the three indicated above polymers prove to be analogous by the shape, but differing by ν_{cl} absolute values [111]. One more general characteristic of amorphous polymers structure local ordering degree is the clusters relative fraction φ_{cl}, which can be determined according to the equation [107]:

$$\varphi_{cl} = Sl_{st}\nu_{cl} \tag{91}$$

where S is cross-sectional area of macromolecule.

One should dwell upon the question of φ_{cl} evaluation for PMMA. Owing to bulk side groups availability this polymer has large S value (approx. 64 Å^2 [85]), that determines unreal values $\varphi_{cl}>1$, defined according to the Eq. (91), for this polymer. It is obvious that the side groups cannot be included in densely packed clusters—this is possible only for the main chain segments. Small ν_{cl} values for PMMA in comparison with PC and PAr are due to steric hindrances to cluster formation owing to bulk side groups' availability. Therefore, the authors [105] used S value for the main chain PMMA, which was evaluated according to X-raying data [107].

For the studied polymers all data in coordinates φ_{cl}-(T_g-T) lie down on one straight line [111]. Hence, as ΔH_d, the local ordering degree in polymers is defined by closeness of T to T_g and this allows one to suppose the possibility of correlation of ΔH_d and ν_{cl} or φ_{cl}. Let us note, that similar by shape dependences of polymers properties on (T_g-T) are often found in literature. The dependences of mechanical properties on (T_g-T) for PMMA, PC and PAr can serve as an example [112]. Treating included in clusters segments as linear defects (analog of dislocations), the mathematical calculus of the dislocations theory can be used for amorphous polymers structure and properties description [107]. So, the dissociation energy in counting per one segment U_{cl} can be expressed as follows [107]:

$$U_{cl} = \frac{Gb_B^2 l_{st}}{4\pi}\ln\left(\frac{r}{r_0}\right), \tag{92}$$

where b_B is Burgers vector, r and r_0 are radii of external and internal power fields of dislocation, respectively.

The value b_B was estimated according to the following empirical equation [107]:

$$b_B = \left(\frac{60.7}{C_\infty} \right)^{1/2}, \text{Å}, \tag{93}$$

and the ratio r/r_0 is usually accepted equal to approx. 10.

Hence, the dissociation energy of densely packed clusters U_{dis} per polymer volume unit can be expressed as follows [105]:

$$U_{dis} = U_{cl} \nu_{cl}. \tag{94}$$

The values U_{dis} and ΔH_d comparison as function of $(T_g - T)$ for the three considered polymers, adduced in Fig. 47, has shown their close correspondence, that allows one to confirm, that ΔH_d value represents the energy, which is necessary for dissociation (decay) of clusters in polymers dissolution process. The last phenomenon means the transition of polymer from metastable nonequilibrium state in quasiequilibrium solution [109].

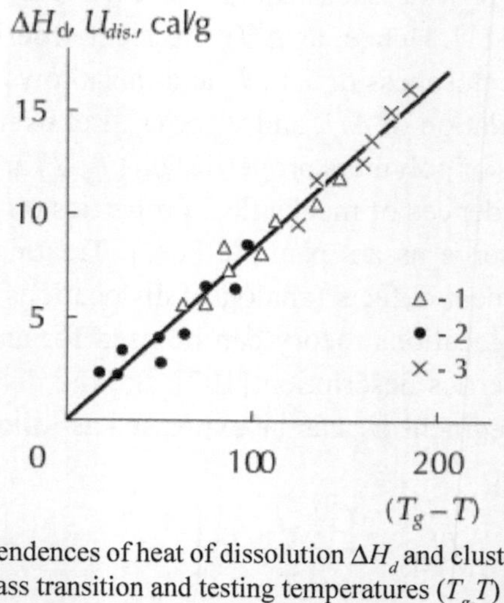

FIGURE 47 The dependences of heat of dissolution ΔH_d and clusters dissociation energy U_{dis} on difference of glass transition and testing temperatures $(T_g - T)$ for PMMA (1), PC (2) and PAr (3). The straight line gives the dependence $\Delta H_d(T_g - T)$ according to the data of Ref. [108].

2.6 THE INTERCOMMUNICATION OF STRUCTURES IN DILUTED SOLUTION AND POLYMERS CONDENSED STATE

At present it is well-known through Ref. [100], that the structure and properties of products, received from polymer solutions, depend essentially on the solvent nature, since concentrated solutions already contain structures, very similar to structures simple types of solid-phase polymers. However, such assumptions remain on the qualitative level and at present there are no analytical relationships, connecting structures of a macromolecular coil in solution and solid-phase polymer between themselves or with the last properties. The development in the last years of modern physical conceptions, similar to fractal analysis, gives prerequisites for this problem solution. As it has been shown above, a macromolecular coil in diluted solution is the fractal, whose dimension D_f and, hence, coil spatial structure is determined by two groups of factors: interactions polymer-solvent and coil elements interactions between themselves. The change of medium, surrounding a macromolecular coil (for example, solvent molecules dimension change, transition in condensed state and so on) leads to the change of its structure, characterized by the value D_f. From the said above it is obvious, that in addition to the indicated in Ref. [100] factors, reflecting the first group interactions, it is necessary to take into account polymer chain structure (its rigidity, side substituents availably and so on), that is, it is necessary to take into consideration the second group of the indicated above factors as well. The authors [113, 114] performed theoretical analysis of intercommunication of macromolecular coil in solution and solid-phase polymer structures and also the last influence on one from the most important polymer properties—glass transition temperature. This analytical intercommunication was obtained within the framework of fractal analysis with the cluster model of polymers amorphous state structure [106, 107] representations participation on the example of amorphous glassy copolymers—aromatic copolyethersulfoneformals (PESF) [47].

As it has been shown above, both solvent type and polymer chain structure will influence on solid-phase polymer structure and, hence, its properties. Both indicated factors have an influence on D_f value and it is obvious, that the chain structure will have main influence on the dimension D_f variation, since the data only for PESF solutions in tetrachloroethane were used. The dependence of intrinsic viscosity [η] on polysulfone fragments contents c_{PSF} in PESF, adduced in Fig. 48, has shown the correlation [η](c_{PSF}) nonadditivity. As it is known through Ref. [115], the value [η] at other equal conditions (as it will be shown further, the dependence of molecular weight *MM* on c_{PSF} is additive practically) means coil sizes increase, that is, its compactness degree reduction. This is expressed analytically by the fractal relationship Eq. (8).

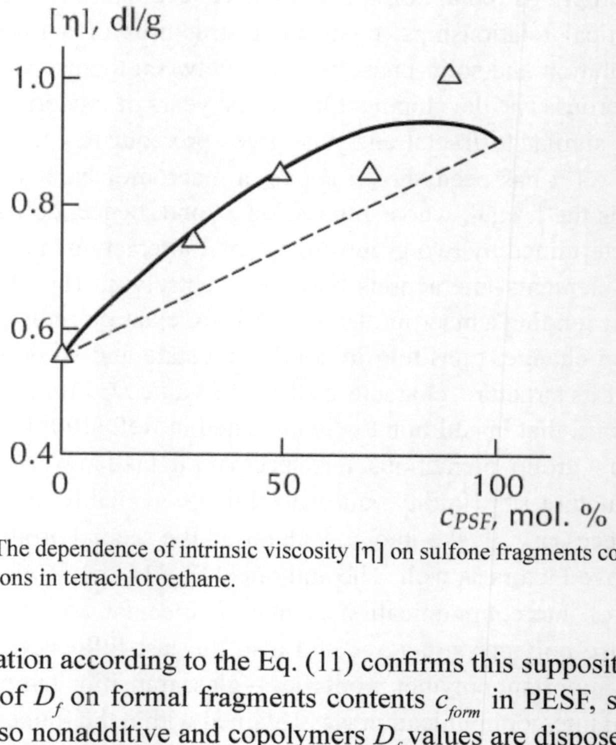

FIGURE 48 The dependence of intrinsic viscosity [η] on sulfone fragments contents c_{PSF} for PESF solutions in tetrachloroethane.

D_f calculation according to the Eq. (11) confirms this supposition. The dependence of D_f on formal fragments contents c_{form} in PESF, shown in Fig. 30, is also nonadditive and copolymers D_f values are disposed below the additive dependence. Thus, the data of Figs. 30 and 48 allow one to trace PESF macromolecular coil structure in tetrachloroethane solution change as a function of either chain fragments contents.

Let us consider now, how these changes influence on the structure of copolymer in solid-phase state. The value D_f with the excluded volume interactions accounting can be determined according to the Eq. (12) of Chapter 1. Since all considered copolymers are linear polymers, then for them d_s=1.0

should be accepted in the indicated equation. The gelation transition or transition to the condensed state is characterized by a macromolecular coil environment change and now it is in similar coils environment instead of low-molecular solvent molecules. This results in its fractal dimension change and for the condensed state its value d_f is determined according to the Eq. (14) of Chapter 1. The combination of the Eqs. (12) and (14) of Chapter 1 at the indicated above conditions d_s=1.0 and d=3 gives for linear polymers [114]:

$$d_f = 1.5D_f . \tag{95}$$

The Eq. (95) demonstrates clearly the genetic intercommunication between structures of reaction products (macromolecular coil in solution) and polymers condensed state.

The structure correct quantitative model is necessary for analytic intercommunication between polymers structure and properties obtaining. As it has been noted above, the cluster model of polymers amorphous state structure will be used with this purpose [106, 107]. The notion of local (short-order) order forms the basis of this model and local order domains (clusters) relative fraction φ_{cl} is connected with glass transition temperature T_g according to the following percolation relationship [107]:

$$\varphi_{cl} = 0.03\left(T_g - T\right)^{0.55}, \tag{96}$$

where T is the testing temperature.

Within the framework of the cluster model glass transition is considered as thermo-fluctuational clusters formation at temperature T_g [110]. Besides, the fractal conceptions are interconnected with the cluster model and these models main characteristics relationship is given as follows [107]:

$$d_f = d - 6\times10^{-6}\left(\frac{\varphi_{cl}}{SC_\infty}\right)^{1/2}, \tag{97}$$

where S is cross-sectional area of macromolecule, expressed in m^2, C_∞ is characteristic ratio, which is polymer chain statistical flexibility indicator [1].

S values for forming PESF fragments of sulfone and formal are accepted equal to 35 and 23 Å^2, respectively [85], and the values S for copolymers are evaluated from the additivity condition according to the mixtures rule. C_∞ value is calculated according to the following equation [107]:

$$C_\infty = \frac{2d_f}{d(d-1)(d-d_f)} + \frac{4}{3}, \tag{98}$$

where d is dimension of Euclidean space, in which a fractal is considered (it is obvious, that in our case d=3).

The Eqs. (14) and (95)–(97) combination allows to calculate T_g value, proceeding only from the structure of macromolecular coil in diluted solution, characterized by dimension D_f, and molecular parameter S. In Fig. 49 the comparison of experimental and calculated according to the proposed model T_g values has shown, from which their good correspondence follows (the discrepancy between them does not exceed of 4%) [113]. Let us note, that Figs. 30 and 49 comparison with using the stated above calculation scheme demonstrates the mode of copolymers with T_g values above additive obtaining—for this D_f values should be less than additive ones. Besides, the reason of T_g increasing at the introduction of bulk side substituents in polymer chain can be pointed out—this results in S essential growth and, hence, φ_{cl} according to the Eq. (97), that leads to T_g increasing, proceeding from the Eq. (96).

The main reason of D_f enhancement at formal fragments contents increase is copolymer chain rigidity reduction, which is due from the chemical point of view to reduction of rigid and very polar diphenylsulfone fragments contents owing to their replacement by methylene groups, increasing polymer chain flexibility. In Fig. 50 the dependence of dimension D_f on groups SO_2 contents is adduced for the considered copolymers, confirming the made above suppositions.

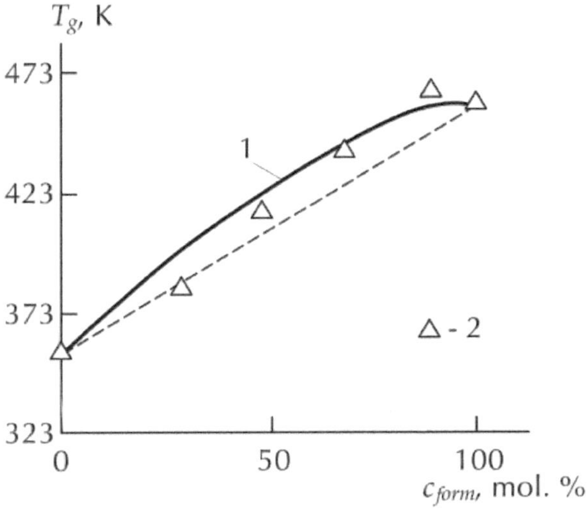

FIGURE 49 The dependence of glass transition temperature T_g on formal fragments contents c_{form}. 1—the experimental data; 2—calculation according to the Eqs. (14) of Chapter 1 and (95)–(97).

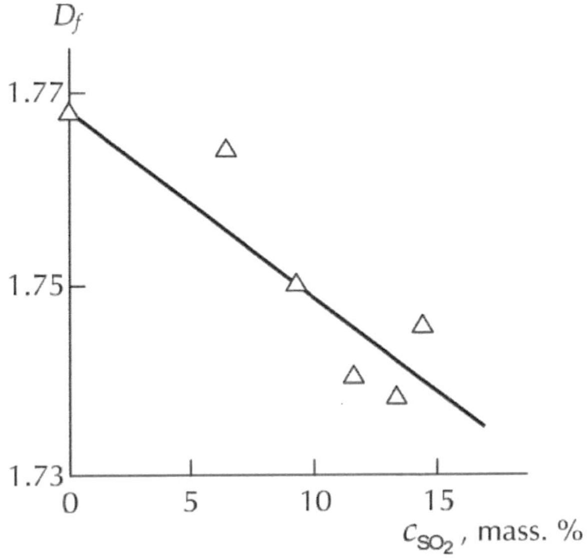

FIGURE 50 The dependence of macromolecular coil fractal dimension D_f for PESF solutions in terachloroethane on groups SO_2 c_{SO_2} contents.

D_f value for the stated above estimations was calculated according to the Eq. (11), that makes necessary solvents choice correctness verification: tetrachloroethane as good solvent and 1,4-dioxane as θ-solvent. As it is known through Ref. [1], this can be made with the aid of the following relationship:

$$\frac{[\eta]}{MM^{1/2}} = f\left(\frac{1}{MM^{1/2}}\right) \tag{99}$$

The indicated correlation should be linear and for good solvent it changes the slope in respect to $MM^{-1/2}$ axis and for θ-solvent it should be parallel one to the indicated axis. In Fig. 51 the correlations, corresponding to the Eq. (99), are adduced for PESF solutions in tetrachloroethane and 1,4-dioxane. As it follows from the given above definition [1], the results of Fig. 51 have pointed out, that for PESF 1,4-dioxane is close to θ-solvent and tetrachloroethane is a good solvent.

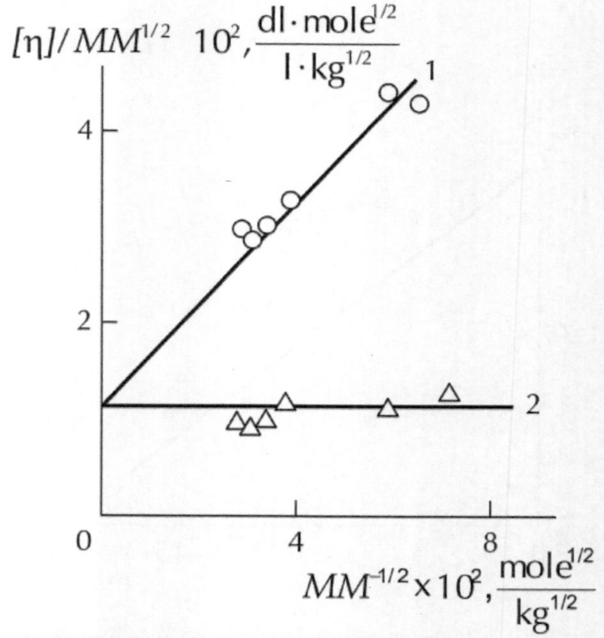

FIGURE 51 The dependences of $[\eta]/MM^{1/2}=f(MM^{-1/2})$, corresponding to the Eq. (99), for PESF solutions in tetrachloroethane (1) and 1,4-dioxane (2).

Hence, the stated above data demonstrated the genetic intercommunication of the structures of synthesis products (macromolecular coil in diluted solution) and polymers condensed state. The technique of the analytical description of intercommunication of the indicated structures and properties of polymers on the example of glass transition temperature, which has shown a good correspondence to the experimental data [113].

The intercommunication of polymers molecular and structural characteristics with their properties in condensed state is known long ago [116]. Kargin formulated this intercommunication scheme most precisely, pointing out, that polymer properties were encoded in molecular level and were realized in supramolecular one [116]. For the purpose of this postulate confirmation the authors [117] used the considered above model [113] for the description of glass transition temperature of copolymers polyarylate-polyarylenesulfonoxide [115, 118, 119].

In Fig. 52 the relationship between D_f and T_g is adduced for five considered copolymers. As it follows from the adduced plot, T_g linear decay at D_f growth is observed, that confirms the cited above postulate about polymers properties encoding on the molecular level. Let us remind, that D_f value characterizes the structure of macromolecular coil in diluted solution, that is, as a matter of fact the structure of an isolated macromolecule.

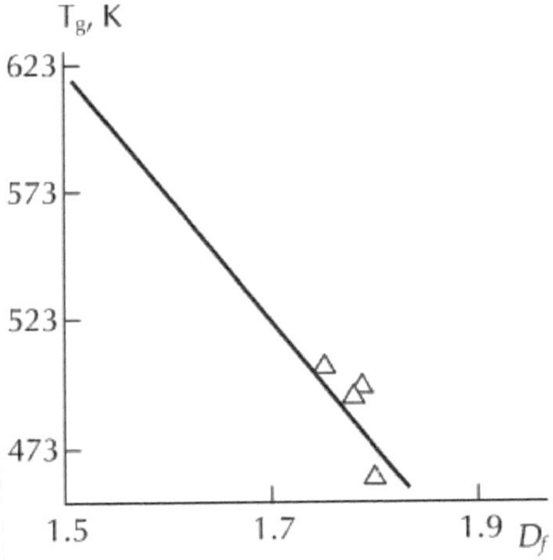

FIGURE 52 The dependence of glass transition temperature T_g on macromolecular coil fractal dimension D_f in tetrachloroethane for copolymers polyarylate-polyarylenesulfonoxide. The points—experimental data; the straight line—calculation according to the Eqs. (14) of Chapter 1 and (95)–(97).

However, the macroscopic properties, such as T_g, are measured for polymers condensed (block) state, but not for their solutions. In Fig. 53 the dependence of T_g on the indicated state dimension d_f, is adduced which finds out again T_g linear decay at d_f growth. Thus, the clear genetic intercommunication of polymer structures in solution and condensed state is traced. Any change of synthesis conditions, effecting on D_f value, results in d_f change and, hence, polymer properties. This rule illustrates clearly properties variation for the same polymer (polyarylate), synthesized by the polycondensation different methods (equilibrium and nonequilibrium ones) and having by virtue of this different a_η values and, hence, different D_f values according to the Eq. (4).

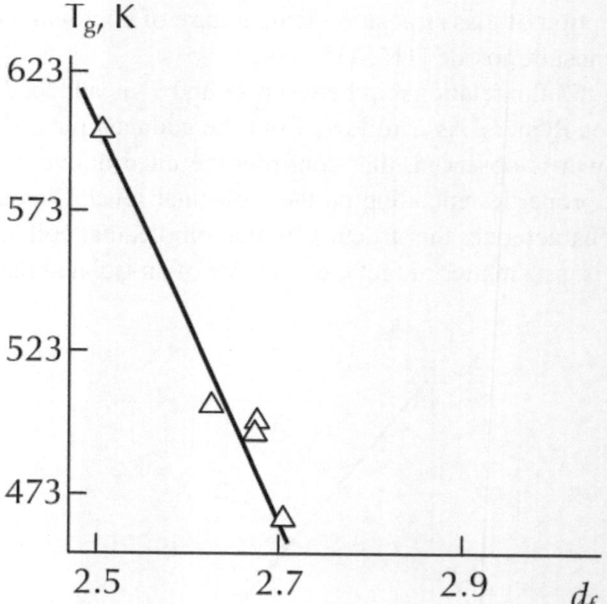

FIGURE 53 The dependence of glass transition temperature T_g on condensed state structure fractal dimension d_f for copolymers polyarylate-polyarylenesulfonoxide. The points—experimental data; the straight line—calculation according to the Eqs. (14) of Chapter 1 and (95)–(97).

Nevertheless, the proposed scheme allows to obtain the relationships, similar to the adduced in Figs. 52 and 53, only for any one polymers type. The model [113] gives a general relationship that allows one to evaluate

T_g value theoretically from the Eqs. (14) and (95)–(97) combination. In Figs. 52 and 53 the theoretical dependences $T_g(D_f)$ and $T_g(d_f)$ are shown by solid lines, which prove to be almost linear and correspond excellently to T_g experimental values. The calculation T_g for polycarbonate at S=32 Å^2 [85] and C_∞=2.4 [69] gave the value T_g=426 K, which is close to the experimental one (≈426 K [50]).

Hence, the considered above scheme of genetic intercommunication of synthesis products and condensed state structures with polymer properties demonstrates clearly, that the foundation of the latter lays (encodes) quite at the synthesis stage and cannot be set by only singular polymer macromolecule chemical structure. The using of the concrete quantitative model of solid polymers structure together with fractal analysis methods gives the possibility of prediction of not only glass transition temperature T_g, but also polymers other important properties number [120].

The stated above model of the dimensions D_f and d_f intercommunication allows one to develop two techniques of D_f value prediction in good solvent on the basis of the known molecular characteristics C_∞ and S [84]. As it has been indicated above, C_∞ and d_f intercommunication is given by the Eq. (98), which for d=3 is simplified to [84]:

$$C_\infty = \frac{d_f}{3(3-d_f)} + \frac{4}{3}. \tag{100}$$

The Eqs. (95) and (100) combination allows to obtain the relationship for the dimension D_f evaluation [84]:

$$C_\infty = \frac{D_f}{3(2-D_f)} + \frac{4}{3} \tag{101}$$

Another variant of obtaining the intercommunication between molecular characteristics C_∞ and S, on the one hand, and D_f, on the other hand, uses the Eq. (97). Between φ_{cl} and C_∞ the following relationship exists [106]:

$$\frac{2}{\varphi_{cl}} = C_\infty^{D_{ch}} \tag{102}$$

where D_{ch} is fractal dimension of macromolecule part between clusters.

TABLE 15 The literary values of parameters and the fractal dimension calculated values D_f^T.

Polymer	Solvent	δ_p, (cal/cm^3)$^{1/2}$	δ_s, (cal/cm^3)$^{1/2}$	C_∞	S, Å^2	D_f	The Eq. (101)		The Eq. (72)	
							D_f^T	Δ, %	D_f^T	Δ, %
PS	Tetrahydrofuran	8.60–9.52	9.53–9.90	8.82–13.80	69.8	1.818	1.914	5.3	1.923	5.8
PET	Chlorophenol	10.0–10.1	11.40	4.10–4.21	20.0	1.809	1.791	1.0	1.700	6.0
PVC	Dichloroethane	9.50–9.66	9.80	6.70–6.80	27.2	1.796	1.882	4.0	1.838	2.3
PSF	Chloroform	10.35–10.64	9.13–9.40	3.60	32.0	1.744	1.744	—	1.719	1.0
PDMS	Benzene	7.30–7.35	9.15–9.22	5.20–7.30	43.6	1.676	1.840	9.8	1.835	9.5
PMMA	Acetone	9.10–9.50	9.62–10.0	7.0–8.70	63.8	1.764	1.889	7.0	1.899	7.7
PC	Methylene chloride	9.68–9.93	9.70–9.88	2.40	30.9	1.648	1.525	7.7	1.575	4.4
PBN	Carbon tetrachloride	8.43	8.60	4.90–6.80	20.7	1.724	1.829	6.0	1.746	1.3
PAr	Tetrachloroethane	10.70–10.78	10.30–10.40	2.30	30.9	1.548	1.488	4.1	1.558	0.6

Footnote: PET—poly(ethylene terephthalate), PBN—butadiene-nitrile rubber.

The minimum value d_f should be used in the Eq. (95) for the minimum (in good solvent) D_f value obtaining. As it has been shown in Ref. [121], at the minimum d_f values D_{ch} magnitude a little less than one and therefore to the first approximation D_{ch}=1.0 can be accepted. At this conditions the Eqs. (97) and (102) combination allows to obtain analytical intercommunication between C_∞, S and D_f in the Eq. (72) form [84].

In Table 15 the ranges of the literary values δ_p [1, 14, 16, 49, 122], δ_s [1, 14, 16], C [1, 69, 70, 123, 124], S [85] and also D_f values, calculated according to the Eq. (4) with literary values a_η using [5, 29–31, 125, 126] are adduced.

For the comparison in the same table the theoretical D_f^T values, calculated according to the Eqs. (101) and (72) and also their discrepancy with D_f Δ are adduced. Besides, the graphic comparison of D_f and D_f^T values is performed in Fig. 54. As it follows from the data of Table 15 and Fig. 54, between theoretical and experimental D_f values a good correspondence was obtained—in 18 considered cases discrepancy does not exceed 10%. Calculation according to the Eq. (72) gave more precise results—the average discrepancy between D_f and D_f^T, calculated according to this equation, makes up 5.4% and according to the Eq. (101)—7%.

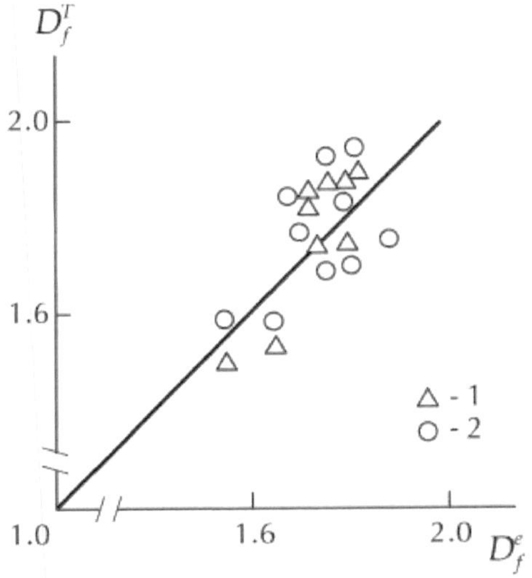

FIGURE 54 The comparison of experimental D_f and calculated according to the Eqs. (101) (1) and (72) (2) values of macromolecular coil fractal dimension in good solvent for the indicated in Table 15 polymers.

Let us consider D_f^T calculations error reasons. In Ref. [84] it has been accepted, that a good solvent for polymer is the one, for which the condition $\delta_s=\delta_p$ is performed [16]. As one can see from the data of Table 15, this condition is not always performed, that is an error source. Besides, the literary sources give a large enough C_∞ variation range. And at last, the usual error of the exponent a_η determination makes up ±0.05 [42], that for D_f value, calculated according to the Eq. (4) in the middle of the range of D_f=1.5–2.0 gives the error of 3.5%. Taking into account all the said, above the adduced in Table 15 discrepancy between D_f and D_f^T should be assumed acceptable.

Hence, the stated above similar techniques allow one to determine the fractal dimension of a macromolecular coil in a good solvent on the basis of the known molecular characteristics of polymer. These techniques give identical results, but D_f^T evaluation precision with two characteristics (C_∞ and S) is somewhat higher, than at one (C_∞) usage.

2.7 THE INFLUENCE OF MACROMOLECULAR COIL STRUCTURE ON THE PROCESSES OF POLYMERS SYNTHESIS IN SOLUTION

Various organic solvents were used as reactionary medium at nonequilibrium polycondensation in solution realization [96]. The solvent type influence on the synthesis reaction main characteristics (conversion degree *Q* and molecular weight *MM*) is well known and is explained usually by solvent various characteristics (dielectric constant, solubility parameter, heat of dissolution and so on) [96]. However, up to now the indicated effects general theoretical explanation is not obtained. Besides, at the solvent type influence analysis its correlation with polycondensation process quantitative characteristics (the same *Q* and *MM*) is usually considered, but any changes of polymer structure or reaction mechanism are not assumed, although the possibility of side reactions is noted repeatedly [96]. The authors [71, 127] studied the solvent influence on the enumerated above characteristics on the example of the rules of chloranhydride of terephthalic acid and phenolfthaleine low-temperature polycondensation (polyarylate F-2), performed in 8 different solvents [128].

The exponent a_η values for F-2 solutions, synthesized in the same solvent, in three different solvents (tetrachloroethane, tetrahydrofuran and 1,4-dioxane) is adduced in Ref. [5]. The using of these a_η values for D_f calculation according to the Eq. (4) has shown, that D_f value is dependent on the used solvent type. Such effect was expected, since at present it is known through Ref. [13], that the dimension D_f of a macromolecular coil in solution is defined by two groups of factors: polymer-solvent interactions and interactions between drawing closer to one another elements of a coil itself. In the first approximation the first from the indicated factors can be taken into account with the aid of the polymer δ_p and solvent δ_s solubility parameters difference $\Delta\delta=|\delta_{p-}\delta_s|$ [16]. This method is an approximate one and has predictive power for only certain sets polymer-solvent [22]. Nevertheless, the authors [71] traced on the plot of Fig. 55 three calculated according to the equation (4) D_f values as a function of $\Delta\delta$ and this dependence continuation used as calibrating curve. As it was to be expected [16], $\Delta\delta$ increase (solvent quality change for the worse in respect to polymer) leads to D_f growth: at $\Delta\delta=0$ F-2 macromolecular coil is permeable ($D_f=1.50$), a good solvent for F-2 (D_f1.667) the solvent with $\Delta\delta\approx0.9$ (cal/cm^3)$^{1/2}$ can be assumed and θ-point for this polymer is reached at $\Delta\delta\approx2.2$ (cal/cm^3)$^{1/2}$. Such gradation supposes, that polyarylate F-2 is a linear polymer [25]. The estimated according to calibrating curve D_f^{cal} values are adduced in Table 16.

Another (more precise) calculation D_f method uses the Eq. (11). Since in Ref. [128] all values of the reduced viscosity η_{red} for polyarylate F-2, synthesized in 8 different solvents, are measured in the same solvent (tetrachloroethane), then for $[\eta]_\theta$ evaluation Mark-Kuhn-Houwink equation can be used [5]:

$$[\eta]_\theta = 0.266\times10^{-4} MM^{a_\eta}, \tag{103}$$

where a_η is equal to characteristic value for θ-solvent, namely, 0.5 [10]. In its turn, the value $[\eta]$ can be calculated according to the Eq. (48).

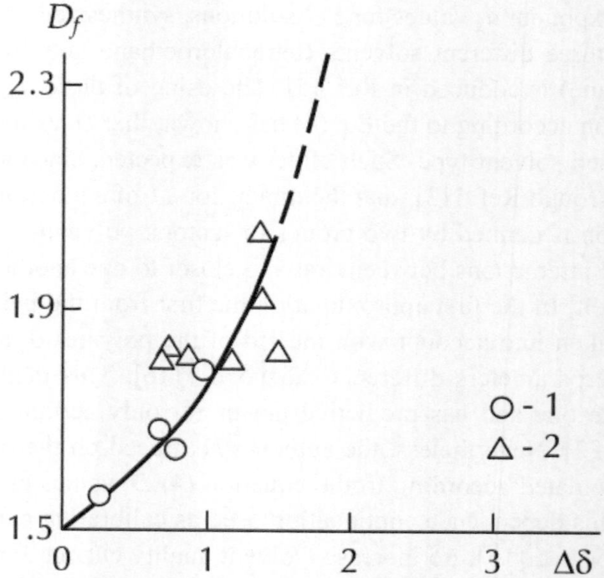

FIGURE 55 The dependence of macromolecular coil fractal dimension D_f on polymer and solvent solubility parameters difference Δδ for polyarylate F-2. 1—calculation according to the Eq. (4); 2—calculation according to the Eq. (11); 3—a calibrating curve.

TABLE 16 The values of fractal dimension of a macromolecular coil of polyarylate F-2, synthesized in different solvents.

Solvent	D_f^{cal}	D_f^T
N, N-dimethylformamide	1.80	2.03
Nitrobenzene	1.70	1.80
Acetone	1.70	1.81
Dichloroethane	1.63	1.80
Chloroform	1.59	1.90
1,2,4-Trichlorobenzene	1.99	1.80
Benzene	2.14	1.81
Hexane	2.36	2.05

The fractal dimension D_f^T values, calculated with the Eqs. (11), (48) and (103) using, are presented as a function of $\Delta\delta$ in Fig. 55, from which their satisfactory correspondence with the calibrating curve follows, excluding D_f^T for F-2, synthesized in hexane. D_f^T values are also adduced in Table 16, from which D_f^T and D_f^T variation tendencies identity follows. Let us note, that the dimensions D_f^{cal} and D_f^T were received for two different groups of polyarylate F-2: the first was received for the same polymer, but was measured in different solvents, the second was received for different polymers actually (synthesized ones in different solvents), but measured in the same solvent. Their identity supposes, that the macromolecular coil formation in synthesis process submits to the same laws, as dissolution, that is, controlled by the same groups of factors and dimension is determined according to the same Eq. (11).

In Fig. 56 the dependence of D_f^T on solubility parameter of using at synthesis solvent δ_s for polyarylate F-2 is adduced. The typical for the similar dependences plots cupola-like shape is observed [16] (compare with Fig. 2), which confirms D_f^T increase (compactness and branching of coil enhancement) with solvent quality change for the worse.

The adduced in Figs. 55 and 56 plots allow one to proceed to F-2 polycondensation mechanism analysis within the framework of irreversible aggregation models [129–131]. First of all it should be noted, that D_f^T values variation exceeds practically the fractal dimension change possible interval in aggregation cluster–cluster models (D_f=1.75–2.11 [131]). As it is known through Ref. [131], the lower limit of fractal dimension (D_f =1.75) corresponds to diffusion-limited aggregation and the upper one (D_f=2.11)—to the so-called chemically limited aggregation, in which only rare clusters collisions lead to their aggregation. In the last model the aggregate dimension D_f is determined by the aggregation probability p and the greatest value D_f corresponds to the smallest p value of order of 0.01 [132]. The aggregation, in which cluster is reformed during its growth, that is, in it two competing processes (aggregation and disaggregation) proceed, in it is another possible variant of the macromolecular coils with large D_f realization [133]. According to the available data to solve this dilemma is impossible, but D_f increase for solvents with the smallest affinity in respect to polymer (N, N-dimethylformamide and hexane, having the greatest values $\Delta\delta$) allows one to confirm, that aggregation mechanism replacement occurs [71].

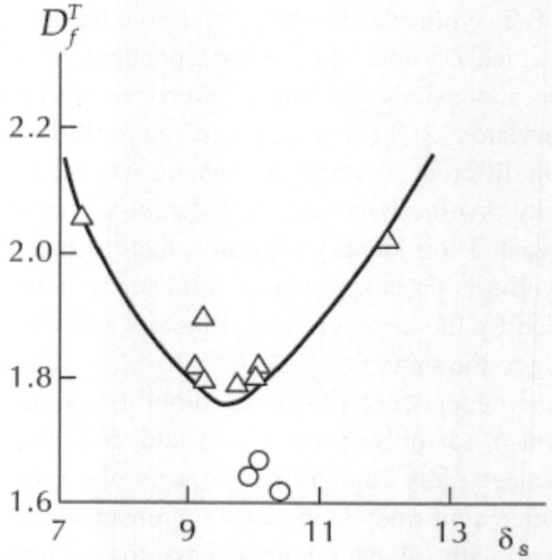

FIGURE 56 The dependence of macromolecular coil fractal dimension D_f^T on solubility parameter of using at synthesis solvent δ_s for polyarylate F-2.

Since [η] values, according to which D_f^T values were estimated according to the Eq. (11) were measured in the same solvent (tetrachloroethane), then this means, that the interaction of macromolecular coil elements between themselves was the only factor, influencing on D_f [13]. This supposes concretely weakening of repulsive forces between themselves and, respectively, attractive interactions intensification. Let us note, that D_f^T increasing means polymer branching degree enhancement. This rule can be demonstrated by calculation of the spectral dimension d_s, characterizing connectivity degree of fractal [45], according to the Eq. (12) of Chapter 1. The evaluation according to this equation gives d_s=1.125 for F-2, synthesized in dichloroethane, and d_s=1.39 for F-2, synthesized in hexane. The theoretical values d_s vary within the following limits: for a linear polymer chain d_s=1.0, for a branched one—d_s=1.33 [45]. Thus, received in N, N-dimethylformamide and hexane polyarylates F-2 by their properties are branched polymers and received in remaining solvents are close to linear ones (both within the framework of fractal analysis and within the framework of Flory mean field theory [25]).

The considered effect is very similar to the received one at the comparison of polyarylates, synthesized by equilibrium and nonequilibrium (interphase) polycondensation [5]. So, for polyarylate F-2, received by the first from the indicated methods, estimation according to the Eq. (4), gives D_f=1.77 and by the second one—D_f=1.55. This distinction was explained by the polyarylates structure distinction, received by the indicated above methods. Hard conditions of equilibrium polycondensation (high temperature, large process duration) can cause the appearance of branched reaction products owing to lacton cycle rupture in phenolphthaleine residues; then the exponent a_η in Mark-Kuhn-Houwink equation should be reduced, since its value is less for a branched polymer, than for a linear one [5]. If it is like that, then D_f value should be increased respectively according to the Eq. (4).

The relationship, based on Flory-Fox theory, can be used also for branched polymer characteristic [5]:

$$g = \frac{\overline{R_\theta}^2}{\overline{R_{l,\theta}}}, \quad (104)$$

where g is a branching factor, $\overline{R_\theta}$ and $\overline{R_{l,\theta}}$ are mean hydro dynamical radius of branched and linear macromolecule coils in θ-solvent, respectively.

Evaluating a_η values according to D_f^T known values according to the Eq. (4) and the coefficient K_η magnitude in Mark-Kuhn-Houwink equation according to the obtained by such mode a_η values (the Eq. (2)), the intrinsic viscosities in θ-solvent $[\eta]_\theta$ and $[\eta]_{l,\theta}$ values can be calculated for F-2 samples, synthesized in hexane and dichloroethane (at the condition a_η=0.5) and then g value can be evaluated according to the relationship [5]:

$$\frac{[\eta]_\theta}{[\eta]_{l,\theta}} = g^{2-a_\eta}, \quad (105)$$

where the exponent in Mark-Kuhn-Houwink equation for a linear polymer is accepted as a_η.

These estimations have shown, that g=0.806, that is close again to the values, characterizing polyarylates F-2 distinction, received by equilibrium and interphase polycondensation [5].

The indicated aspect of polymers synthesis in different solvents has an important applied significance. As it is known through Ref. [5], mechanical properties of polyarylates of comparable molecular weight, but received by methods of equilibrium and nonequilibrium polycondensation, differ strongly, in addition the first from the indicated methods gives a better complex of mechanical properties.

Let us also note, that the structure of a macromolecular coil in synthesis process defines to a considerable extent the received polymer molecular weight. In Fig. 57 the dependence of mean-viscous molecular weight M_η on D_f value for polyarylate F-2, synthesized in different solvents is adduced. M_η values were received with the aid of Mark-Kuhn-Houwink equation for solutions in tetrachloroethane [5]. As one can see, the systematic M_η reduction at D_f growth, obtained by both considered above methods, is observed. D_f increase means coil larger compactness and p smaller values. Both these factors should result in formation of aggregates with smaller sizes large number [131], that is, in M_η reduction.

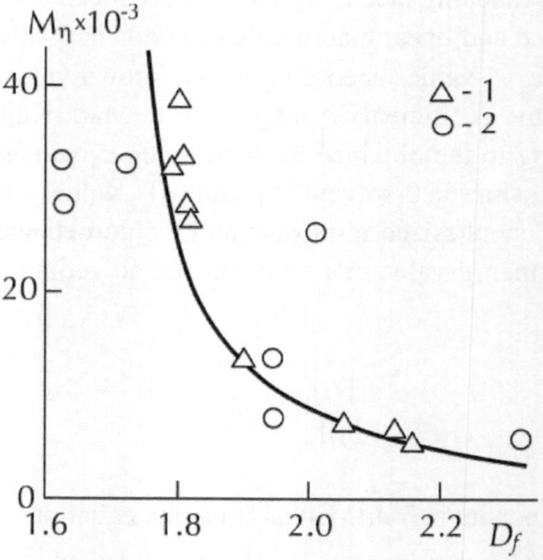

FIGURE 57 The dependence of mean-viscous molecular weight M_η on macromolecular coil fractal dimensions D_f^T (1) and D_f^T (2) for polyarylate F-2.

Hence, the performed above analysis has shown that different solvents using in low-temperature nonequilibrium polycondensation process can result not only in synthesized polymer quantitative characteristics change, but also in reaction mechanism and polymer chain structure change. This effect is comparable with the observed one at the same polymer receiving by methods of equilibrium and nonequilibrium polycondensation. Let us note, that the fractal analysis and irreversible aggregation models allow in principle to predict synthesized polymer properties as a function of a solvent, used in synthesis process. The stated above results confirm Alexandrowicz's conclusion [134] about the fact that kinetics of branched polymers formation effects on their topological structures distribution and macromolecules mean shape.

It is accepted to call fractal reactions either fractal objects reactions or reactions in fractal spaces [135]. The characteristic sign of such reactions is autodeceleration, that is, reaction rate k_r reduction with its proceeding duration t [136]. Let us note, that for Euclidean reactions the linear kinetics and respectively the condition k_r=const are typical [137]. The fractal reactions in wide sense of this term are very often found in practice (synthesis reactions, sorption processes, stress-strain curves and so on) [74]. The following relationship is the simplest and clearest for the indicated effect description [136]:

$$k_r \sim t^{-h}, \tag{106}$$

where h is the heterogeneity exponent ($0 \leq h \leq 1$), turning into zero in case of classical behavior (of reaction in homogeneous or Euclidean mediums) and k_r=const, as it was noted above.

The Eq. (106) allows one to give the following treatment of physical significance of a reactionary medium heterogeneity: the higher this heterogeneity (fractality) degree is the stronger k_r decay is at reaction proceeding. It is obvious, that the incommunication of heterogeneity degree and reactionary medium structural state is necessary to establish for this postulate practical application [138].

The general fractal relationship for the description of reaction in solutions kinetics looks like [87]:

$$Q \sim c_0 \eta_0 t^{(d-D_f)/2}, \tag{107}$$

where Q is the conversion degree, c_0 is reagents initial concentration, η_0 is reactionary medium initial viscosity, t is reaction duration, d is the dimension of space, in which reaction proceeds (further it was accepted, that the indicated space was Euclidean one with dimension d=3), D_f is fractal dimension of a macromolecular coil in solution.

The reaction rate k_r can be received by the Eq. (107) differentiation by time t [138]:

$$k_r = \frac{dQ}{dt} \sim c_0 \eta_0 t^{(1-D_f)/2} \sim t^{-(D_f-1)/2} \tag{108}$$

The relationships (106) and (108) comparison allows to receive the following equation [138]:

$$h = \frac{D_f - 1}{2}. \tag{109}$$

Let us estimate the values of fractal dimension of a macromolecular coil D_f in solution for two cases: h=0 (D_f^0) and h=1.0 (D_f^1). From the Eq. (109) it follows: D_f^0=1.0 and D_f^1=3.0. The physical significance of the received limiting D_f values in this case is obvious. For Euclidean (homogeneous) reactionary medium (h=0) k_r is the greatest for all t>1 and D_f^0=1.0 means, that the macromolecule represents a completely stretched chain, but not a macromolecular coil. If the latter is a fractal object, for which internal regions are screeched by the surface and accessible for reaction proceeding [139], then for the stretched chain all reactive centers are accessible and steric hindrances are absent. Therefore, k_r value is the greatest and constant. The value D_f^1=3.0 means, that macromolecule is rolled in a compact globule, for which a small surface part is accessible for reaction. Therefore, the value k_r is minimum and fast approaches to zero [138].

The reaction rate choice defines a concrete type of this reaction. So, in polymers synthesis reactions a high rate is desirable and in thermooxidative degradation process—a low one. As it follows from the Eq. (106), k_r value is defined by the heterogeneity degree. Besides, from the Eq. (108) it follows, that the parameters c_0 and η_0 will also influence on reaction rate.

Hence, the proposed analysis allows connecting macromolecular coil structure and medium heterogeneity degree for polymer solutions. In its turn, this intercommunication knowledge and ability to control a macromolecular coil structure allows one to regulate the chemical reaction rate in desirable direction [138].

It has been noted above, that for polyarylates (PAr) series, received by high-temperature (equilibrium) and interphase polycondensation, different values of the exponent a_η in Mark-Kuhn-Houwink equation for PAr of the same chemical composition, but received by different methods, are observed. In addition, the value a_η for PAr, received by interphase polycondensation, systematically exceeds the corresponding exponent for PAr, received by high-temperature polycondensation [5]. In Ref. [5] this distinction was explained by different structure of PAr, synthesized by the indicated polycondensation modes. Hard conditions of high-temperature polycondensation (high temperature, large duration of process) can result in appearance of branched reaction products owing to lacton cycle rupture in phenolphthaleine residues that leads to branched polymer chains appearance and corresponding a_η reduction. Such explanation is an individual one: a_η similar distinction is observed and for PAr on the basis of diane, which do not have phenolphthaleine in their composition [5]. Therefore, it should be supposed, that the explanation should be sought in distinctions of polycondensation methods, by which the considered PAr were synthesized. Proceeding from this, the authors [140, 141] proposed the general treatment of the indicated above distinctions of a_η values and, hence, D_f (see the Eq. (4)) for PAr, received by different polycondensation methods, with the fractal analysis methods using a_η values for seven polyarylates of different chemical structure, received by high-temperature (equilibrium) and interphase polycondensation and determined for three different solvents (tetrachloroethane, tetrahydrofuran and 1,4-dioxane) are accepted according to the data of Ref. [5]. The experimental values of fractal dimension of a macromolecular coil D_f^e in the indicated solvents are determined according to the Eq. (4). The solubility parameter δ_s values for these solvents are taken from the literary sources [1, 14, 16]. The fractal dimension of low-molecular solvents molecules totality ("swarm") is determined according to the Eq. (82).

As it has been shown in Chapter 1, several methods of macromolecular coils fractal dimension determination exist. So, phantom (taking into no account the excluded volume effects) fractal dimension D_f^{ph} is determined according to the Eq. (11) of Chapter 1 and swollen (taking into account the indicated effects) dimension D_f according to the Eq. (12) of Chapter 1. The last equation was obtained for the case, when solvent molecules have zero-dimensional (point) structure. If the solvent molecules structure is more complex, then D_f value is determined according to the Eqs. (80) and (81). From the comparison of Eqs. (12) of Chapter 1 and (80) it follows, that for δ_f>0 the last equation gives higher values D_f (smaller values a_η), than the first ones. Such relation supposes the formal possibility of the description of macromolecular coil structure, obtained in the high-temperature polycondensation process, with the aid of the Eq. (80) and by the interphase one—the Eq. (12) of Chapter 1. The same treatment the indicated polycondensation modes mechanisms distinction supposes: if high-temperature polycondensation proceeds in solution, then the interphase one—on boundary of phases division. This means that at the first from the indicated polycondensation modes the structure of solvent molecules influences actively on forming structure of synthesized macromolecular coil, screening repulsive interactions between elements of coil, whereas in the second from the indicated modes such screening is absent. Therefore, received in the solvent with δ_f>0 availability the coil structure is more compact, than obtained without solvent influence. This effect of structural memory is maintained at polymers dissolution in the same solvent, that determines the exponent a_η different values for polyarylates of the same chemical structure, but received by the indicated above polycondensation modes. For this hypothesis verification in Fig. 58 the theoretical dependences $D_f(d_s)$, calculated according to the Eqs. (12) of Chapter 1 and (80), with experimentally obtained D_f^e values comparison is performed. The values d_s were determined according to the Eq. (80) at known D_f^e values (the Eq. (4)) and δ_f (the Eq. (82)). As it follows from this comparison, D_f^e values for PAr, received by high-temperature polycondensation, correspond to the Eq. (80) and received by interphase polycondensation—to the Eq. (12). Thus, the data of Fig. 58 confirm the made above supposition about a_η values (or D_f) distinction reasons for PAr, received by different polycondensation modes. It is important to note, that this correspondence is equally correct for both linear (d_s=1.0) and branched (d_s=1.20–1.33) PAr, that is, the supposition about chain branching influence on a_η values distinction, offered in Ref. [5], in the given case is not valid [141].

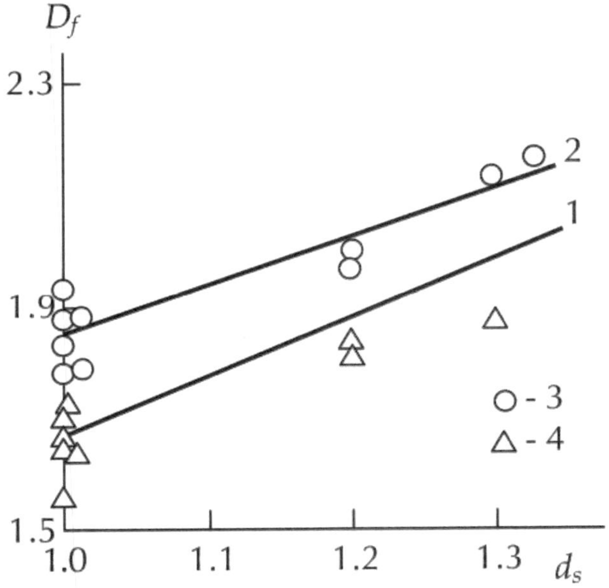

FIGURE 58 The dependences of macromolecular coil fractal dimension D_f on spectral dimension d_s for polyarylates. 1—calculation according to the Eq. (12) of Chapter 1; 2—calculation according to the Eq. (80) at δ_f=0.60; 3, 4—calculation according to the Eq. (4) for PAr, received by interphase (3) and high-temperature (4) polycondensation.

In Table 17 the comparison of different parameters for PAr on the basis of phenolphthaleine (F-2 according to the designation, accepted in Ref. [5]) for different solvents, in which Mark-Kuhn-Houwink parameters were determined. Firstly, δ_f calculation according to the Eq. (18) of Chapter 1 gave these dimension values of 0.36–0.84 for F-2, received by equilibrium polycondensation, and $\delta_f \leq 0$—for received by interphase polycondensation. This means that the solvent molecules structure does not influenced (does not screened repulsive interactions) on F-2 macromolecular coil structure, received by interphase polycondensation, and influences on the similar factor for F-2, received by high-temperature polycondensation. In other words, if the solvent influences on formation of coil structure in synthesis process, then it will be influenced and on this polymer formed coil structure at subsequent dissolution as well and vice versa. Secondly, δ_f values calculated according to the Eq. (82) and Eq. (18) of Chapter 1 correspond satisfactorily to one another [142]. Thirdly, if for polyarylate F-2, received by interphase polycondensation, D_f values are close to dimension of macromolecular coil in good solvent (approx. 1.667 [10]), then for F-2, received by high-temperature polycondensation, they grow systematically at δ_f increase within the range of D_f^e =1.77–1.94. And at last, D_f^e values for F-2, received by high-temperature and interphase polycondensation, correspond to D_f values, calculated according to the Eq. (80) and Eq. (12) of Chapter 1, respectively, that confirms the made above supposition.

TABLE 17 The characteristics of polyarylate F-2 macromolecular coil in different solvents.

Solvent	δ_f, the Eq. (18) of Chapter 1		δ_f, the Eq. (82)	D_f^e		D_f, the Eq. (80)	D_f, the Eq. (12) of Chapter 1
	F-2e*	F-2i**		F-2e	F-2i		
Tetrachloroethane	0.36	0	0.60–0.62	1.77	1.55	1.85	1.67
Tetrahydrofuran	0.68	0	0.40–0.55	1.88	1.64	1.83	1.67
1,4-dioxane	0.84	0	0.50–0.72	1.94	1.67	1.89	1.67

*F-2 was synthesized by high-temperature (equilibrium) polycondensation.
**F-2 was synthesized by interphase polycondensation.

Let us note, that the value D_f of a macromolecular coil in solution defines the polymer structure in condensed state (see Section 2.6). For instance, higher values D_f for PAr, received by high-temperature polycondensation, define higher fractal dimension of solid-phase PAr [114] and, hence, larger values of deformation up to failure [143], that is confirmed experimentally [5].

Hence, the stated above results have shown that a common reason of different structure of a macromolecular coil of PAr of the same chemical structure, but received by different polycondensation modes, is participation (or participation absence) of solvent with dimension δ_f>0 in this coil structure formation in synthesis process. The obtained in this process macromolecular structure is maintained and at subsequent polymer dissolution and solvents influence in this case is similar to their influence in the polymer synthesis process. The fractal analysis methods give mathematical calculus for this problem quantitative description [140, 141].

In Ref. [144] it has been shown that polyurethanearylates on the basis of dianurethane and dichlorandhydride of terephthalic acid (PUAr), received by interphase and acceptor-catalytic polycondensation, have glass transition temperatures T_g, differing on about 50 K. This distinction was explained by different conformations of macromolecules and, as consequence, by different supramolecular structure: fibrillar and globular, respectively [5, 144]. Forming in the interphase polycondensation process fibrillar structure possesses higher ordering, than forming at acceptor-catalytic polycondensation globular structure, that defines properties distinction of the indicated PUAr at comparable molecular weight. This conclusion was confirmed by an indirect evidences number [5, 144]. Nevertheless, the quantitative estimation of different kinds factors influence, defining the indicated distinctions, on conformation, structure and properties of polymers were not performed. The authors [145] have shown the possibility of such prediction on the example of PAUr, received by interphase and acceptor-catalytic polycondensation, in the framework of fractal analysis [146] and cluster model of polymers amorphous state structure [106, 107].

Chlorobenzene in case of interphase polycondensation ($PUAr^{if}$) and acetone in case of acceptor-catalytic ($PUAr^{ac}$) served as organic mediums at synthesis of PUAr on the basis of dianurethane and dichloranhydride of terephthalic acid. It is obvious, that the change of organic medium, in which synthesis proceeds, should result in the change of interactions polymer-solvent and, respectively, in the change of PUAr macromolecular coil dimension D_f, which can be determined according to the Eq. (13). D_f values, obtained by the indicated method, are adduced in Table 18. As it was to be expected from the stated above considerations, $PUAr^{ac}$ macromolecular coil proves to be much more compact than $PUAr^{if}$ coil [145].

TABLE 18 The characteristics of polyurethanearylates (PUAr), received by different polycondensation modes.

Polymer	T_g^T, K	D_f	d_f	C_∞	φ_{cl}	T_g^T, K	Δ, %
PUAr^{ac}	395	1.876	2.814	6.38	0.313	364	7.8
PUAr^{if}	445	1.801	2.702	4.36	0.459	436	2.0

As it has been shown in Section 2.6, D_f value defines polymer structure in condensed state, characterized by the fractal dimension d_f. For linear polymers dimensions d_f and D_f are linked by the Eq. (95) and obtained according to the indicated relationship d_f values are adduced in Table 18. Further the value of characteristic value C_∞ can be determined, which is the indicator of polymer chain statistical flexibility, according to the fractal Eq. (18). From the adduced in Table 18 C_∞ values it follows, that polycondensation mode change leads to essential variation of chain statistical flexibility: in case of PUAr^{ac} the chain is more flexible than for PUAr^{if}. The local order domains (clusters) relative fraction φ_{cl}, which characterizes ordering degree of amorphous polymer structure [107], can be evaluated from the Eq. (102) at the condition D_{ch}=1.0. From the adduced in Table 18 φ_{cl} values PUAr^{if} structure higher ordering degree in comparison with PUAr^{ac} follows, that corresponds to the treatment, offered in Ref. [144]. And at last, the theoretical value of glass transition temperature T_g (T_g^T) can be determined within the framework of percolation theory according to the Eq. (96). T_g^T values, calculated by the indicated method, are adduced in Table 18. The comparison of these values and experimental magnitudes T_g^e has shown that the stated in the present section model describes adequately glass transition temperature variation owing to structural changes, which due to PUAr synthesis different modes. The experimental and theoretical T_g values discrepancy does not exceed of 8%, which is quite acceptable for preliminary estimations.

Hence, the stated above results confirmed just on qualitative level the explanation of polyurethanarylates structure and properties distinctions, proposed in chapters [5, 144]. The glass transition temperature increase for polymers, received by the interphase polycondensation, in comparison with synthesized by acceptor-catalytic polycondensation is due to different solvents choice for synthesis realization by the change owing to it for a macromolecular coil conformation, fixed in the synthesis process and, as consequence, by the change of ordering degree of structure in condensed state. However, the offered treatment allows one the indicated changes quantitative description that in its turn, gives the possibility of polyurethanarylates properties variation prediction on the example of their glass transition temperature [145].

As it is known through Ref. [147], one from the copolymers important characteristics is forming macromolecules structure regularity or their compositional heterogeneity, characterized quantitatively by the microheterogeneity coefficient K_M. For two homopolymers mixture K_M=0, for strictly alternating copolymer K_M=2.0. Between these extreme cases, corresponding completely to the ordered links arrangement all possible values K_M are located, in addition to completely statistical (i.e., the most disordered) distribution K_M=1.0. By K_M deviation from one absolute value one can judge quantitatively about distribution ordering degree of links in copolymer and by the fact, to which side this deviation is observed—about monomers (oligomers) tendency to either alternation in chains (K_M>1) or to both comonomers long blocks formation (K_M<1) [29].

Comonomers functional groups activity χ change is one from the important factors, effecting on K_M value [147]. This change is defined by the ratio of reaction rate constants of the first k_1 and the second k_2 functional groups after the first group introduction in reaction. In Ref. [147] K_M increase at χ growth has been shown. The authors of Refs. [148, 149] found the dependence of χ and, hence, K_M on the structure of a copolymers macromolecular coil in solution, characterized by its fractal dimension D_f, on the example of three copolymers: PESF, di-block-copolymers of oligoformal 2,2-di-(4-oxiphenyl)-propane and oligosulfone phenolphthaleine (CP-OPD-10/OSP-10) and block-copolymers of oligoformal 2,2-di-(4-oxiphenyl)-propane, phenolphthaleine and dichloranhydride of isophthalic acid (CP-OPD-10/P-1) [47], for which K_M values were determined with the aid of Gordon-Tailor-Wood equation [150]:

$$T_g = K_M\left[\left(T_{g_1} - T_{g_2}\left(\frac{W}{1-W}\right)\right)\right] + T_{g_1}, \tag{110}$$

where T_g is glass transition temperature of copolymer, T_{g_1} and T_{g_2} is glass transition temperatures of homopolymers, W is molar fraction of comonomer.

In Fig. 59 the dependences of T_g on composition are adduced for three considered copolymers. As one can see, the dependences of T_g on composition for three copolymers have different character. For PESF T_g are disposed above the additive glass transition temperature T_g^{ad}, that is, this copolymer has tendency to links regular alternation, for CP-OPD-10/P-1 $T_g<T_g^{ad}$, that is, this copolymer has tendency to the same links long sequences formation, and for CP-OPD-10/OSP-10 the transition from the second of the indicated copolymers types to the first type at c_{form}>30 mol. % is observed.

In Fig. 60 the dependences of D_f, calculated according to the equation (11) on copolymers composition are adduced, in respect to which the following should be said. All D_f values were determined for copolymers solution in chloroform, though copolymers synthesis was performed in other solvents: PESF—in dimethyl sulfoxide and two remaining copolymers—in methylene chloride. Since the interactions polymer-solvent fix macromolecular coil structure at its synthesis [141], then either D_f determination for the indicated solvents or D_f value recalculation in chloroform solution on D_f values of copolymers in solvents, used in copolymerization process are necessary. The authors [148] used the second from the indicated methods with the Eq. (13) application. D_f values, recalculated by the indicated method, were used for the plot of Fig. 60 construction.

As one can see from the plots of Figs. 59 and 60 comparison, the dependences $T_g(c_{form})$ and $D_f(c_{form})$ find out clearly the expressed similarity in respect to additive values (the stroked lines). In other words, D_f increase means T_g growth and corresponding K_M enhancement. Within the framework of fractal analysis the reaction rate constant k_r can be determined according to the Eq. (108), from which χ definition follows as the ratio k_1 and k_2 [148]:

$$\chi = \frac{k_1}{k_2} \sim t^{\left(D_{f_1} - D_{f_2}\right)/2}, \qquad (111)$$

where t is copolymerization reaction duration.

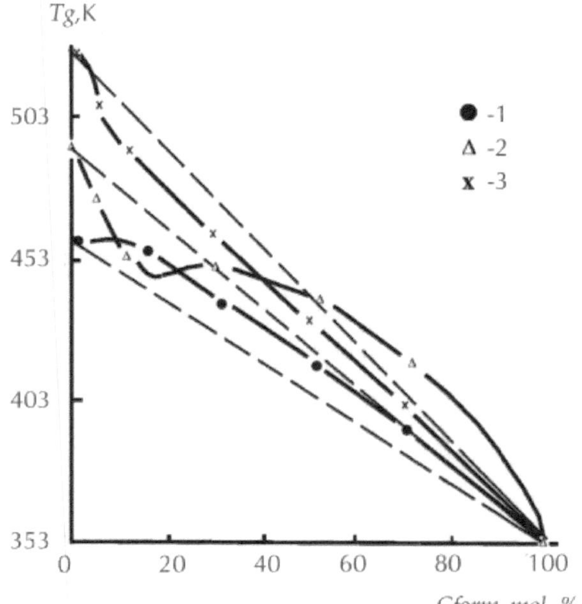

FIGURE 59 The dependences of glass transition temperature T_g on formal contents c_{form} for PESF (1), CP-OPD-10/OSP-10 (2) and CP-OPD-10/P-1 (3).

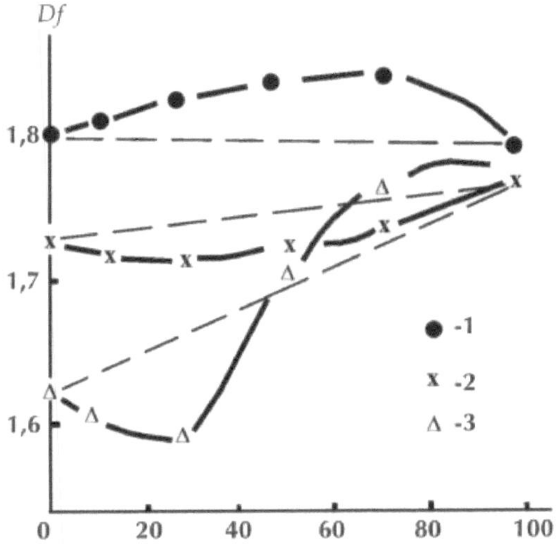

FIGURE 60 The dependences of macromolecular coil fractal dimension D_f, calculated according to the Eq. (13), on formal contents c_{form} for PESF (1), CP-OPD-10/OSP-10 (2) and CP-OPD-10/P-1.

Let us consider the dimensions D_{f_1} and D_{f_2} physical significance. As it is known through Ref. [147], the value K_M=1 corresponds to χ=1 or, proceeding from the Eq. (111), to $D_{f_1}=D_{f_1}$. Since K_M=1 means completely statistical copolymer formation, then this supposes the equality of D_f values for copolymer and the additive dependence of D_f in respect to homopolymers D_f values (the stroked lines in Fig. 60). Proceeding from the said above, D_{f_1} value in the Eq. (111) is the fractal dimension of a copolymer macromolecular coil and D_{f_2}—this dimension additive value. If $D_{f_1}>D_{f_1}$, then K_M>1 and copolymer finds out the tendency to alternating links formation, at $D_{f_1}<D_{f_2}$ K_M<1 and copolymer has the tendency to the same links long sequences formation and at $D_{f_1}=D_{f_2}$ K_M=1 and purely statistical copolymer is formed. The adduced in Fig. 61 dependence K_M(ln χ), calculated according to the Eq. (111) at the condition t=1200s, confirms the estimations made above. This correlation is similar to the dependence $K_M(\chi)$, adduced in Ref. [147], for which χ value was calculated according to the purely kinetic data.

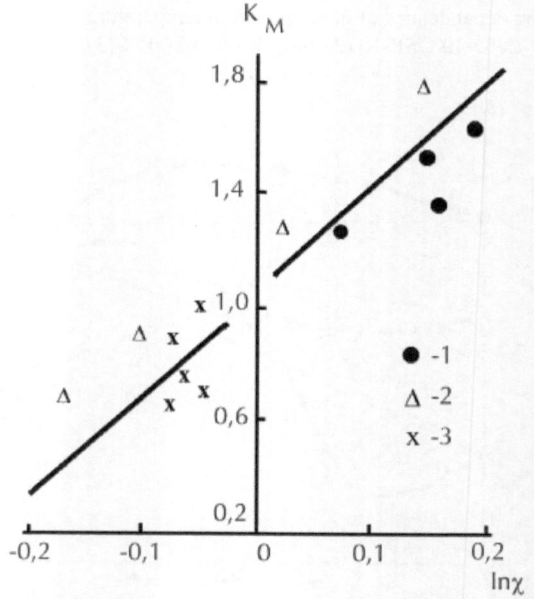

FIGURE 61 The relation between microheterogeneity coefficient K_M and functional groups activity χ in logarithmic coordinates for PESF (1), CP-OPD-10/OSP-10 (2) and CP-OPD-10/P-1.

Hence, the adduced above results have shown the decisive influence of a macromolecular coil structure, characterized by the fractal dimension, on the forming copolymer type. The cause of this is obvious enough—irrespective of active centers number and their activity their accessibility is defined by a macromolecular coil structure, namely, the higher D_f is, the more compact its structure is and the stronger coil internal regions are screened, that decelerates copolymerization reaction proceeding. Let us note one more important aspect—the Eq. (111) introduces the temporal dependence of K_M. It is obvious, that the larger reaction duration t is, the larger K_M is at $D_{f_1}>D_{f_2}$ and the smaller it is at $D_{f_1}<D_{f_2}$. The reaction duration does not influence on the copolymer type only at the condition $D_{f_1}=D_{f_2}$, that is, at completely statistical copolymer formation [148, 149].

In Ref. [150] it has been shown that epoxy polymer curing can proceed in both Euclidean three-dimensional space and in fractal one. In the last case the space dimension is equal to dimension D_f of the formed in curing process microgels. The main difference of kinetic curves conversion degree-reaction duration (*Q-t*) in the last case is practically linear dependence $Q(t)$ almost up to gelation point and variation (increase) of D_f value on this part of curve $Q(t)$. The authors [151–155] performed the further study of epoxy polymers curing in fractal space, in particular the reaction rate constant k_r and microgels self-diffusivity D_{sd} changing character on the example of haloid-containing oligomer on the basis of hexachlorobenzene curing [156].

The kinetics of curing of haloid-containing oligomer on the basis of hexachlorobenzene (conditional designation EPS-1) was studied. This oligomer was cured by 4,4′-diaminodiphenylmethane (DDM) at stoichiometric ratio of DDM: EPS-1 [156]. As it has been shown in Ref. [150], the value of reaction rate constant k_r within the range of Q=0–0.70 for curing reaction in Euclidean space is constant. The relation between k_r, Q and D_f has the following form [157]:

$$t^{(D_f-1)}=\frac{c_1}{k_r(1-Q)}, \tag{112}$$

where c_1 is a constant.

For the system EPS-1/DDM the average value k_r=0.97×10^{-3} moll/s was determined by method of inverse gas chromatography [156]. From the equation (112) the value c_1 can be determined at the average values of the parameters included into it: t=1.5×10^{-3} s, D=1.99 and Q=0.35. In this case c_1=0.0244 mol·l/s. As the calculations have shown, k_r reduction from 4.16×10^{-3} up to 0.76×10^{-3} mol·l/s was observed within the range of t=(0.5–2.5)×10^{-3} s. The range of the microgels fractal dimension values in this case makes up D_f=1.61–2.38 [150], that assumes the microgels formation according to the cluster–cluster mechanism, that is, large microgel formation from smaller ones [158]. In this case the microgels molecular weight *MM* value is determined according to the following scaling relationship [159]:

$$MM \sim Q^{2/(3-D_f)} \tag{113}$$

The microgel gyration radius R_g is connected with *MM* according to the following relationship [158]:

$$R_g \sim MM^{1/D_f} \sim Q^{2/D_f(3-D_f)} \tag{114}$$

The obtained results allow one to perform the system EPS-1/DDM curing kinetic analysis within the framework of irreversible aggregation models [160]. In general case the relationship between k_r and R_g can be written as follows [160]:

$$k_r \sim R_g^{2\omega} \tag{115}$$

In its turn, the exponent ω is defined by the parameters describing clusters (microgels) motion in space and their structure. This intercommunication has the following form [160]:

$$2\omega = -\gamma + d - D_w \tag{116}$$

where γ characterizes the dependence of microgels self-diffusivity D_{sd} on their sizes (D_{sd}–$R_g^{-\gamma}$), d is the dimension of space, in which curing reaction occurs, D_w is the dimension of microgels random walk trajectory.

For the reactions in Euclidean space $d=3$, $D_w=2$ (Brownian motion of microgels), $\gamma=-1$ and then $\omega=0$. This means, that in the given case the condition should be fulfilled [151]:

$$k_r = \text{const} \tag{117}$$

The condition of Eq. (117) is confirmed experimentally (k_r value does not change at R_g increasing) [156]. For the curing reaction proceeding in fractal space the situation differs completely from the described above. This aspect attains special meaning within the framework of nanochemistry [161], therefore deserves consideration in more detail.

As it is known through Ref. [162], in nanochemistry there are two fundamental notions—nanoparticle and nanoreactor: the first characterizes dimensional parameter while the second defines nano-object function. Thus, iron cluster loses almost completely its specific properties (ionization energy, magnetism) and approaches metallic iron at a number of atoms in cluster $n=15$. At $n>15$ it remains a nano-object in dimensional sense, but loses "nanoreactor" qualities, for which properties become a size function. In Fig. 62 the dependence of curing rate constant k_r on microgels diameter $2R_g$ is adduced, which has a very specific shape. Within the range of microgels (although the term "nanogel" is a more precise one) diameters less than 100 nm, k_r value is clearly expressed rapidly decreasing function of diameter $2R_g$ and at $2R_g>100$ nm the indicated dependence is practically absent. Let us note that the size 100 nm is assumed as an upper limit for nanoworld objects (although conditional enough) [162]. Hence, the data in Fig. 62 demonstrate clearly that microgel at $2R_g<100$ nm is a nanoreactor in which reaction (curing) rate is a strong function of its size, and at $2R_g\geq 100$ nm microgel loses this function and as a matter of fact becomes a chemically inert enough particle. Let us note that the indicated transition nanoreactor—nanoparticle is only possible in the fractal space. In Euclidean space these notions do not differ (k_r=const). In Fig. 63 the dependence $k_r(R_g)$ for the system EPS-1/DDM in double logarithmic coordinates is shown, which is approximated well by a straight line. From the slope of this straight line the value $2\omega=-0.58$ can be determined. As it has been noted above, the dimension of the space, in which the curing reaction occurs, is equal to D_f and the value D_w can be determined according to Aarony-Stauffer rule [163]:

$$D_w = D_f + 1 \tag{118}$$

$$D_w = D_f + 1.$$

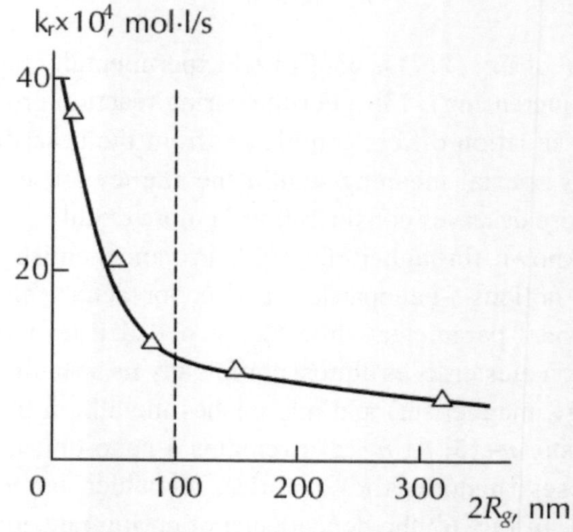

FIGURE 62 The dependence of reaction rate constant k_r on microgels diameter $2R_g$ for system EPS-1/DDM.

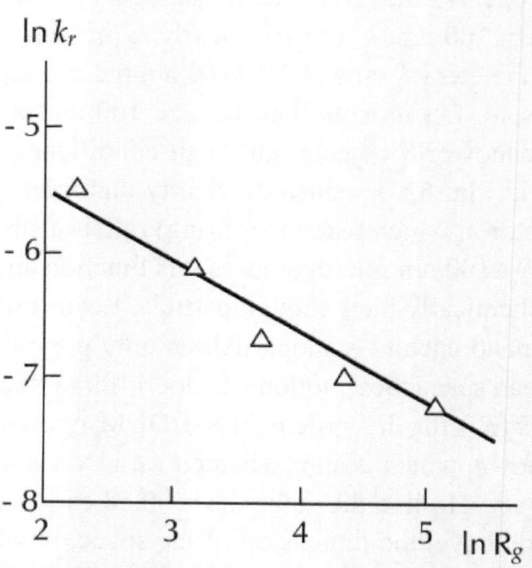

FIGURE 63 The dependence of reaction rate constant k_r on microgels gyration radius R_g in double logarithmic coordinates for system EPS-1/DDM.

Hence, the Eq. (116) for the considered case can be rewritten as follows (for any D_f value) [152]:

$$2\omega = -\gamma - 1 \tag{119}$$

Then according to the Eq. (119) γ=-0.42 can be obtained. This means that the self-diffusivity value D_{sd} reduces with R_g growth much slower (D_{sd}–$R^{-0.42}$) in comparison with the reaction in Euclidean space ($D_{sd} - R_g^{-1}$). The indicated distinction is demonstrated in Fig. 64 in a diagram form.

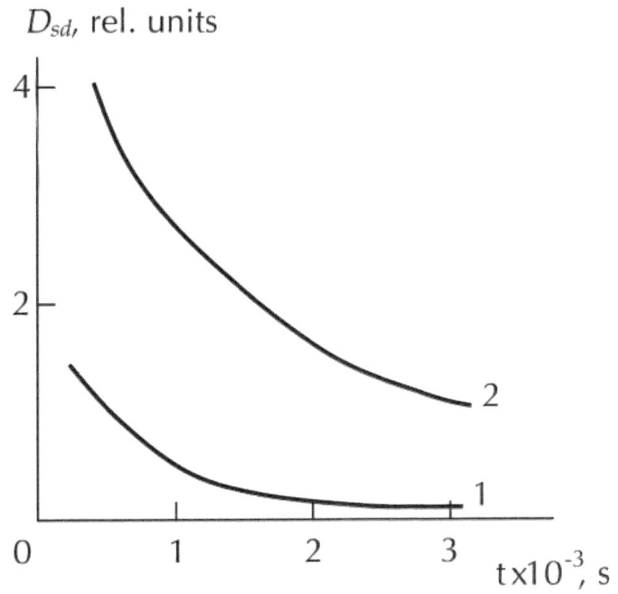

FIGURE 64 The dependences of nicrogels self-diffusivity D_{sd} on curing reaction duration t in Euclidean (1) and fractal (2) spaces for system EPS-1/DDM.

Molecular weight *MM* of microgels depends on curing duration t as follows [161]:

$$MM \sim t^{D_f/(D_f - 2\omega)} \tag{120}$$

As it has been noted above, in Euclidean space 2ω=0 and the exponent in the Eq. (120) is equal to one. This assumes *MM–t*. For the reaction in fractal space $2\omega<0$ and the exponent in the Eq. (120) is less than one. This means that in fractal space *MM* value grows slower than in Euclidean one. This relation for the system EPS-1/DDM is shown in Fig. 65 in a diagram form. Since *Q* value in the second case is larger than in the first then this means that the curing reaction in fractal space gives a larger number of small clusters (microgels).

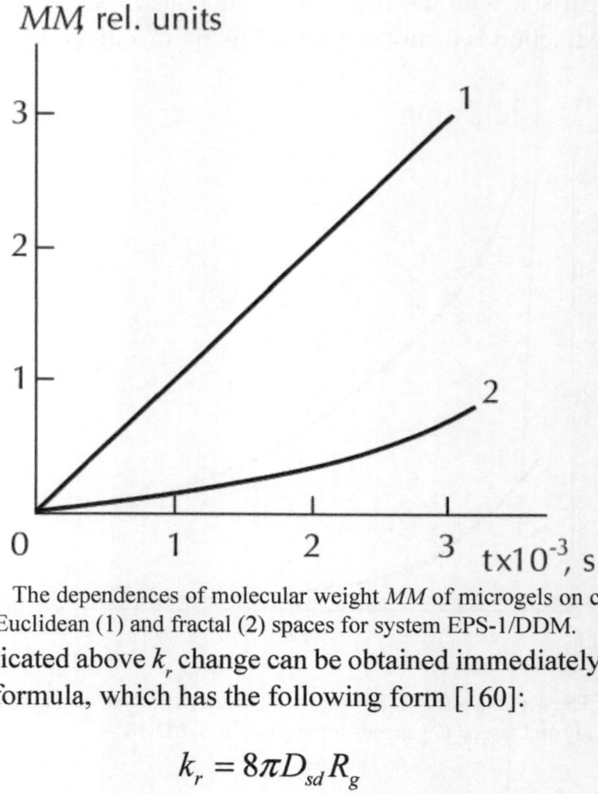

FIGURE 65 The dependences of molecular weight *MM* of microgels on curing reaction duration t in Euclidean (1) and fractal (2) spaces for system EPS-1/DDM.

The indicated above k_r change can be obtained immediately from Smoluchovski formula, which has the following form [160]:

$$k_r = 8\pi D_{sd} R_g \tag{121}$$

For the reaction in Euclidean space $D_{sd} - R_g^{-1}$ and k_r=const, for the reaction in fractal space for the system EPS-1/DDM D_{sd}–$R^{-0.42}$ and $k_r - R_g^{0.58}$, that is, k_r reduction at the curing reaction proceeding (R_g or *MM* growth) is supposed.

Let us note in conclusion the strong dependence of k_r on the microgels structure, characterized by the fractal dimension D_f (Fig. 66). As it follows from the plot adduced in Fig. 66, k_r sharp decay is observed for D_f increase at D_f<2 and the attainment of the values k_r on asymptotic branch at D_f>2. As it is known through Ref. [160], within the framework of irreversible aggregation models the following relationship is valid:

$$k_r \sim D_{sd} R_g^{d-2} \tag{122}$$

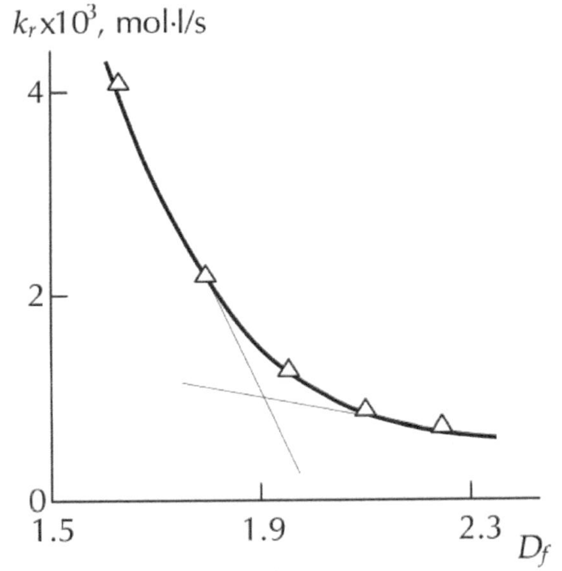

FIGURE 66 The dependence of reaction rate constant k_r on microgels fractal dimension D_f for system EPS-1/DDM.

If for Euclidean space, for the example, with d=3, the exponent in the Eq. (122) is constant and equal to one, then for fractal space with variable value D_f the situation will be essentially different. For D_f<2 the exponent in the Eq. (122) is less than zero and R_g growth results in k_r reduction under other equal conditions. At D_f=2 k_r is independent on R_g. And at last, at D_f >2 k_r value should be increased at R_g enhancement. This is expressed by k_r sharp decay at D_f<2, since both D_{sd} and $R_g^{D_f-2}$ reduce this parameter at *MM* microgels growth. At D_f>2 D_{sd} reduction is compensated to a certain extent by $R_g^{D_f-2}$ growth and k_r decay at *MM* increasing is decelerated. It is easy to see that at D_f=2.48 the condition k_r=const is fulfilled for the system EPS-1/DDM. Analytically the correlation $k_r(D_f)$ can be presented as follows [155]:

$$\lg k_r = -1.56 - 1.26\left(D_f - 1\right) \tag{123}$$

Thus, the stated above results have shown that for curing reaction proceeding in fractal space the reaction rate constant reduction is typical. The formation of a large number of microgels with smaller molecular weight in comparison with reaction in Euclidean space at the same conversion degree is also typical for such reaction. The dimensional border between nanoreactor and nanoparticle for the considered curing reaction has been obtained.

As it is known through Ref. [164], the kinetic curves conversion degree—polymerization duration (Q–t) in the radical polymerization case have a sigmoid shape and are divided into three parts, usually called as an initial stage, autoacceleration and finish polymerization stages. Q values at transition from one stage to a subsequent one can be designated as Q_1 and Q_2, respectively, and the final conversion degree—as Q_f. It has been shown earlier [165], that at structures self-organization from previous structure instability point to the subsequent one the adaptation universal algorithm is realized:

$$A_m = \frac{Z_i}{Z_{i+1}} = \Delta_i^{1/m}, \tag{124}$$

where Z_i and Z_{i+1} are critical values of governing parameter, controlling structure formation; their ratio defines a system adaptability measure A_m to structural reformation; m is reformations number; Δ_i is structure stability measure, maintaining constant at m change from m=1 up to m=m_{max}.

Assuming, that in radical polymerization process, including polymerization of the considered in Ref. [166] dimethyldiallyl ammonium chloride (DMDAAC) the conversion degree is a governing parameter, A_m value for the two last from the indicated stages can be determined as the boundary values Q ratio, that is, A_m^{aut}=Q_1/Q_2 and A_m^{aut}=Q_2/Q_f for the autoacceleration and finish polymerization stages, respectively. The calculations have shown large distinction of these parameters: A_m^{aut}=0.232 and A_m^{fp}=0.915. For the estimation of Δ_i and m values determined by gold proportion law evolving systems self-similarity constants can be used [165]. The tabulated data allow one to receive Δ_i=0.232 and m=1 for the autoacceleration stage and $_i$=0.232 and m=16 for the finish polymerization stage. Thus, the system stability at both stages is the same, but their adaptability measure changes sharply at the expense of possible reformations number m variation [166].

It can be supposed, that DMDAAC macromolecular coil adaptability to polymerization reaction realization will be defined by its structure, characterized by the fractal dimension D_f: the less D_f is, the easier and more intensively the indicated reaction proceeds and the lower system (macromolecular coil) adaptability A_m is. In Fig. 67 the dependences $A_m(D_f)$ for three concentrations of the initial reagents c_0 is adduced. As one can see, the linear correlation is obtained, which is extrapolated to A_m=0 at D_f^{tr}=1.5 (transparent coil [10]) and to A_m=1.0 at D_f=3.0 (compact globule). In other words, the factor, defining macromolecular coil adaptability to reactionary medium, is purely physical parameter: its fractal dimension D_f [166]. The indicated extrapolation limiting points were expected from the most general considerations of fractal analysis: at D_f^{tr}=1.50 the permeable coils are transparent for one another, that gives the greatest reaction effectiveness, and for compact globule a chemical reactions are impossible. Let us note that a reagents chemical parameters change does not alter this picture: the limiting values Q_1, Q_2 and Q_f can be changed only, that is observed at c_0 variation [164]. Analytically the relationship between A_m and D_f can be expressed as follows [166]:

$$A_m = 0.667\left(D_f - D_f^{tr}\right) \tag{125}$$

As it is known through Ref. [157], within the framework of fractal analysis the polymerization kinetics is described by the general Eq. (79). The combination of the relationships (79) and (125) allows to obtain radical polymerization kinetics description within the framework of synergetics [166]:

$$Q \sim t^{0.75(1-A_m)} \tag{126}$$

As it was to be expected, at A_m=0 reaction proceeds most intensively (Q–$t^{0.75}$) and at A_m=1.0 reaction does not realize at all (Q~const). Let us note, that the known condition for a reaction in polymer solutions D_f≤2.28 [157] imposes restrictions on m value: m≤2, that is, it restricts structure adaptability.

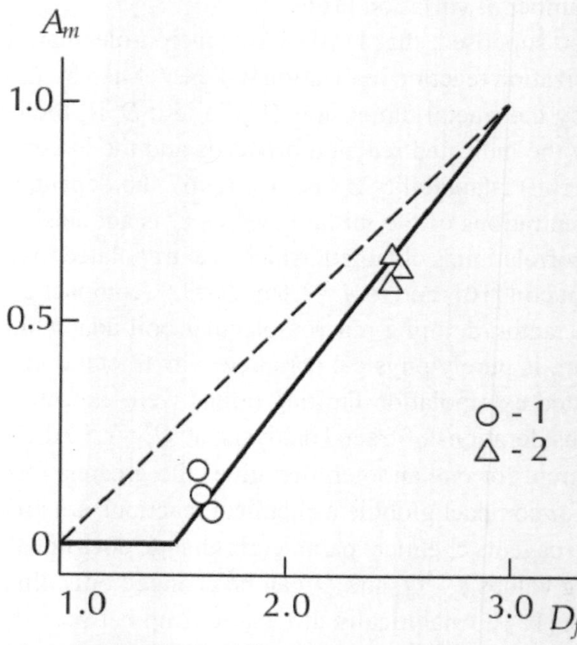

FIGURE 67 The dependences of adaptability measure A_m on macromolecular coil fractal dimension D_f for DMDAAC. The autoacceleration (1) and finish polymerization (2) stages.

In Fig. 67 the hypothetical dependence $A_m(D_f)$ within the framework of the Eq. (79), assuming A_m0 at D_f^{tr}=1.50 (A_m=0 at D_f=1.0 and A_m=1.0 at D_f=3.0), is shown by a stroked line. The discrepancy the dependences $A_m(D_f)$, received within the frameworks of synergetics and fractal analysis, supposes that self-organizing structures are macromolecular coils (or their totalities) only, but not monomer molecules, the reaction of which proceeds within the range of D_f=1.0–1.50 [166].

2.8 THE INTERCOMMUNICATION OF MACROMOLECULE BRANCHING AND MACROMOLECULAR COIL STRUCTURE IN DILUTED SOLUTIONS

The branching degree (and/or bulk side substituents availability) is important property of polymer chain, to a great extent defining polymers behavior in both solutions and condensed state. It has been shown [5, 51], that side substituents molecular weight and branching degree increase leads to systematic reduction of the exponent a_η in Mark-Kuhn-Houwink equation, that at other equal conditions means a macromolecular coil gyration radius decreasing, that is, its compactness degree enhancement. The side substituents molecular weight increasing in case of solid-phase polymers results to their glass transition temperature growth, reducing simultaneously their macroscopic plasticity [96]. The enumeration of the indicated factors influence on polymers structure and properties can be continued. The authors [167, 168] studied a factors, influenced on D_f value of macromolecular coil of branched polymers in diluted solutions and defined D_f variation tendencies at these factors change for two groups of polymers: statistically branched polyphenylquinoxalines (PPQX) [51] and bromide-containing aromatic copolyethersulfones (B-PES) [169]. The latter are interesting by the fact, that within the narrow range of decabromide diphenyloxide residues contents (3–10 mol. %) the intrinsic viscosity [η] in chloroform sharp decrease is observed at practically constant molecular weight [169], supposing the polymer chain branching degree increasing [51].

The branching factor *g*, determined according to the Eq. (105), was used for polymer branching degree quantitative evaluation. For copolymers B-PES the aromatic polysulfone, not containing side substituents, was accepted as their linear analog [169].

In Fig. 68, the dependences of the macromolecular coil fractal dimension D_f for PPQX in N-methylpyrrolydone and chloroform and B-PES in chloroform on branching factor *g* (points) are adduced. As one can see, in any case branching degree increasing, characterized by *g* reduction [5, 51], leads to D_f growth, that is, to macromolecular coil compactness degree enhancement. One more factor, influencing on D_f value, is solvent thermo-dynamical quality in respect to polymer: poor solvent for PPQX (N-methylpyrrolydone) results in essentially higher D_f values in comparison with good solvent for both considered polymers (chloroform). Let us note that for these polymers linear analogs (at g=1) the data do not locate on the dependence for branched polymers, in addition any objective regularity for relation of branched and linear polymer D_f values are absent. So, for solution in N-methylpyrrolydone D_f for linear analog is less than D_f for branched polymer and in chloroform—larger, D_f for B-PES has the dependence of D_f on g, common for both types.

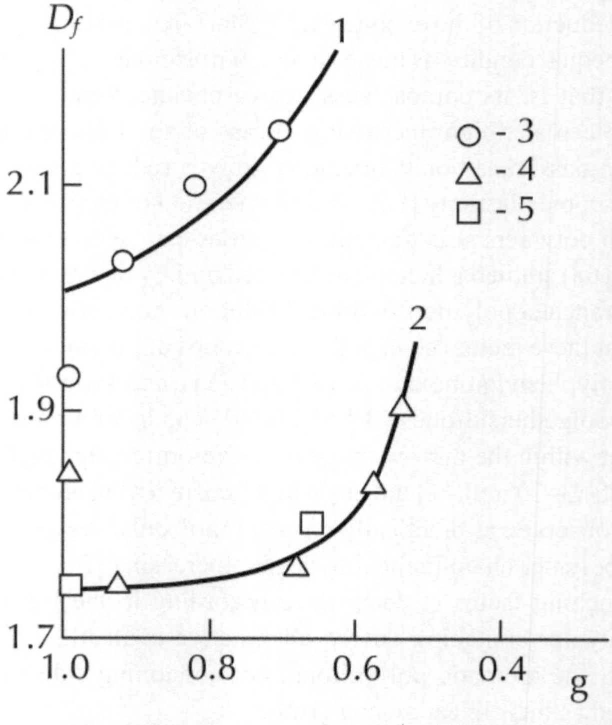

FIGURE 68 The dependences of macromolecular coil fractal dimension D_f on branching factor g for PPQX (1–4) and B-PES (5). The solvents: N-methylpyrrolydone (1, 3) and chloroform (2, 4, 5). 1, 2—calculation according to the Eq. (127); 3–5—the experimental data.

These results allow one to make a number of conclusions. Firstly, as it was to be expected [13], D_f value of branched polymers is controlled by two factors: interactions polymer-solvent and interactions of macromolecular coil elements between themselves. Secondly, the branching factor *g* is a dominant parameter at the second factor definition—any correspondence to linear analogs is not observed. For the evaluation of interactions character of macromolecular coil elements between themselves the parameter of volume interactions ε, which result in coil shape deviation from an ideal Gaussian one, can be used. ε value can be calculated according to the Eq. (22). In Fig. 69 the dependences of ε on *g* are shown for the considered polymers, from which it follows, that polymer chain branching degree increasing weakens repulsive interactions between coil elements, that is, defines positive values ε reduction. For PPQX macromolecular coils in N-methylpyrrolydone the attractive interactions (negative ε values) are predominated. Thus, the data of Figs. 68 and 69 suppose, that at the same *g* values a good solvent (chloroform) screens effectively attractive interactions between branches and/or side substituents that results in smaller D_f values of macromolecular coils of branched polymers in such solvent in comparison with a poor solvent.

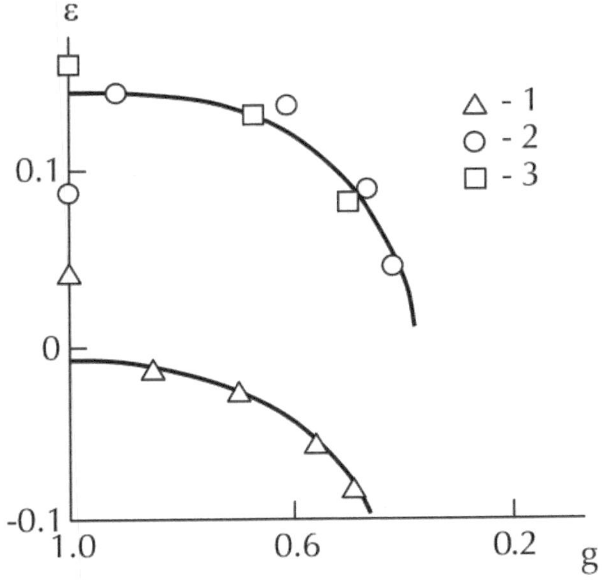

FIGURE 69 The dependences of volume interaction parameter ε on branching factor *g* for PPQX (1, 2) and B-PES (3). The solvents: N-methylpyrrolydone (1) and chloroform (2, 3).

The Eqs. (4) and (105) can be used for the analytical relationship between D_f and g obtaining. Their combination allows to receive the following relationship [167]:

$$D_f = \frac{3\ln g}{3\ln g - \left(\ln[\eta]_b - \ln[\eta]_l\right)}, \tag{127}$$

where $[\eta]_b$ and $[\eta]_l$ are intrinsic viscosities of branched polymer and its linear analog, respectively.

In Fig. 68 the comparison of the dependences $D_f(g)$, received according to the Eqs. (4) and (105) (points) and to the Eq. (127) (the solid curves) is adduced. From this comparison a good correspondence of the dependences $D_f(g)$, obtained by both indicated methods, follows and, as consequence, the possibility of the Eq. (127) application for D_f calculation according to the known g values or, on the contrary, g values according to the known D_f values. The solvent type influence in the Eq. (127) is taken into account by the intrinsic viscosities $[\eta]_b$ and $[\eta]_l$ logarithms difference, adduced in brackets—the better solvent thermo-dynamical quality in respect to polymer is the larger this difference is and, respectively, the smaller D_f value is. Let us note an interesting feature of the Eq. (127)—D_f increasing at polymer branching degree enhancement is not defined by g change, but by the change of the indicated logarithms difference $[\eta]_b$ and $[\eta]_l$ value. As the evaluations have shown, for hypothetical polymer with the indicated difference constant value the dependence $D_f(g)$, opposite to the shown in Fig. 68, would be observed, that is, polymer chain branching degree enhancement (g reduction) would not result in D_f decrease, but in its growth [167, 168].

Thus, the stated above results supposed the specific nature of the dependence of the fractal dimension of a macromolecular coil in diluted solution for branched polymers on polymer chain molecular characteristics—the indicated parameter depends on the chain branching degree only, that does not allow any extrapolations to g=1.0 for the estimation of their linear analog D_f value. Side branches appearance results in attractive interactions between macromolecular coil elements intensification, that is the cause of D_f increasing, that is, a coil compactness degree enhancement. Although this conclusion was made on the example of the considered polymers limited number, the community high enough degree of this conclusion should be supposed. In favor of such supposition speak the results for polyarylates series, obtained by equilibrium and nonequilibrium polycondensation, for the example, in case of the first method application the formation of branched chains is possible, that always leads to reduction of the exponent a_η in Mark-Kuhn-Houwink equation (Table 5, p. 120 in Ref. [5]) or to D_f increasing according to the Eq. (4).

The spectral (fraction) dimension d_s, which characterize macromolecular coil connectivity degree, are its structure important characteristic (particularly for the branched polymers). For linear polymer chains d_s=1.0 and for the branched ones d_s value is varied within the limits of 1.0–1.33 [45]. The dimension d_s are a decisive one at D_f calculation for fractals of different types [25]. So, phantom (without accounting for the excluded volume effects) dimension of macromolecular coil D_f^{ph} is given by the Eq. (11) of Chapter 1 and swollen (accounting for the indicated effects) dimension D_f—by the Eq. (12) of Chapter 1. The Eq. (12) of Chapter 1 is correct for determination of D_f value in solvents, whose molecules can be simulated by zero-dimensional objects, that is, in case, when molecule dimension δ_f=0. If a molecule (molecules totality, see Section 2.4) of solvent has a more complex structure, then D_f value is determined according to the Eqs. (18) or (80). From the comparison of the Eqs. (12) of Chapter 1 and (80) it is easy to see, that at δ_f>0 D_f value, determined according to the Eq. (80), exceeds the corresponding parameter, determined according to the Eq. (12) of Chapter 1 at other equal conditions, namely, at the same d_s. This means that solvent with δ_f>0 screens repulsive interactions of a macromolecular coil the stronger, the larger δ_f is. It is obvious, that in this case the Eq. (12) used for d_s evaluation can be resulted in essential errors. Therefore, the authors [170] developed the technique for d_s value correct determination for branched and linear polymer chains in diluted solutions. With this purpose the data of Ref. [51] for PPQX solutions with its macromolecules different branching degree in N-methylpyrrolydone was used. δ_f values for different solvents were determined according to the Eq. (82). Besides, D_f values for polyarylate F-2 on the basis of phenolphthaleine, synthesized in different solvents (Table 19) [5], determined according to the Eq. (11).

TABLE 19 The fractal dimension of a macromolecular coil of polyarylate F-2, synthesized in different solvents.

Solvent	D_f, the Eq. (11)	D_f, the Eq. (80)
N, N-dimethylformamide	2.03	2.0
Nitrobenzene	1.80	1.82
Acetone	1.81	1.82
Dichloroethane	1.80	1.81
Chloroform	1.90	1.77
1,2,4-Trichlorobenzene	1.80	1.85
Benzene	1.81	1.86
Hexane	2.05	1.67

In Fig. 70 the dependence of spectral dimension d_s on the fraction of functional branching component γ, characterizing macromolecules branching degree, is adduced for PPQX, where d_s value was determined according to the Eq. (12) of Chapter 1. As one can see, d_s value for a linear chain (γ=0) is equal to approx. 1.24 and for γ>0.005 exceeds the limiting value d_s^{lim}=1.33 (a stroked horizontal line), that contradicts to the known data [45], cited above. The similar dependence $d_s(\gamma)$, calculated according to the Eq. (80) with δ_f=0.76, adduced also in Fig. 70, gives the results, corresponding completely to the data [45]. For PPQX linear chain (γ=0) d_s=1.0 and at γ=0.04 d_s value attains the asymptotic value d_s=1.33 for very branched polymer chain. Thus, the results, adduced in Fig. 70, demonstrate clearly, that for macromolecular coils in solvents with molecules (molecules totality) complex enough structure, that is, for δ_f>0, the Eqs. (80) (or (18) of Chapter 1) application is necessary for coil spectral dimension d_s correct determination.

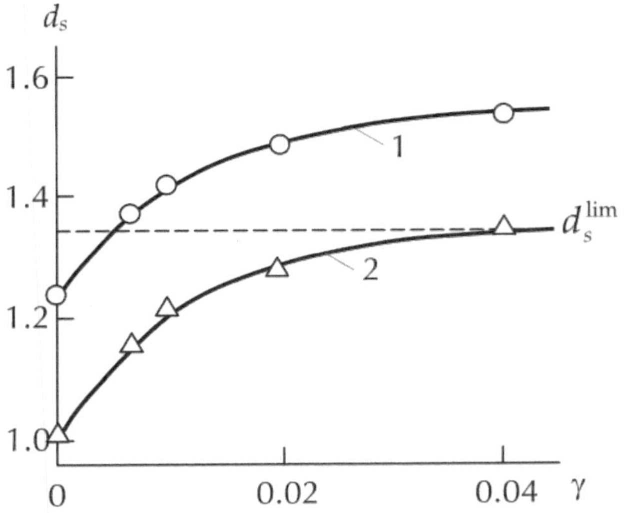

FIGURE 70 The dependences of spectral dimension d_s, calculated according to the Eqs. (12) (1) and (80) (2), on fraction of polyfunctional branching component γ for PPQX. The horizontal stroked line indicates the greatest theoretical value $d_s^{\lim}$.

The stated above technique allows the correct determination of the macromolecule type (linear or branched), which is formed in polymers synthesis process. Such estimation was performed on the example of polyarylate F-2 synthesis in 8 solvents (Table 19). D_f values comparison, calculated according to the Eqs. (11) and (80), has shown a good correspondence for all solvents, excluding hexane. Calculation according to the Eq. (80) was performed in F-2 linear chain supposition, that is, at d_s=1.0. The mean discrepancy of D_f values, determined by the indicated methods, makes up approx. 2.3% for seven solvents and for hexane—23 %, that is, larger on one order of magnitude. Therefore, the assumption was made, that in hexane F-2 branched macromolecule is formed with value d_s=1.33. Then calculation according to the Eq. (12) of Chapter 1 gives D_f2.0, which coincides with the estimation according to the Eq. (11) within the limits of 2.5%. Thus, the adduced above estimations suppose, that in F-2 synthesis process in seven solvents a linear polymer is formed and in hexane—a very branched one. This supposition corresponds to the conclusions of Ref. [171], received by independent methods.

Let us consider the dependence of d_s on the chain of PPQX branching degree more strictly. The number of branching nodes m was used for branching degree estimation [1]. *M* value can be determined by two modes. The first from them was considered in Ref. [51]:

$$m = \frac{N\gamma}{(2+\gamma)}, \tag{128}$$

where *N* is a polymerization degree.

The second mode uses the formula for trifunctional branching nodes [1, 51]:

$$g = \left[\left(1+\frac{m}{7}\right)^{0.5} + \frac{4m}{9\pi}\right]^{-0.5}, \tag{129}$$

where *g* is a branching factor, determined according to the Eq. (105).

In Fig. 71 the dependence $d_s(m)$ for PPQX is adduced. As it was to be expected from the most general considerations, d_s value grows monotonously at *m* increase, approaching to the asymptotic value d_s=1.33 at $m4$. This dependence is approximately quadratic and it can be linearized, by writing d_s as a function of $m^{1/2}$. The dependence $d_s(m)$ analytic form will look like [170]:

$$d_s = 1 + 0.17m^{0.5},\ m \le 4, \tag{130}$$

$$d_s = 1.33,\ m > 4.$$

The Eq. (130) of Chapter 1 supposes, that a chain branching degree reaches saturation very fast and further γ or *m* growth does not change macromolecular coil structure, that is, dimensions d_s and D_f.

Hence, the stated above results have shown that for correct evaluation of the spectral dimension value of a macromolecular coil in solvents, molecules (molecules totalities) of which have complex enough structure (with nonzero dimension) it is necessary to account for this structure characteristics. The proposed technique also allows simple and precise determination of chains branching availability.

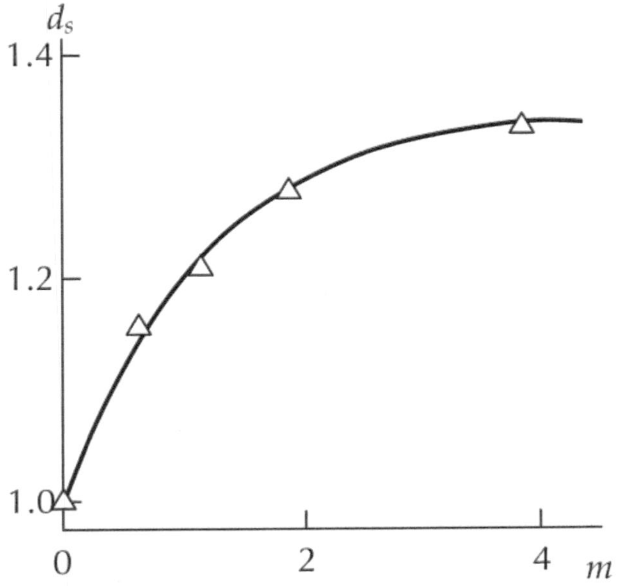

FIGURE 71 The dependence of spectral dimension d_s, calculated according to the Eq. (80), on branching nodes number m for PPQX.

The authors [172, 173] continued to study a macromolecular coil structure for branched polymers within the framework of fractal analysis [174] on the example of branched PPQX with branching different degree in two solvents [51]. As it has been noted above, the Eq. (12) of Chapter 1 gives the correct D_f value in the solvents, consisting of point (zero-dimensional) molecules, whose fractal dimension δ_f=0 [25]. From the Eq. (82) it follows, that the condition δ_f=0 is reached for $\delta_s \leq 8.0$ $(cal/cm^3)^{1/2}$. Since δ_s values for the used solvents exceed the indicated magnitude, then for them δ_f>0. From the Eq. (82) δ_f values can be estimated as approx. 0.19 for chloroform and approx. 0.91 for N-methylpyrrolydone. In this case D_f value is determined according to the Eq. (18) of Chapter 1.

In Fig. 72 the dependences of D_f and D_f^{ph} on d_s are adduced for all considered cases. The dependences $D_f(d_s)$ for real polymers should be located between curves $D_f^{ph}(d_s)$ and $D_f(d_s)$, where D_f value is determined according to the Eq. (12) of Chapter 1 (the curves 1 and 2, respectively). For PPQX macromolecular coils in N-methylpyrrolydone this condition is fulfilled actually and, what is more, the experimental D_f values (the Eq. (4)) correspond excellently to D_f calculation according to the Eq. (18) of Chapter 1, where δ_f =0.91 and D_f^{ph} value was calculated according to the Eq. (11) of Chapter 1.

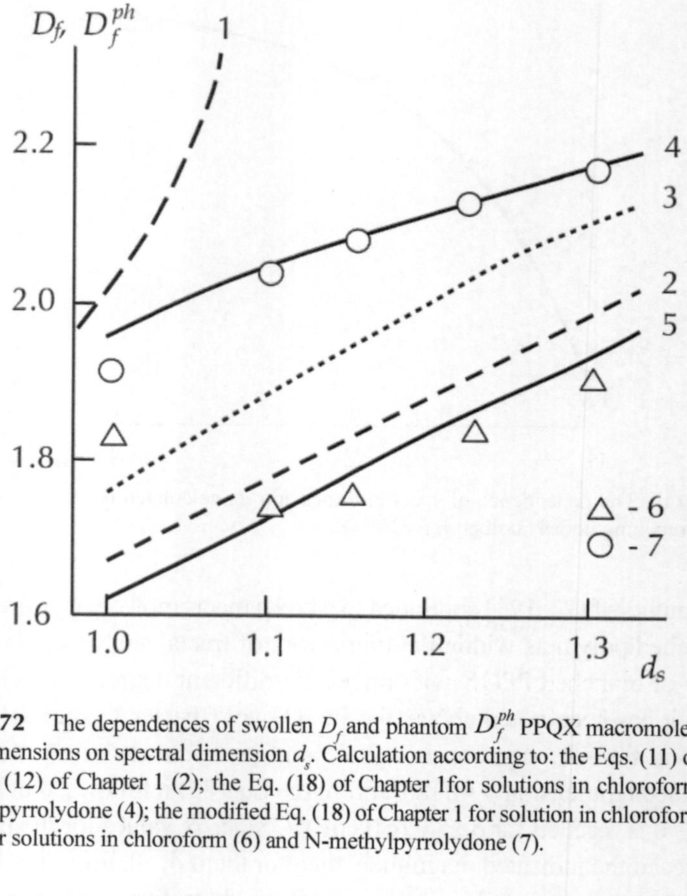

FIGURE 72 The dependences of swollen D_f and phantom D_f^{ph} PPQX macromolecular coil fractal dimensions on spectral dimension d_s. Calculation according to: the Eqs. (11) of Chapter 1 (1) and (12) of Chapter 1 (2); the Eq. (18) of Chapter 1for solutions in chloroform (3) and N-methylpyrrolydone (4); the modified Eq. (18) of Chapter 1 for solution in chloroform (5); the Eq. (4) for solutions in chloroform (6) and N-methylpyrrolydone (7).

At the same time the data for PPQX macromolecular coil in chloroform, estimated according to the Eq. (4), are disposed lower than the dependence $D_f(d_s)$ for swollen coils, although the Eq. (12) of Chapter 1 supposes minimum D_f value for them (δ_f=0). The dependence $D_f(d_s)$ for swollen coils (the curve 2 in Fig. 72) does not correspond to the experimental data. Let us consider the Eq. (18) of Chapter 1 derivation, adduced in Ref. [25], for this discrepancy explanation. It is supposed, that the fractal solvent molecules screen the excluded volume interactions, that is, reduce the level of repulsive interactions between macromolecule elements. Therefore, δ_f increase leads to D_f enhancement. Further the parameter α, determined according to the Eq. (81), was introduced. The screening interactions were approximated by two bodies interaction [25]:

$$U^{(2)} \sim (\vartheta / N) MM^2 / R_g^d = \left(\vartheta / r^{\delta_f}\right) MM^2 / R_g^d, \tag{131}$$

where $U^{(2)}$ is the energy of two bodies repulsive interaction, ϑ is the excluded volume parameter, N is polymerization degree, MM is molecular weight, r and R_g are gyration radii of solvent molecule and macromolecular coil, respectively.

Then the value $U^{(2)}$ was minimized, owing to that the relationship was obtained [25]:

$$\frac{F}{kT} = \left(\frac{R_g}{R_{g0}}\right)^2 + \frac{\vartheta MM^{2-\alpha}}{R_g^d}, \tag{132}$$

where F is free energy, K is Boltzmann constant, T is temperature, R_{g0} is gyration radius of coil in ideal conditions.

The Eq. (80) is a final result of this treatment. It is easy to see, that in this derivation the possibility of attractive interactions between macromolecule elements is not taken into account. Such effect should result in coil compactness degree reduction and D_f value decrease, that is observed for PPQX coils in chloroform (Fig. 72). The most simple mode of this effect appreciation is replacement of sign "plus" before δ_f in the Eq. (18) of Chapter 1 by "minus" or replacement of sign "minus" before in the Eq. (80) by "plus." D_f calculation according to the modified thus mode Eqs. (18) of Chapter 1 and (80) like that gave an excellent correspondence to the experiment (curve 5, Fig. 72). This allows one to suppose that chloroform molecules screen attractive interactions between PPQX macromolecule side branches, but do not screen them for polymer main chain (D_f experimental value for linear PPQX corresponds well to calculation according to the Eqs. (18) of Chapter 1 or (80) at d_s=1.0, Fig. 72). If this is correct, then molecular weight MM increase, proportional to the increase of side branches in a macromolecular coil number, should intensify this effect, expressed by the excluded volume factor α^3, which is determined according to the Eq. (6). In Fig. 73 the dependence $\alpha^3(MM)$ is adduced, plotted according to the data of Ref. [51], which confirms the present supposition. Let us note that the value α^3 is linked unequivocally with the macromolecular coil fractal dimension D_f that follows from the equation (39). The indicated equation allows to estimate the excluded volume factor for each solvent (unlike the Eq. (6)). The dependence of α^3 on branching component fraction γ for PPQX is adduced in Fig. 74. As it was expected, the extreme change of $\alpha^3(\gamma)$ for PPQX solutions in chloroform and monotonous reduction α^3 at γ growth for these polymer solutions in N-methylpyrrolydone is observed.

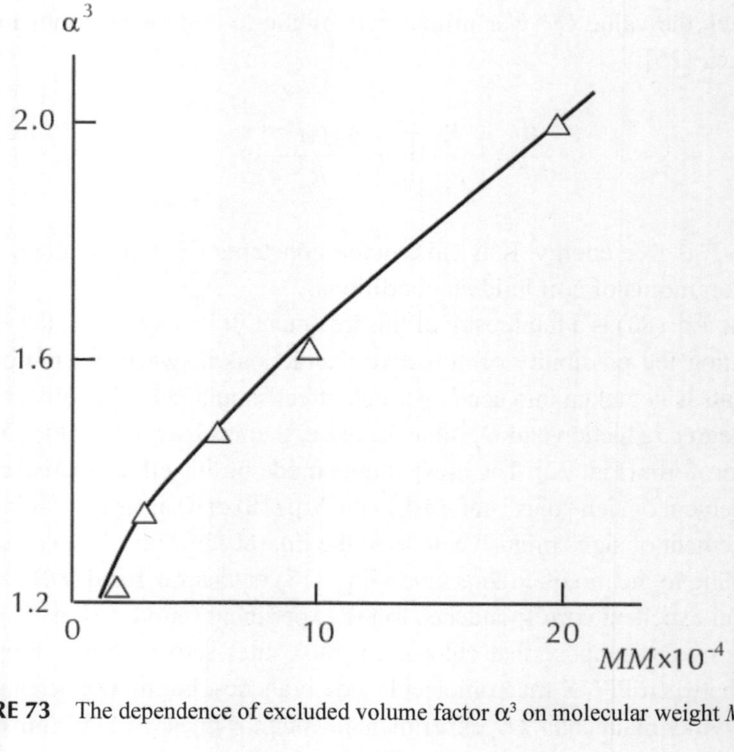

FIGURE 73 The dependence of excluded volume factor α^3 on molecular weight *MM* for PPQX.

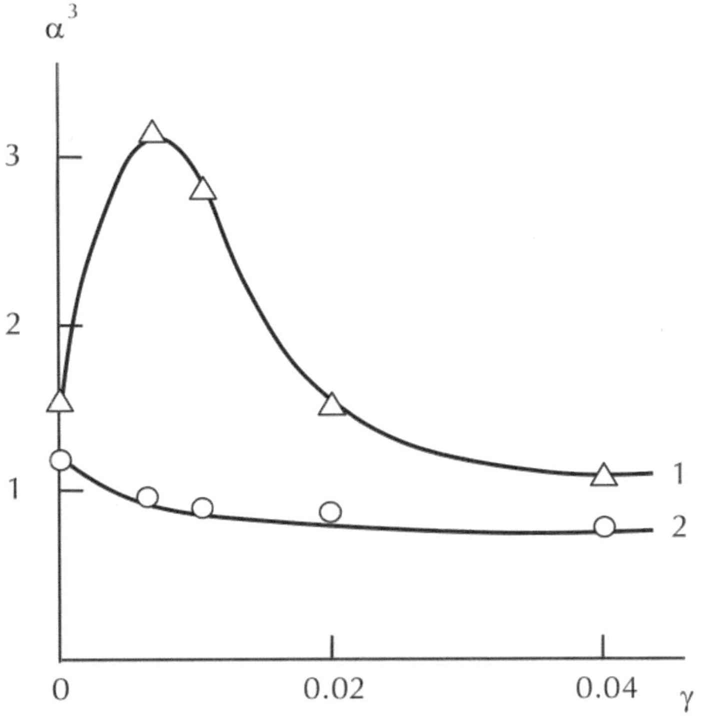

FIGURE 74 The dependence of excluded volume factor α^3 on branching component fraction γ for PPQX in solution in chloroform (1) and N-methylpyrrolydone (2).

Maximum dependence $^3(\gamma)$ in Ref. [51] was explained only qualitatively. It has been assumed, that the first part of this dependence (up to the maximum) is due to links density increase for a coil in good solvent that should increase α^3.

The second part of the curve is defined by polymer chain rigidity enhancement with branching degree increasing, linked with internodal distances decrease. Within the framework of fractal analysis the general quantitative treatment of this effect can be obtained. In Fig. 75 the dependences $\alpha^3(D_f)$ comparison is adduced, where α^3 values were determined theoretically according to the Eq. (39) and experimentally according to the Eq. (6). As one can see, a good correspondence between these dependences was received.

Hence, the stated above results demonstrated once more the fractal analysis methods applicability for the quantitative description of branched polymer macromolecular coil structure. The type of interactions between macromolecule elements (attraction or repulsion) is an important aspect of the proposed treatment. The indicated macromolecule structure, characterized by its fractal dimension, is one more important factor.

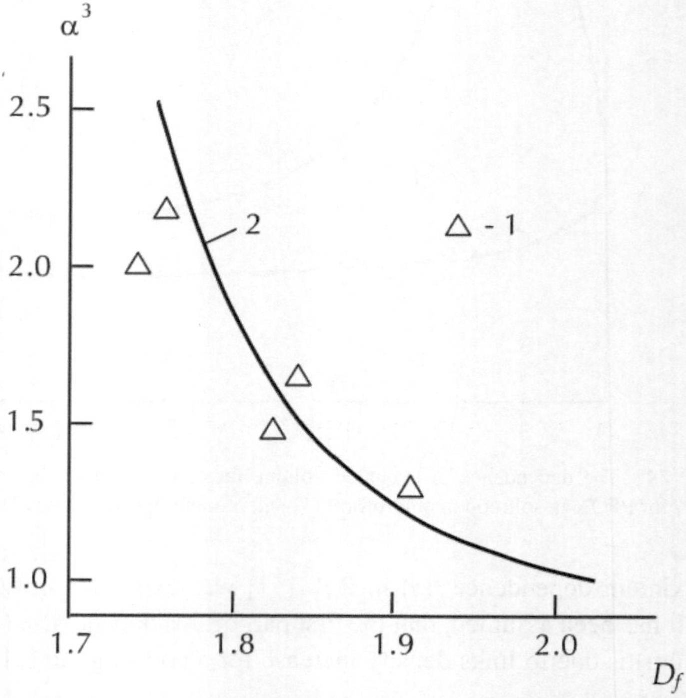

FIGURE 75 Comparison of the dependences of excluded volume factor α^3 on macromolecular coil fractal dimension D_f, calculated according to the Eqs. (6) (1) and (39) (2) for PPQX at MM=2×10^5.

As it has been shown above, the interactions of macromolecular coil elements between themselves play an important role for branched polymers: it has been shown [167, 168] that the chain branching degree increase (branching factor g decrease) results in a macromolecular coil compactization and in corresponding D_f growth. In Refs. [167, 168] the Eq. (127), expressed the intercommunication of D_f and g for branched polymers, was obtained. The authors [131] obtained a more strict analytical intercommunication of g and D_f:

$$g = CN^{3/D_f - 3/2} \tag{133}$$

where C is constant, N is polymerization degree.

The authors of Ref. [175] considered the possibility of the Eq. (133) usage for D_f value estimation on the example of polymers two groups: polyhydroxiether (PHE), received at different synthesis temperatures, and branched PPQX [51]. The experimental values of fractal dimension D_f for PHE were determined with the aid of the Eq. (79) by the slope of corresponding curves Q-t [176] and for PPQX—according to the Eq. (4). The experimental values of a branching factor g for PHE were determined as follows [175]. According to the experimental data the dependences of molecular weight MM on synthesis duration t were constructed at synthesis different temperatures in double logarithmic coordinates, corresponding to the relationship [134]:

$$MM \sim t^{\gamma_t}, \tag{134}$$

from the slope of which the exponent γ_t was determined. Then the branching factor g was determined [134]:

$$\gamma_t^{-1} = 1 - g \tag{135}$$

For PPQX the value g was determined according to the Eq. (105). The necessary for further calculations values a_η, $[\eta]$, N, D_f, g and K_η for the indicated polymers and solvents are listed in Tables 20–22.

TABLE 20 Characteristics of macromolecular coils of polyhydroxiether, synthesized at different temperatures T_{syn}, in chloroform.

T_{syn}, K	a	K_η	$[\eta]$, dl/g	N	D_f	g	D_f^T	C_∞	S, Å²	g^{cor}	D_{f1}^{cor}
333	0.515	1.4×10^{-3}	0.352	158	1.96	0.394	1.67	12.4	128.0	0.086	1.98
338	0.587	6.0×10^{-4}	0.432	250	1.89	0.478	1.64	6.8	57.1	0.190	1.81

TABLE 20 *(Continued)*

T_{syn}, K	a	K_η	[η], dl/g	N	D_f	g	D_f^T	C_∞	S, Å²	g^{cor}	D_{f1}^{cor}
343	0.775	7.0×10⁻⁴	0.528	349	1.69	0.774	1.59	3.5	28.6	0.595	1.63
348	0.923	1.3×10⁻⁵	0.612	399	1.56	1.0	1.56	2.7	23.1	1.0	1.56

The constant C in the Eq. (133) can be determined from the following considerations. For a linear polymer g=1 and substituting this value, its fractal dimension D_f and N value in the Eq. (133), C value can be estimated for each from polymers studied series. Then according to the same equation at the condition C=const theoretical values of fractal dimension D_f^T can be calculated for branched polymers. These D_f^T values are also listed in Tables 20–22. As one can see, between dimensions D_f and D_f^T a qualitative correspondence is observed, namely, their increase at polymer chain branching degree growth or branching factor g reduction. However, a quantitative correspondence is much worse—D_f and D_f^T absolute values discrepancy is large enough.

TABLE 21 Characteristics of macromolecular coils of polyphenylquinoxaline, synthesized with different contents of polyfunctional component γ, in N-methylpyrrolydone solution.

γ	a_η	K_η×10⁴	[η], dl/g	N	D_f	g	D_f^T	C_∞	S, Å²	g^{cor}	D_{f1}^{cor}	MM	D_{f2}^{cor}
0	0.57	12.9	0.60	212	1.91	1.0	1.91	7.8	67.7	1.0	1.91	47470	1.91
0.0067	0.47	25.0	0.51	314	2.04	0.892	1.94	9.8	91.5	0.710	1.99	83080	2.03
0.01	0.44	32.0	0.42	285	2.08	0.779	1.97	14.5	162	0.420	2.12	64280	2.16
0.02	0.41	43.3	0.35	241	2.13	0.686	2.0	17.0	207	0.315	2.20	45130	2.03
0.04	0.38	55.4	0.30	114	2.17	0.616	2.03	19.5	258	0.250	2.33	36260	2.28

TABLE 22 Characteristics of macromolecular coils of polyphenylquinoxaline, synthesized with different contents of polyfunctional component γ, in chloroform solution.

γ	a_η	$K_\eta\times10^4$	[η], dl/g	N	D_f	g	D_f^T	C_∞	S, Å²	g^{cor}	D_{f1}^{cor}	MM	D_{f2}^{cor}
0	0.64	7.94	0.92	212	1.83	1.0	1.83	5.2	42.0	1.0	1.83	61500	1.83
0.0067	0.72	2.30	0.85	314	1.74	0.943	1.85	5.7	46.5	0.86	1.87	91000	1.89
0.01	0.71	2.51	0.54	285	1.75	0.674	1.92	8.3	73.3	0.42	2.03	50000	1.96
0.02	0.63	4.91	0.42	241	1.84	0.560	1.96	12.4	128	0.23	2.19	45940	2.02
0.04	0.57	7.40	0.38	114	1.91	0.520	1.97	15.8	185	0.17	2.31	56800	1.94

The reasons of the indicated disparity are obvious enough. As it is known through Ref. [1], g value is determined correctly according to the Eq. (105) only in the case, when a branched polymer and its linear analog have the same molecular weight and chain rigidity. At the first approximation the last factor can be taken into account as follows [1]:

$$g^{cor} = g\frac{l_{st}^{l}}{l_{st}^{b}}, \tag{136}$$

where g^{cor} is the corrected g value, l_{st}^{l} and l_{st}^{b} are statistical segments length for linear and branched polymers, respectively.

l_{st} value is calculated according to the Eq. (71) and, believing l_0=const, the Eq. (136) can be rewritten as follows [175]:

$$g^{cor} = g\frac{C_\infty^{l}}{C_\infty^{b}}, \tag{137}$$

where C_∞^{l} and C_∞^{b} are characteristic ratios for linear and branched polymers, respectively.

C_∞^{b} variation can be estimated with the aid of the approximate equation (72). It is necessary to note, that chain branching degree increase causes macromolecule cross-sectional area S enhancement [177]. S and C values are connected between themselves as follows [175]:

$$S = (3 + 0.67C_\infty)^2 \qquad (138)$$

C_∞ and S values, calculated according to the Eqs. (72) and (138) with dimension D_f using are also listed in Tables 20–22. Further the fractal dimension D_{f1}^{cor} can be calculated according to the Eqs. (133) and (137) with correction by statistical flexibility change of branched polymer chain. The adduced in Table 20 results for PHE have shown a clear improvement of D_f and D_{f1}^{cor} correspondence in comparison with D_f and D_f^T—the average discrepancy reduces from approx. 11.6 up to 2.7%. The results adduced in Tables 21 and 22 have shown that in case of N-methylpyrrolydone using as solvent the calculated D_{f1}^{cor} and D_f values correspondence improves, where as for chloroform this correspondence changes for the worse. Thus, the mean discrepancy between D_f and D_f^T makes up approx. 5.6% for N-methylpyrrolydone and approx. 6% for chloroform and mean discrepancy between D_f and D_{f1}^{cor} is approx. 3.6 and 13.5% for the indicated solvents, respectively [175].

The two reasons of experimental and theoretical D_f values discrepancy are possible. The first from them consists of approximate character of the equation (72), which gives C_∞ overstated values [84]. The second reason consists of the usage of polymers with different molecular weights, whereas correct application of the Eq. (105) for g evaluation requires their equality. The adduced in Tables 21 and 22 *MM* values for PPQX have shown its large enough variation (almost in 3 times). Therefore, g values for PPQX were recalculated for the same MM=5×10^4. With this purpose the values $[\eta]_l$ and $[\eta]_b$ for the indicated *MM* were calculated according to Mark-Kuhn-Houwink equation (the indicated equation parameters K_η and a_η are listed in Tables 21 and 22) and then the corrected values $g(g^{cor})$ were calculated according to the Eq. (105). D_{f2}^{cor} values, calculated by the indicated mode (with correction on C_∞ and *MM*) are also adduced for PPQX in Tables 21 and 22. As one can see, for PPQX solutions in N-methylpyrrolydone D_f and D_{f2}^{cor} mean discrepancy does not change practically in comparison with the similar value for D_f and D_{f1}^{cor}—approx. 3.6 and approx. 3.4%, respectively. For PPQX solutions in chloroform the indicated mean discrepancy is decreased almost twice—from approx. 13.5% up to approx. 7.3%. For PHE correction on *MM* was not fulfilled because of two reasons: firstly, for this polymer g determination takes into account MM variation (the Eqs. (134) and (135)) and, secondly, the mean D_f and D_{f1}^{cor} discrepancy is insignificant—approx. 2.7%.

Thus, the adduced above results have shown the unequivocal intercommunication of polymer chain branching degree, expressed by the branching factor g, and macromolecular coil fractal dimension D_f. Certain difficulty in case of D_f determination by g value (or vice versa) are different values of constant C in the Eq. (133) for different polymers and solvents. However, the values D_f and N knowledge for linear analog allows this constant simple estimation. The necessity of keeping the constant molecular weight and chain rigidity conditions is observed at g determination according to the Eq. (105). If these conditions are not performed, then the authors [175] offered the value g correction techniques by the indicated parameters.

2.9 THE FRACTAL ANALYSIS OF FLOCCULATING ABILITY OF POLY(DIMETHYL DIALLYL AMMONIUM CHLORIDE)

As it has been shown above both macromolecular coils [10] and polymer chains sections [40] in polymers are fractal objects. This means that the indicated objects properties (including diffusive ones [178]) will differ essentially from the supposed ones in classical space and their description within the framework of Euclidean geometry can be only approximation, the authenticity degree of which is defined by the approaching degree of the corresponding static geometrical exponents (dimensions) of fractal and Euclidean objects [44]. The possibility of the object structure quantitative description with the aid of its fractal dimension D_f an unquestionable dignity of fractal analysis. This circumstance will be used in the present section for the description of admixtures absorption ability (flocculation) by polymer on the example of poly(dimethyl diallyl ammonium chloride) (PDMDAAC) [179].

As it is known through Ref. [180], D_f value characterizes "accessibility" (looseness) degree of object structure—the smaller D_f at the fixed values of surrounding Euclidean space d, is the smaller object compactness degree is. The fractal object "looseness" degree can be expressed quantitatively by its density ρ, determined as follows [180]:

$$\rho = \rho_0 R_g^{D_f - d}, \tag{139}$$

where R_g is a fractal object gyration radius, ρ_0 is Euclidean object density of the same material as fractal object.

For a macromolecular coil the relation between R_g and D_f is given by the Eq. (8), where polymerization degree N can be received by polymer molecular weight MM division by monomer link molecular weight m_0 (for PDMDAAC m_0=176 g/mole [34]) and let us obtain finally [181]:

$$\rho \sim \left(\frac{MM}{m_0} \right)^{-(d-D_f)/D_f}, \tag{140}$$

where d in the considered case is obviously equal to 3.

The fractal dimension of PDMDAAC macromolecular coil D_f was determined according to the Eq. (4) and Mark-Kuhn-Houwink equation for the indicated polymer looks like [34]:

$$[\eta] = 1.12 \times 10^{-4} MM^{0.82}. \tag{141}$$

With the Eqs. (4) and (141) using the following results were received for PDMDAAC: in water solution D_f=1.44 and in NaCl solution D_f=1.65. The indicated D_f values define macromolecular coil conformation for linear polymers [10]. The first from the indicated values corresponds to permeable coil fractal dimension, the second one—to D_f value for a macromolecular coil in good solvent. D_f decreasing means more unfolded conformation of polymer chain (D_f=1.0 corresponds to completely stretched chain and to transition to Euclidean behavior [182]). Thus, the main influence of medium on polymer macromolecule is its conformation change (and, hence, D_f value) at variation of solvent quality in respect to polymer [56].

As it follows from the Eq. (140), for compact (Euclidean) object (D_f =d) the value ρ=1 (in relative units). One from the most important properties of fractal objects is their density reduction (loosening) at the size increasing (R_g or *MM*) even at the constant D_f value. It is obvious, that flocculation degree will be a function of not only active centers number (centers of flocculator and admixtures interaction), but also their accessibility degree. Within the framework of fractal analysis the latter factor can be characterized by the macromolecular coil density ρ—the less , is the more possible adsorbed substance capture (in the considered case—phosphatides [183]) over all coil volume is. As a matter of fact the similar explanation of the observed effect was given in Ref. [183], where the flocculator (PDMDAAC) concentration reduction at [η] growth was also connected with macromolecular coil geometrical features, although this was made on purely qualitative level. In Fig. 76 the dependence ρ(*MM*) for PDMDAAC in Nach solution is adduced (D_f=1.65), which has illustrative character and demonstrates two aspects, important for fractal objects. Firstly, this plot has shown, as far as density ρ of fractal object with D_f=1.65 is lower than the corresponding value for Euclidean object (D_f=*d*=3) (let us remind, that for the last $\rho=\rho_0=1.0$).

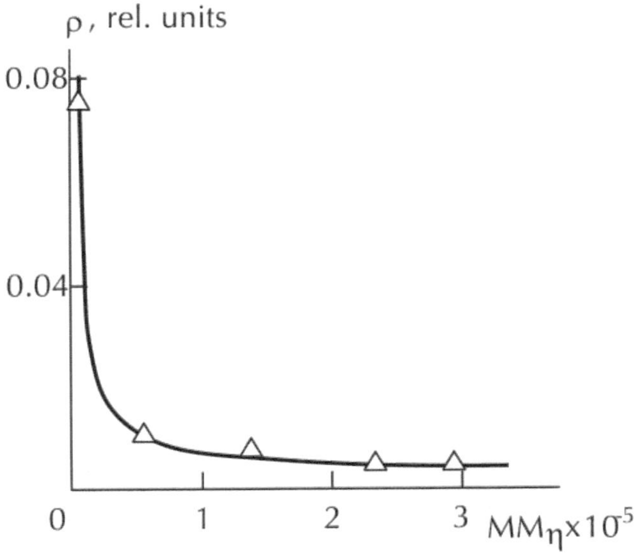

FIGURE 76 The dependence of macromolecular coil density ρ (in relative units) on average viscous molecular weight MM_η for PDMDAAC (D_f=1.65).

Secondly, this plot shows very strong dependence of ρ on *MM*, that in its turn explains macromolecular coil accessibility degree strong enhancement in respect to admixtures (phosphatides) penetration in its internal regions and, hence, the strong dependence of flocculator (PDMDAAC) optimal concentration c_f on *MM* [183].

In Fig. 77 the relationship $c_f(MM^{-1})$ in double logarithmic coordinates is adduced, which was constructed according to the data of Ref. [183]. As one can see, this relationship can be described by the following scaling expression [181]:

$$\tilde{n}_f \sim MM^{-n}, \tag{142}$$

where the exponent n0.80, that is very close by absolute value to the exponent $(d\text{-}D_f)/D_f$ in the Eq. (140), which for D_f=1.65 is equal to 0.818.

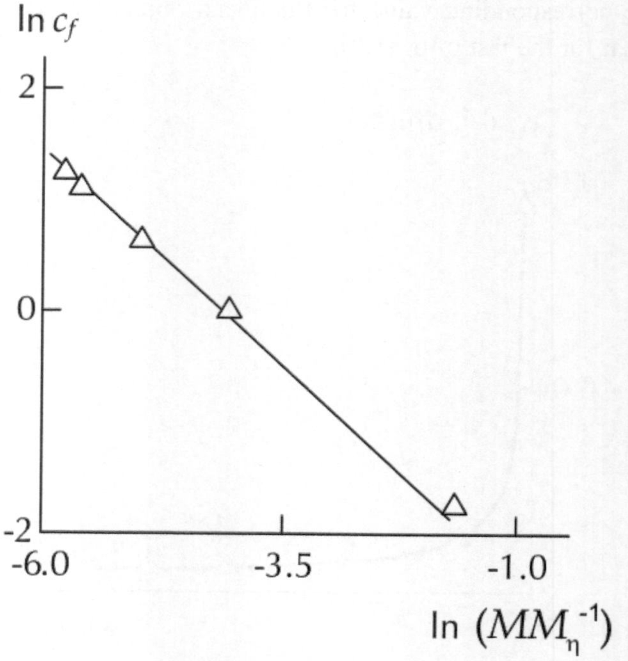

FIGURE 77 The scaling relationship between PDMDAAC concentration c_f and its average viscous molecular weight MM_η in double logarithmic coordinates.

The dependence (142) physical significance can be understood easily within the framework of modern representations about fractal clusters [135, 160]. The particle penetration in cluster depth Δr is determined as follows [160]:

$$\Delta r \sim N^{q}\,. \tag{143}$$

The exponents ratio q/D_f for three-dimensional Euclidean space is approximately equal to 0.8 [160]. For D_f=1.65 and with the condition q–n^{-1} appreciation let us obtain n≈0.76, that is close to n value, evaluated from the plot of Fig. 77. Thus, the larger penetration depth of phosphatides in PDMDAAC macromolecular coil is, the lower flocculator optimal concentration is.

It is obvious, that the macromolecular coil fractal dimension D_f, which can be changed, for example, by solvent, type of polymer variation, so forth, is one more factor, influencing on flocculation process effectiveness. Thus, from the Eq. (142) it follows, that the value c_f can be maintained on previous level by MM variation at D_f change. The authors [181] compared this possibility for PDMDAAC in two solvents: water and NaCl water solutions (the value D_f=1.44 and 1.65, respectively) and obtained coefficient k_c, which shows, in how many times a polymer MM should be reduced at transition from D_f=1.65 to D_f=1.44 for the previous value c_f preservation. The dependence of k_c on MM is adduced in Fig. 78, from which the strong dependence of c_f on a macromolecular coil fractal dimension follows or, in other words, on coil accessibility degree to particles penetration. Thus, at D_f reduction from 1.65 up to 1.44 for the initial MM=2.99×10 at previous c_f value preservation MM decrease almost on one order is allowed [181].

Hence, the stated above results have shown that the molecular weight, macromolecular coil structure, characterized by its fractal dimension, and medium, in which flocculation occurs, are included in a factors number, influencing on adsorption of admixtures (flocculation) effectiveness by a polymer flocculator PDMDAAC. The medium cans influence on flocculation process by two ways: by a macromolecular coil fractal dimension change and viscosity variation. The latter factor can influence on diffusive processes intensity. The sole from the enumerated above factors, which will influence equally on fractal and compact objects properties, is a medium viscosity variation and three remaining factors influence is defined exclusively by the fractal nature of polymer flocculator macromolecular coil [179].

Let us note one more interesting dependence, adduced in Ref. [183], namely, the extreme change of phosphatides removal degree as PDMDAAC concentration function for each molecular weight. Because of its applied and theoretical importance this question requires special consideration. Therefore, in Ref. [184] this effect theoretical description for PDMDAAC was given within the framework of fractal analysis.

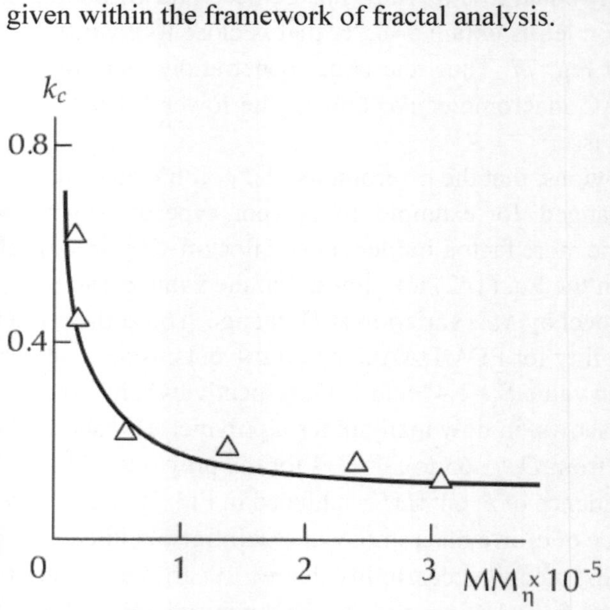

FIGURE 78 The dependence of average viscous molecular weight MM_η reduction coefficient k_c at transition of D_f from 1.65 to 1.44 for PDMDAAC.

The increase of phosphatides removal degree Q up to some polymer flocculator concentration c_{max}, at which the greatest value Q (Q_{max}) is reached, is due to a simple increase of "active centers" (capture sites [183]) number of flocculation owing to a macromolecular coils (polymer concentration) number growth in solution. At the concentration c_{max} macromolecular coils touching occurs, that corresponds to Q_{max}. At c_f growth coils interpenetration is realized and the indicated transition scale r_c is determined by the criterion [131]:

$$r_c \sim c_f^{-1/\left(d-D_f\right)} \qquad (144)$$

r_c value at the condition of coils touching is accepted equal to $2R_g$ [184]. In Fig. 79 the calculated according to the Eq. (144) theoretical values c_{max} (c^T_{max}) and this parameter experimental values comparison is adduced [183]. For all molecular weights, excluding the lowest from the used ones ([η]=0.10 dl/g or MM=4×10³) a good correspondence of theory and experiment was obtained, that confirms the made above supposition in respect to physical significance of c_{max} value. As for PDMDAAC with $MM_\eta \approx 4\times10^3$, then c^T_{max} value proves to be approximately in 8 times larger than c_{max}, that is, low-molecular flocculator effectiveness turns out to be much higher than an expected one. This important aspect deserves special consideration.

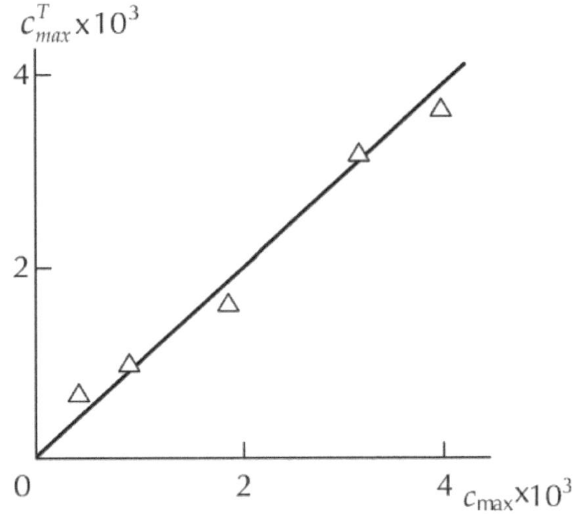

FIGURE 79 The relationship between calculated according to the Eq. (144) c^T_{max} and obtained experimentally c_{max} PDMDAAC concentrations, at which the greatest degree of phosphatides removal is reached. Here and in Fig. 80 c_f values are given in mass % to oil mass.

By its molecular weight PDMDAAC with $MM_\eta \approx 4\times10^3$ is closer to oligomer than to polymer. Within the framework of irreversible aggregation model [160] it has been shown that at small N small rodlike aggregates are formed, that supposes for them smaller D_f value than the obtained according to the Eq. (4) at a_η=0.82 for high-molecular polymer [130]. Besides, it is known through Ref. [185], that D_f value is the monotonously growing N function. Thus, all reasons are to be expected, that for low-molecular ($MM_\eta \approx 4\times10^3$) PDMDAAC D_f value will be essentially lower than the value 1.65, obtained for high-molecular polymers. From the condition $c^T_{max}=c_{max}$ and with the relationships (4), (8) and (144) D_f value for low-molecular PDMDAAC was estimated, which turns out to be equal to approx. 1.26, that corresponds completely to the made above suppositions and the results of Refs. [130, 185]. This example demonstrates strong influence of macromolecule spatial geometrical structure, characterized by the fractal (Hausdorff) dimension D_f, which controls the accessibility of admixtures (in our case—phosphatides) into flocculator internal regions in flocculation process. As it has been noted above, this factor can be expressed quantitatively with the aid of the fractal density ρ, determined according to the Eq. (139).

The power dependence of ρ on D_f defines the strong dependence of Q on D_f. Thus, for twofold increase of Q value for PDMDAAC with $MM=3\times10^5$, D_f reduction from 1.65 up to 1.50 is required at constant flocculator concentration. As it is known through Ref. [1], the exponent a value in Mark-Kuhn-Houwink equation depends on a type of the solvent, used at measurements, therefore macromolecular coil conformation can be changed by the last suitable choice and this way to control its ability to flocculate admixtures. This ability control can be performed by the exponent a_η variation and measurement—as it follows from the Eq. (4), a_η increasing means D_f reduction.

Let us consider now the causes of the extreme dependence $Q(c_f)$, obtained experimentally in Ref. [183] for each [η] value. As it follows from the mentioned above considerations, Q value should depend, as minimum, on two parameters [184]:

$$Q = k_0 \frac{c_f}{\rho}, \qquad (145)$$

where k_0 is a constant, is the macromolecular coil density, determined according to the Eq. (139).

For $c_f \leq c_{max}$ coils do not come into contact with one another, ρ value for them will be constant (at *MM*=const) and *Q* as a matter of fact is only a function of c_f. For $c_f > c_{max}$ macromolecular coils overlapping by their external regions occurs, that is, as it they flocculate one another, that reduces their ability to flocculate admixtures out of solution. Let us remind that fractal aggregates have one common feature—for large enough *N* particles (or particles small clusters) addition to such aggregate can occur only over narrow enough external "active zone" and their internal regions are screened by the indicated zone and cannot added particles [186]. Therefore, between effective (unscreened) macromolecular coil gyration radius R_{ef} and polymer concentration c_f the following relationship exists [186]:

$$c_f \sim R_{ef}^{D_f-d}, \tag{146}$$

from which R_{ef} reduction at c_f growth follows at $c_f > c_{max}$. Using the value R_{ef}, received by such mode, the macromolecular coil effective density ρ_{ef} can be evaluated according to the Eq. (139) and then, calculating the constant k_0 in the Eq. (145) from the condition of *Q* theoretical and experimental values at the maximum point for PDMDAAC with different *MM*, the theoretical dependence $Q(c_f)$ can be estimated. The theoretical (curves) and experimental (points) dependences $Q(c_f)$ for PDMDAAC with viscosity of 0.9, 1.8 and 3.45 dl/g comparison are adduced in Fig. 80. As one can see, the proposed technique reflects precisely qualitatively *Q* change at c_f variation, although a quantitative correspondence is worse, especially at $c_f < c_{max}$. The authors [184] supposed that this discrepancy was impossible to attribute only at the expense of theoretical calculation imperfection, although its simplicity allows certain possibilities for improvement. Thus, the obvious extrapolation of *Q* to zero at $c_f=0$ should be expected, although experimental data do not always imply (for example, for [η]=1.8 dl/g [183]). Nevertheless, one from possible improvements of the proposed calculation method can be pointed out [184]. All cited up to now equations suppose that aggregate (macromolecular coil) consists of zero-dimensional objects (points). It is obvious, that a real polymer coil consists of monomer links with finite sizes (for example, with length *a*) and in this case dimensional characteristics (scale) introduction in the corresponding relationships is necessary. For example, the Eq. (8) will look like [187]:

$$\left(\frac{R_g}{a}\right)^{D_f} \sim N \tag{147}$$

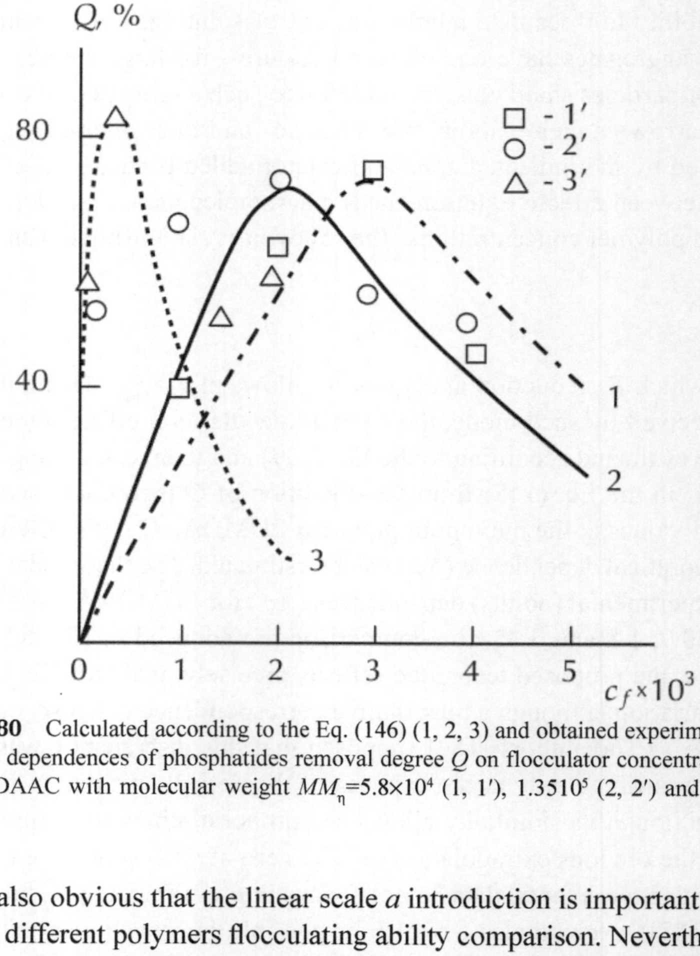

FIGURE 80 Calculated according to the Eq. (146) (1, 2, 3) and obtained experimentally (1,' 2,' 3′) dependences of phosphatides removal degree Q on flocculator concentration c_f for PDMDAAC with molecular weight MM_η=5.8×10⁴ (1, 1′), 1.3510⁵ (2, 2′) and 3.010⁵ (3, 3′).

It is also obvious that the linear scale *a* introduction is important especially at different polymers flocculating ability comparison. Nevertheless, the authors [179, 184] do not try to attain complete quantitative correspondence of theory and experiment, but want to demonstrate the principal possibility of such complex phenomenon as flocculation description within the framework of fractal analysis and irreversible aggregation models.

2.10 THE MECHANISMS OF FILMS FORMATION FROM POLYMER SOLUTIONS

Polymers are often used as films, received from different kinds of solvents [188]. As it is known through Ref. [189], such films of the same polymer properties depend to a considerable extent from the solvent used for their production. Several conceptions, describing polymer films structure and used solvent type exist (for example, [189, 190]). Nevertheless, at present the technique of polymer films structure and properties prediction depending on the used polymer and solvent type is absent. Therefore, the authors [191] performed the quantitative description of structure of polymer films, prepared from different solvents, within the framework of fractal analysis and irreversible aggregation models on the example of amorphous glassy polyblock copolymer polyarylatesulfone (PASF). With this purpose PASF film samples were prepared from solutions in 9 solvents (N, N-dimethylformamide, chlorobenzene, 1,2-dichloroethane, chloroform, N, N-dimethylacetamide, tetrachloroethane, methylene chloride, 1,4-dioxane and tetrahydrofuran). The indicated samples tests on uniaxial tension were performed at temperature 293°K. Besides, for molecular weight MM_{bh} of a chain part between traditional macromolecular entanglements (macromolecular "binary hooking" [70]) nodes determination the tension tests of the same samples were performed above their glass transition temperature T_g ($T_g \approx 473$ K for PASF [192]) at T=523 K and elasticity modulus E_r in rubber-like state was determined. MM_{bh} value was estimated according to the equation [70]:

$$MM_{bh} = \frac{\rho_p RT}{E_r}, \tag{148}$$

where ρ_p is the polymer density, R is universal gas constant, T is testing temperature.

The macromolecular "binary hooking" network density ν_{bh} was determined as follows [70]:

$$\nu_{bh} = \frac{\rho_p N_A}{MM_{bh}}, \tag{149}$$

where N_A is Avogadro number.

Macromolecular coil fractal dimension D_f was calculated according to the Eq. (11), where intrinsic viscosity of polymer in θ-solvent was determined according to Mark-Kuhn-Houwink equation [5]:

$$[\eta]_\theta = K_\theta \overline{MM}_w^{0.5}, \tag{150}$$

where K_θ=10^{-3} [5], $\overline{MM}_w$ is average weight molecular weight of polymer.

The polymer structure fractal dimension d_f is determined as follows [88]:

$$d_f = (d-1)(1+\nu), \tag{151}$$

where d is the dimension of Euclidean space, in which a fractal is considered (it is obvious, in our case d=3), ν is Poisson's ratio, evaluated by the mechanical tests results with the aid of the relationship [37]:

$$\frac{\sigma_Y}{E} = \frac{(1-2\nu)}{6(1+\nu)}, \tag{152}$$

where σ_Y is yield stress, E is elasticity modulus.

According to the representations, available at present [188], polymer samples structure safety is due to formation in them a macromolecular binary hooking network, which spreads stress, acting on a sample, from one macromolecular coil to another. The structure formation at the used films preparation method occurs at solvent evaporation, macromolecular coils interpenetration and macromolecular binary hooking's network formation, ensuring the whole sample connectivity, which can be considered as the gelation transition [193]. The relation between dimensions of macromolecular coil structure D_f and polymer solid-phase structure d_f for linear polymers are defined by the Eq. (95).

Let us consider further the influence of solvent, used for films preparation, on the structure of a macromolecular coil is solution, characterized by the dimension D_f. As it has been noted above, in this case D_f value of a macromolecular coil is defined by two factors: interactions of coil itself elements between themselves and interactions polymer-solvent. Since in Ref. [191] only one polymer (PASF) was used, then D_f value change at solvent variation should be attributed to the influence of the second group interactions. In Fig. 81 the dependence $[\eta](\delta_s)$, where δ_s is solubility parameter of the used solvents [14, 16], is adduced, which has a typical cupola-like shape [16]. $[\eta]$ reduction means macromolecular coil compactization that supposes corresponding D_f increasing [10]. The shown in Fig. 82 dependence $D_f(\delta_s)$ confirms this supposition. D_f calculation according to the equation (95) has shown a similar shape of the dependence $d_f(\delta_s)$ (Fig. 83). Thus, less compact shape of a macromolecular coil in solvent with better thermo-dynamical affinity in respect to polymer defines less D_f values and, respectively, d_f. The latter means higher local order degree of glassy polymers structure [107] and their properties corresponding change [190].

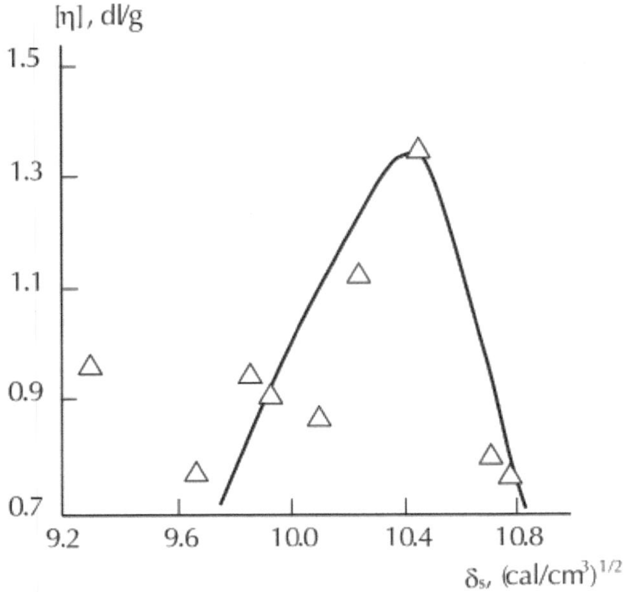

FIGURE 81 The dependence of intrinsic viscosity $[\eta]$ on solvent solubility parameter δ_s for PASF.

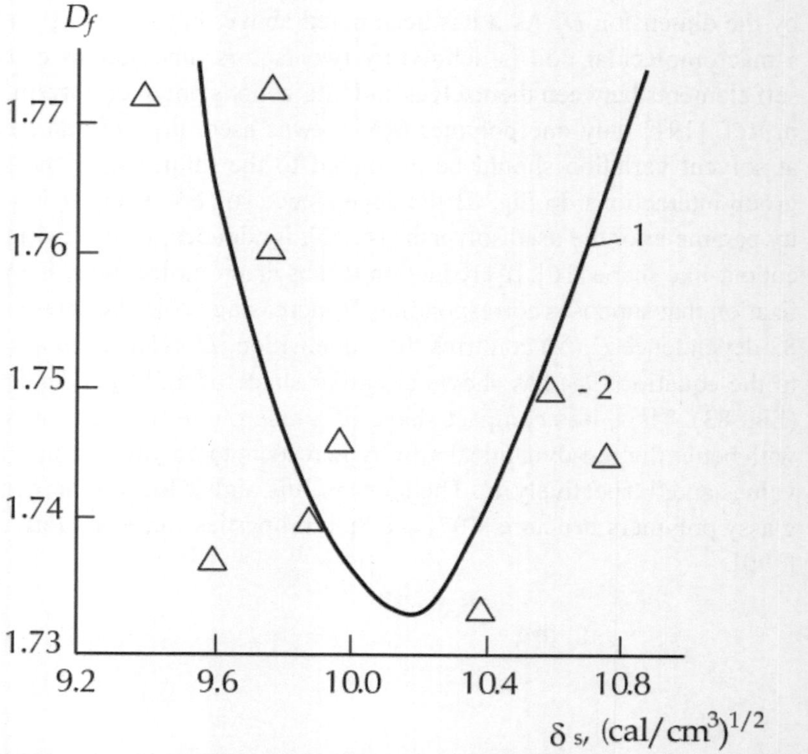

FIGURE 82 The dependences of macromolecular coil fractal dimension D_f on solvent solubility parameter δ_s for PASF. 1—calculation according to the Eq. (11); 2—calculation according to the Eq. (95).

Let us consider the parameter d_f estimation within the framework of Witten-Sander irreversible aggregation model [194]. As it has been shown in Refs. [195, 196], the amorphous glassy polymers structure can be simulated as totality of Witten-Sander clusters (WS clusters) large number, having radius R_{cl}, which is determined as follows [197]:

$$R_{cl} = c_0^{-d}, \tag{153}$$

where c_0 is "seeds" number, around which WS clusters are grown.

The "seeds" role at amorphous polymers structure formation is played by nodes of macromolecular binary hooking network with density ν_{bh}. Then at "seeds" high concentration it can be written [197]:

$$R_{cl}^{d_f} \sim \frac{c}{c_0}, \quad (154)$$

where c is concentration of particles, forming WS cluster.

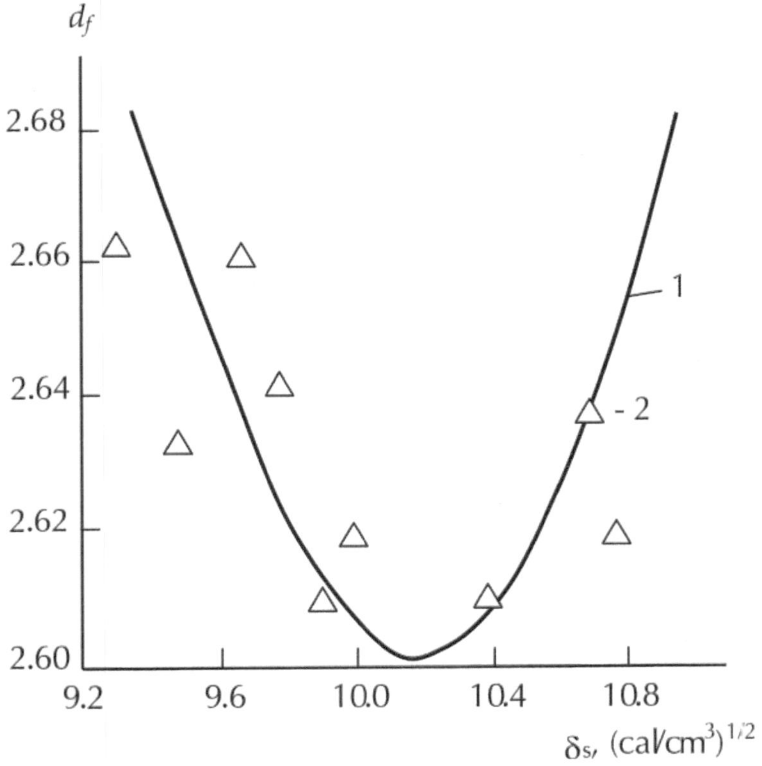

FIGURE 83 The dependences of polymer films structure fractal dimension d_f on solvent solubility parameter δ_s for PASF. 1—calculation according to the Eq. (95); 2—calculation according to the Eq. (154).

In the considered case the statistical segment—macromolecular "stiffness section" with mutually dependent orientation of position in space, is accepted as such particle [49]. The statistical segment length l_{st} is determined according to the Eq. (71). The total length of macromolecules L per polymer volume unit can be determined as follows [107]:

$$L = S^{-1},\tag{155}$$

where S is cross-sectional area of macromolecule, which is equal to 30.7 Å^2 for PASF [190].

Thus, the statistical segments number per polymer volume unit is equal to [191]:

$$c = \frac{L}{l_s} = \frac{1}{S l_0 C_\infty}\tag{156}$$

The fractal dimensions d_f comparison, calculated according to the Eqs. (95) and (154) (at the condition $c_0=\nu_{bh}$), is adduced in Fig. 83, from which their good correspondence follows. This confirms correctness of the description of amorphous polymers structure as WS clusters totality and indicates factors, controlling it. Thus, the polymer chain molecular characteristics (l_0, C_∞, S) and macromolecular binary hooking's network nodes density ($c_0=\nu_{bh}$) influence strongly on polymer structure. In Fig. 82 D_f values, estimated according to the Eqs. (11) and (95) comparison is shown (in the last equation d_f estimation according to WS model was used). As one can see, the comparable values D_f good enough correspondence is obtained again. And let us consider at last the factors, defining the value ν_{bh} or c_0. As it is known through Ref. [25], a fractals intersections number (in our case—macromolecular coils) N_{int} is determined according to the Eq. (13) of Chapter 1. In its turn, between R_g and $\overline{MM}_w$ the fractal Eq. (144) exists. The Eqs. (13) of Chapter 1 and (144) combination gives [191]:

$$c = \frac{L}{l_{st}} = \frac{1}{S l_0 C_\infty}\tag{157}$$

In Fig. 84 the dependence $\nu_{bh}(N_{int})$ at MM_w=const is adduced, which proves to be linear. Hence, ν_{bh} value is also defined by macromolecular coil fractal dimension D_f and polymer molecular weight (the last is a well-known fact [198]).

Hence, the results adduced above have shown correctness of the structure simulation, formed from different solvents, for amorphous PASF as totality of WS clusters. This model parameters are defined by polymer molecular characteristics and interactions polymer-solvent. These results give a strict physical basis for prediction of structure and properties of polymer films, prepared from different solvents.

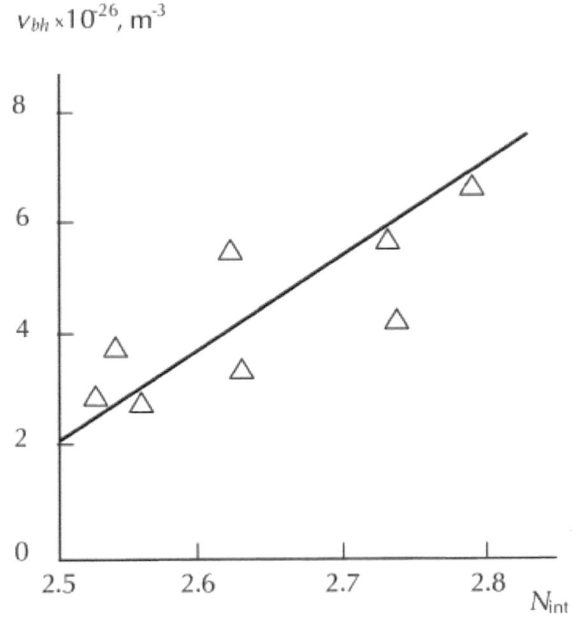

FIGURE 84 The dependence of macromolecular binary hooking's network density ν_{bh} on macromolecular coil intersections number N_{int} for PASF.

Polymers mechanical properties are some from the most important, since even for polymers of different special purpose functions this properties certain level is required [199]. However, polymers structure complexity and due to this such structure quantitative model absence make it difficult to predict polymers mechanical properties on the whole diagram stress-strain (σ–ε) length—from elasticity section up to failure. Nevertheless, the development in the last years of fractal analysis methods in respect to polymeric materials [200] and the cluster model of polymers amorphous state structure [106, 107], operating by the local order notion, allows one to solve this problem with precision, sufficient for practical applications [201].

The mentioned above models are the complement of one another, keeping certain niches in simulation theory [202]. Thus, the fractal analysis operates by very general parameters (fractal dimensions), which do not take into account substance structure specific features, but fix an object elements distribution in space only [86, 203]. This allows establishing the relations between different polymer states (for example, in solution and condensed state, see Section 2.6). On the other hand, the cluster model allows concretization of these general characteristics with polymer specific molecular and structural parameters appreciation [106, 107]. With this reason the indicated models combination was chosen for prediction of the mechanical parameters (elasticity modulus E, yield stress σ_Y, failure strain ε_f), characterizing separate sections of the curve σ–ε (elasticity, yielding and failure, respectively) [201]. The prediction technique verification will be performed on the example of film samples of amorphous glassy PASF, prepared from different solvents [190, 191].

The elasticity modulus E value is connected with polymer molecular characteristics by the following empirical equation [201]:

$$E \approx 0.7\left(\frac{S}{C_\infty}\right)^{1/2}, \text{GPa}, \tag{158}$$

where S is cross-sectional area of macromolecule, equal to 30.7 Å^2 for PASF, C_∞ is characteristic ratio, determined according to the Eqs. (11), (95) and (100).

In Fig. 85 the comparison of experimental E and calculated according to the Eq. (158) E^T elasticity modulus values is adduced for the considered PASF samples. This comparison shows a good correspondence of theory and experiment (the average discrepancy of E and E^T makes up 4%).

The yield stress σ_Y value can be calculated within the framework of polymers yielding dislocation conception [204]:

$$\sigma_Y = \frac{1.73Gb_B}{2\pi}\sqrt{\rho_d}, \tag{159}$$

where G is a shear modulus, calculated according to the Eq. (89), b_B is Burgers vector, estimated according to the Eq. (93), ρ_d is the structure linear defects density (analog of dislocations density for crystalline solid bodies).

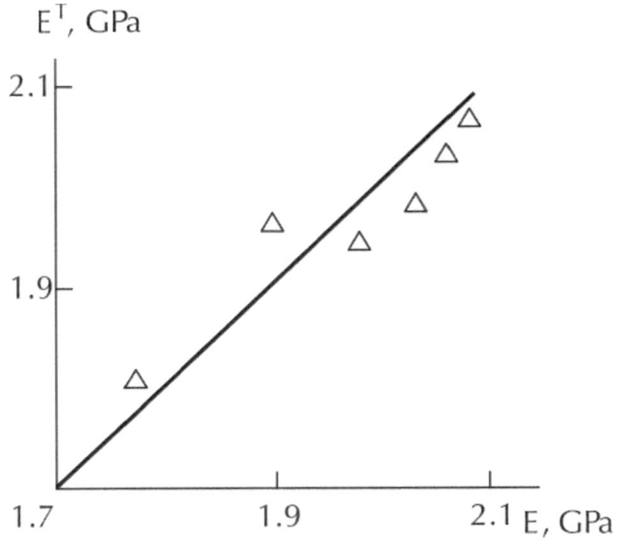

FIGURE 85 The relationship between experimental E and calculated according to the Eq. (158) E^T elasticity modulus values for PASF film samples, prepared from different solvents.

For amorphous polymers ρ_d value is accepted equal to relative total length of polymer segments, included in local order domains (clusters) and estimated as follows [204]:

$$\rho_d = \frac{\varphi_{cl}}{S}, \tag{160}$$

where the ratio (φ_{cl}/S) is calculated directly from the Eq. (97).

The comparison of experimental σ_Y and calculated by the indicated method σ_Y^T yield stress values for PASF film samples, prepared from different solvents, is adduced in Fig. 86. A good correspondence of calculated and experimental data is obtained again (the average discrepancy of σ_Y and σ_Y^T makes up less than 10%).

The strain at fracture ε_f of film samples can be calculated within the framework of a cluster model of polymers amorphous state structure [205]:

$$\frac{1}{\lambda_f} = \frac{\varphi_{cl}}{f} + \frac{(1-\varphi_{cl})^{1/2}}{n_{st}^{1/2}}, \qquad (161)$$

where λ_f is draw ratio at failure, equal to $1+\varepsilon_f$ [201], f is times number, which macromolecule passes thought local order domain (for amorphous polymers f=1 [205]), n_{st} is equivalent statistical links between macromolecular entanglements nodes (binary hooking) in melt.

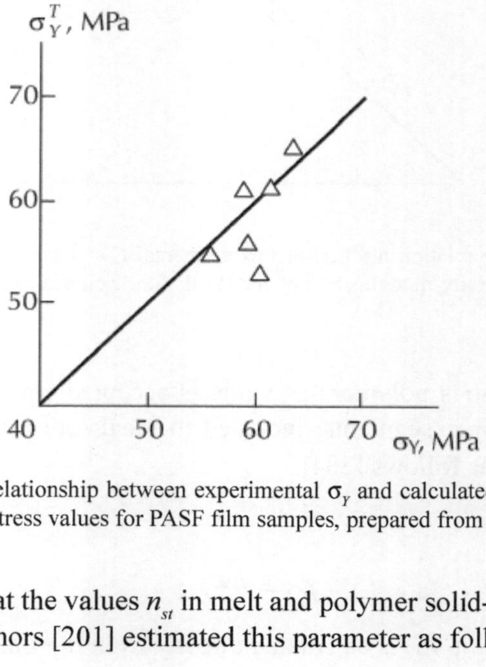

FIGURE 86 The relationship between experimental σ_Y and calculated according to the Eq. (159) σ_Y^T yield stress values for PASF film samples, prepared from different solvents.

Assuming, that the values n_{st} in melt and polymer solid-phase state are the same, the authors [201] estimated this parameter as follows [107]:

$$n_{st} = \frac{2}{\varphi_{cl}} \qquad (162)$$

In Fig. 87 the comparison of experimental ε_f and calculated by the indicated mode ε_f^T strain up to failure values is adduced for PASF film samples, prepared from different solvents. A good enough correspondence of calculation and experiment is obtained again (the average discrepancy of ε_f and ε_f^T does not exceed 20%). Let us note, that on the whole theoretical values ε_f^T exceed experimental ones that were to be expected. This is due to the known fact—the calculated results do not taken into account availability in real polymer of different kinds of defects, reducing ε_f value [143].

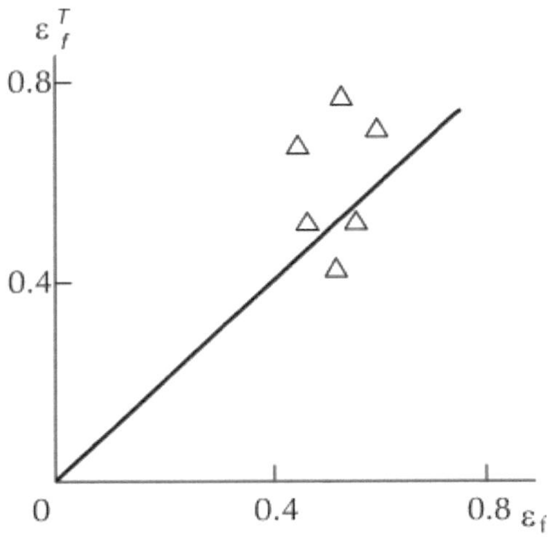

FIGURE 87 The relationship between experimental ε_f and calculated according to the Eq. (161) ε_f^T strain up to failure values for PASF film samples, prepared from different solvents.

Let us consider the same mechanical characteristics prediction technique at testing temperature T variation. For the dependences $E(T)$ estimation the authors [201] modified the Eq. (158) by the simplest mode, assuming that the coefficient in this equation is a function of testing temperature of general view C/T, where C is constant, equal to 205°K, if T is given in K. In Fig. 88 comparison of experimental and calculated by the indicated mode dependences $E(T)$ for two PASF film samples, prepared from solutions in chloroform and methylene chloride, is adduced. This comparison shows applicability of the stated above approximation for prediction of the elasticity modulus temperature dependence.

A more precise method of the dependence $E(T)$ calculation exists. As it is known through Ref. [107], by virtue of clusters thermofluctuational origin their relative fraction φ_{cl} is the reducing function of temperature (the Eq. (96)). This allows calculation d_f value according to the Eq. (97) and subsequent C_∞ estimation according to the Eq. (100). Then the Eq. (158) can be used directly for the dependence $E(T)$ evaluation at the obvious condition S=const. Both methods give coordinated results, namely, E reduction on about 30% at T growth within the range of 293–398 K.

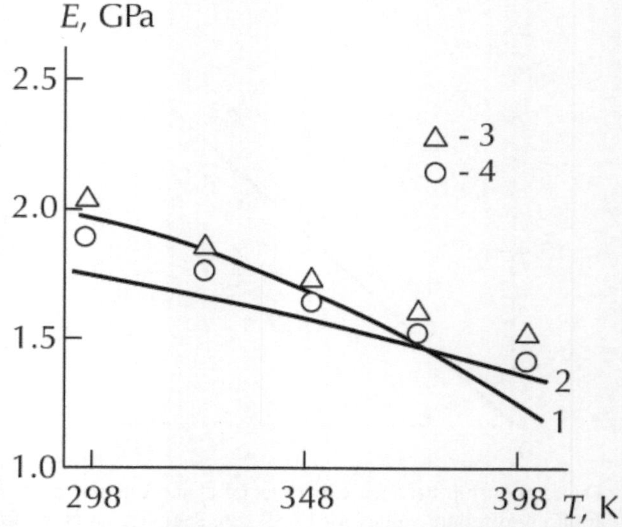

FIGURE 88 Comparison of the experimental (1, 2) and theoretical (3, 4) dependences of elasticity modulus E on testing temperature T for PASF film samples, prepared from solutions in chloroform (1, 3) and methylene chloride (2, 4).

The dependence $_{Y}(T)$ was calculated according to the Eq. (159), where G value was estimated according to the Eq. (89) with the theoretical dependences $E(T)$ using. The comparison of the experimental and theoretical dependences $\sigma_Y(T)$ for PASF samples, prepared from solutions in chloroform and N, N-dimethylformamide, has shown their good enough correspondence (Fig. 89).

The temperature dependence ε_f was evaluated according to the equation (161). φ_{cl} temperature variation knowledge is necessary for such evaluation, which was obtained according to the Eq. (96). Since T_g value is a function of φ_{cl} [206], then at first this parameter was calculated at T=293 K according to the known φ_{cl} values (the Eq. (97)) and then the dependence $\varphi_{cl}(T)$ was estimated, proceeding from the value T_g, obtained by such mode. In Fig. 90 comparison of experimental and calculated by the indicated mode dependences $\varepsilon_f(T)$ is adduced for PASF film samples, prepared from solutions in chloroform and tetrahydrofuran. For the first series of the indicated samples a good correspondence of theory and experiment was obtained and for the films, prepared from PASF solution in tetrahydrofuran, ε_f experimental values at elevated temperatures are essentially lower than theoretical ones by the indicated above reasons.

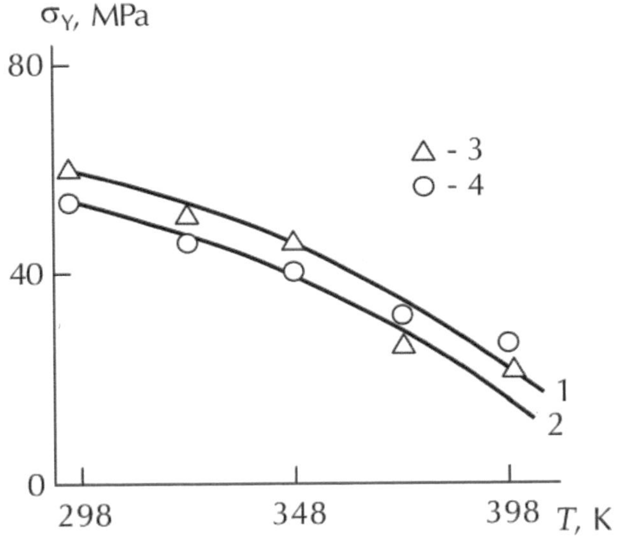

FIGURE 89 Comparison of the experimental (1, 2) and theoretical (3, 4) dependences of yield stress σ_Y on testing temperature T for PASF film samples, prepared from solutions in chloroform (1, 3) and N, N-dimethylformamide (2, 4).

Hence, the results stated above demonstrated that the cluster model of polymers amorphous state structure and fractal analysis allowed quantitative prediction of mechanical properties for polymers film samples, prepared from different solvents. Let us note, that the properties prediction over the entire length of the diagram σ–ε was performed within the framework of one approach and with precision, sufficient for practical applications. This approach is based on strict physical substantiation of the analytical intercommunication between structures of a macromolecular coil in solution and polymers condensed state [201].

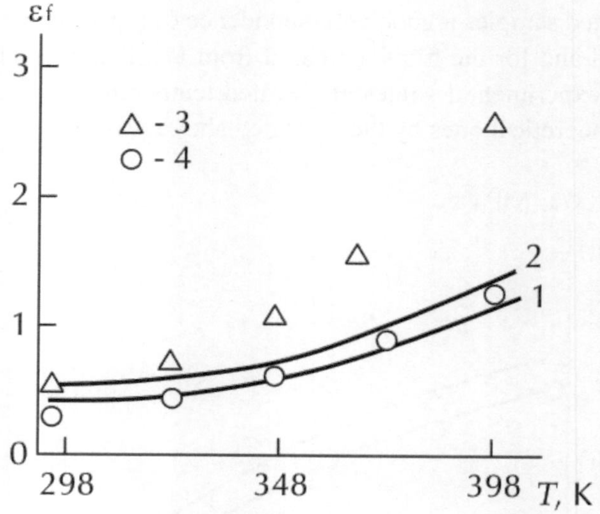

FIGURE 90 Comparison of the experimental (1, 2) and theoretical (3, 4) dependences of failure strain ε_f on testing temperature T for PASF film samples, prepared from solutions in chloroform (1, 3) and tetrahydrofuran (2, 4).

2.11 THE THEORETICAL DESCRIPTION OF POLYMERS MOLECULAR WEIGHT DISTRIBUTION

For high-molecular compounds notions of a molecule and molecular weight *MM* have their features connected with qualitative difference of polymers from low-molecular substances [207]. This difference is linked with the fact that polymers are always a mixture of macromolecules with different molecular weights (mixture of polymer homologs). For such mixture *MM* value, measured by one way or another, is some mean value, which depends on the degree of polydispersity, the type of function of molecular weight distribution (MWD) and the method of experimental determination of *MM* [1]. Therefore, the complete description of the molecular weight characteristics requires knowledge of its MWD.

As it has been noted above, the macromolecular coils in a solution are fractals, whose structure or, more precisely, the spatial distribution of coil elements, is characterized by its fractal dimension D_f. However, despite its unquestionable importance, D_f value gives only limited information about an aggregation (polymerization) process. Firstly, it is a static value and does not describe the dynamics of an aggregation process. Secondly, D_f value characterizes geometrical properties of only one cluster (macromolecular coil) and cannot be used for the description of a set of clusters [208]. Thus, both classical and fractal approaches indicate the need for MWD analysis for the characteristic completeness of both polymers and polymerization process.

D_f value range for macromolecular coils in a solution [10] assumes that the polymerization process proceeds according to the mechanism of irreversible cluster–cluster aggregation. The function of distribution for the indicated mechanism was studied in the scientific chapters [208–210]. The authors [211–213] proposed theoretical description of MWD functions within the framework of theoretical treatment [210] and studied the factors influencing the shape of these functions on the example of polydimethyl diallyl ammonium chloride (PDMDAAC) [34].

In Ref. [210] theoretical treatment of a cluster–cluster aggregation process was proposed, accounting for the existence of adding particle or cluster (monomer or macromolecular coil) to cluster in actual polymerization processes, and their disconnection (destruction) as well. Macromolecules are in a random environment, influencing the processes of aggregation or destruction in dilute polymer solution that is described by the stochastic Eq. (8) of Chapter 1. Following Shiyan's theory [210], the authors [211–213] believe that the reference times of random effects on the fractal aggregation process of macromolecules are much less than the reference times of the aggregation itself and, consequently, it is possible to present ξ_t as white noise of intensity $\langle \xi_t^2 \rangle = \sigma^2$.

The exponent a in the Eq. (8) of Chapter 1 in case of cluster–cluster aggregation is linked with fractal dimension D_f of a macromolecular coil according to the following relationship [210]:

$$a = \frac{2D_f - d}{D_f}, \tag{163}$$

where d is dimension of Euclidean space, in which a fractal is considered (it is obvious, in our case d=3).

The Eq. (8) of Chapter 1 allows one to obtain density of probability $P_s(N)$, which for a stationary solution looks like [210]:

$$P_s(N) = AN^{-a}\exp\left\{2\lambda N^{1-a}\left[1-(1-a)N^{b-a}/\lambda(b+1-2a)\right]/(1-a)\sigma^2\right\}$$

for a≠1, (164)

where A is the normalization constant.

As in the Eq. (164) value has no effect on $P_s(N)$, then the indicated relationship supposes three basic parameters influencing on the distribution $P_s(N)$: a, b and σ^2. Each of the indicated parameters characterizes a definite feature of the polymerization process. This exponent a is as a matter of fact determined by the structure of the macromolecular coil, that follows directly from the equation (163). The value b characterizes the type and intensity of the destructive processes. Parameter σ^2 is determined by the stochastic contribution to a polymerization process and it is possible to assume dependence of σ^2 on c_0: the greater the initial concentration of a monomer is, the larger the probability of random collisions is. All experimental MWD curves for PDMDAAC have the unimodal shape that supposes macromolecular coils low mobility in solution or $\sigma^2<2/3$ [210].

Let us consider first of all general aspects of influence of the three indicated parameters on the shape of MWD curves. For this purpose theoretical dependences $P_s(N)$ were constructed with a serial variation when only one of the indicated parameters varies whereas the other two are constant. In Fig. 91 the curves $P_s(N)$ are shown for the case when the fractal dimension D_f is variable, whereas for a coil (or variable a) both ^{2} and b are constant. In Figs. 91–93 the normalization constant A was chosen so that the maximal magnitude P_s was equal to about 0.4 for all curves $P_s(N)$. As it follows from the data of Fig. 91, the increase in D_f results in displacement of a maximum of distribution in the higher N side. This tendency is most strongly expressed for D_f=2.0, corresponding to θ-conditions [10]. For experimental MWD curves the greatest value of N, corresponding to a maximum of MWD distribution, was obtained for c_0=4.0 mol/l where N=550. As it may be observed from $P_s(N)$ curves, shown in Fig. 91, the indicated value of N corresponds to D_f=1.65, which agrees completely with D_f calculation according to the equation (4). This result shows that the coil contraction up to θ-conditions at high enough c_0 is not realized in practice.

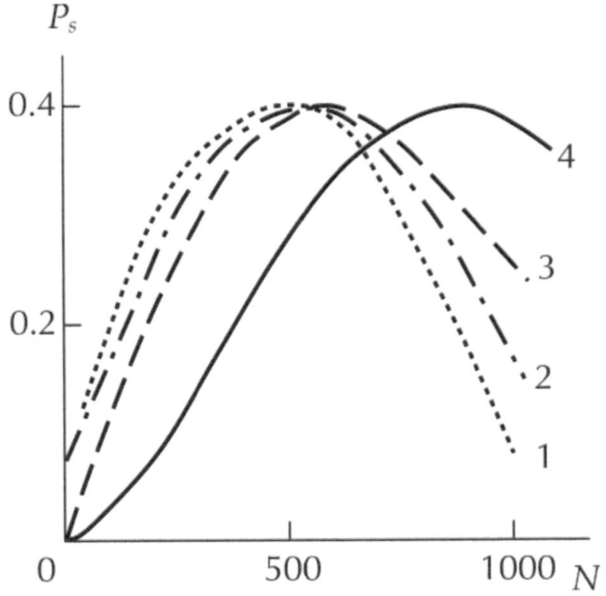

FIGURE 91 Simulation of MWD curves according to the Eq. (164) for D_f=1.50 (1), 1.65 (2), 1.80 (3) and 2.0 (4), b=1.0, ^{2}=0.25.

In Fig. 92 theoretical $P_s(N)$ curves are shown when σ^2 are variable and both D_f and b are constant. As it follows from the plots of this Figure, the value σ^2 exerts primary influence on the width of MWD. Strictly monodisperse distribution can be obtained only at σ^2=0. The increase of the stochastic contribution in polymerization intensity results in symmetrical broadening of MWD relative to its maximum.

And at last, the theoretical curves $P_s(N)$ is shown in Fig. 93, where b is variable and both D_f and σ^2 are constant. The value b=1 means availability of the canal of disintegration for each bond in a cluster. When the greatest part of bonds is fixed (b=0.2), the maximum in the investigated interval N=0–1000 is not reached at all. At b=0.5 (half of the bonds is fixed) $P_s(N)$ curve has maximum, but with obvious asymmetry at large N. Further b growth means, as should be expected, a decrease in the fraction of clusters with large N (high-molecular fraction).

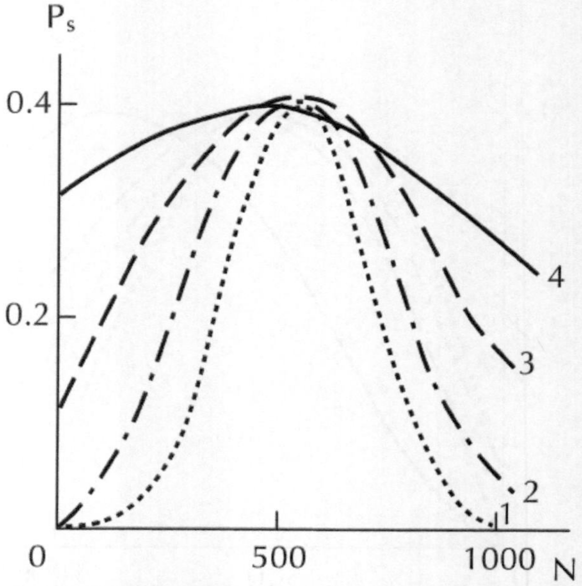

FIGURE 92 Simulation of MWD curves according to the Eq. (164) for σ^2=0.05 (1), 0.10 (2), 0.25 (3) and 0.60 (4). b=1.0, D_f=1.65.

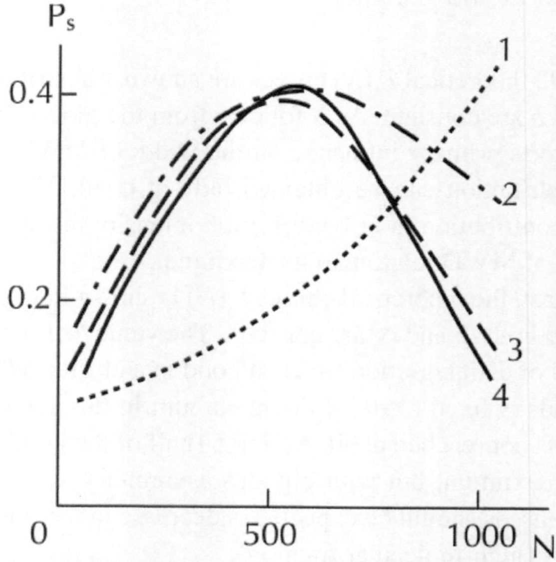

FIGURE 93 Simulation of MWD curves according to the Eq. (164) for b=0.2 (1), 0.5 (2), 1.0 (3) and 2.0 (4). D_f=1.65, ^{2}=0.25.

The comparison of experimental and theoretical MWD curves has shown that these curves for PDMDAAC, synthesized at c_0=4.0 and 5.0 mol/l, are simulated directly at the following parameters of a $P_s(N)$ curve: D_f=1.65 (a≈0.182), σ^2=0.25 and b=1. Such simulation for c_0=4.0 mol/l is adduced in Fig. 94 (curve 3), from which a good conformity of theory and experiment may be observed. For MWD curves at c_0=1.0 and 2.5 mol/l such direct simulation proves to be unsatisfactory, since for them the position of a maximum on MWD curve corresponds to N_{max}=175 and 375, but for the theoretical curves such low values of N_{max} are not attained (Figs. 91–93). Nevertheless, to achieve agreement between theory and the experiment by the use of renormalization is possible for these curves, that is attributed to the automodality of $P_s(N)$ distribution on the variable $\widetilde{N}=N/N_0$ [210]. Simultaneously it is required to perform the renormalization of white noise of intensity σ^2 as follows [210]:

$$\sigma^2 \rightarrow \sigma_0^2 = \sigma^2 N_0^{2a-1-b} \tag{165}$$

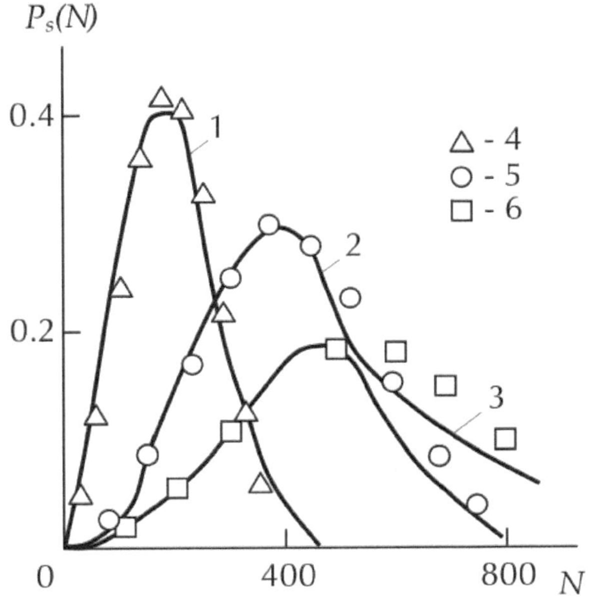

FIGURE 94 Comparison of experimental (1–3) and calculated according to the Eq. (164) (4–6) MWD curves for PDMDAAC at c_0=1.0 (1, 4), 2.5 (2, 5) and 4.0 (3, 6) mol/l.

The indicated renormalization is performed by definition of N_0 under the conditions at which $\widetilde{N}$ is equivalent to the theoretical value N_{max} and N is the experimental value of this parameter. Fig. 94 shows an excellent agreement between experimental and theoretical MWD curves for PD-MDAAC, synthesized at c_0=1.0 and 2.5 mol/l, thus confirming the correctness of the indicated renormalization. In its turn, this normalization reflects one of the basic properties of fractals—their automodality [86].

It was previously established [209] that MWD curve for cluster–cluster aggregation depended on the diffusive characteristics of a system in many respects, that is, on mobility of clusters, expressed by their rate ϑ. The value ϑ can be expressed by the following simple equation [209]:

$$\vartheta = m^{\beta}, \tag{166}$$

where m is mass of the cluster.

In its turn, the value of an exponent β defines the position of maximum in MWD curve [209]:

$$N_{max} = \frac{-\beta}{1-\beta} \tag{167}$$

One should expect that the mobility of clusters ϑ will be higher concomitantly with less viscosity of the initial monomer solution η_0, for which values for PDMDAAC are adduced in Ref. [215]. This supposition was confirmed by the relationship $\vartheta(\eta_0^{-2})$ shown in Fig. 95, which is linear and passes through coordinates origin.

As it was mentioned above, one should expect an increase of the stochastic contribution in polymerization intensity or white noise σ^2 in the growth process of monomer initial concentration c_0. This supposition is confirmed by the dependence $\sigma^2(\eta_0^{-2})$ shown in Fig. 96, which is also linear and passes through coordinates origin. The data of Figs. 95 and 96 demonstrate that parameters of the Eq. (164) can be expressed through technological characteristics of the polymerization process, for example, c_0 and η_0.

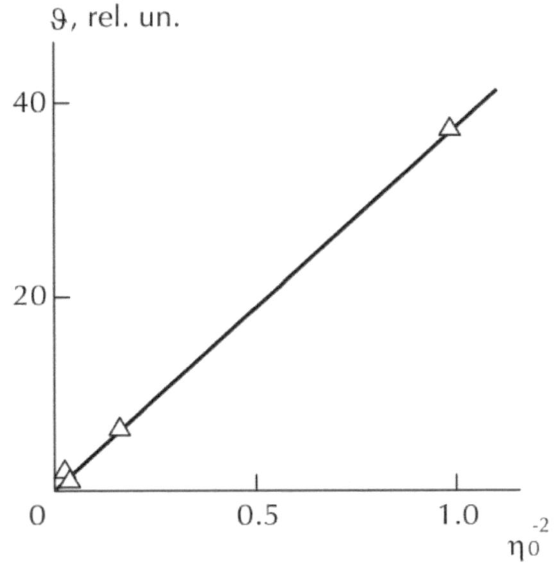

FIGURE 95 The relationship between mobility of macromolecular coil ϑ and initial viscosity of monomer solution η_0 for PDMDAAC.

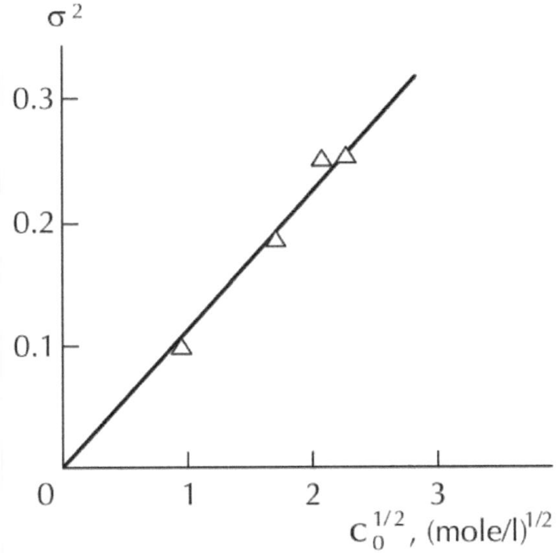

FIGURE 96 The relationship between the white noise of polymerization intensity σ^2 and initial concentration of monomer c_0 for PDMDAAC.

The authors [216] performed the similar analysis for polyarylate on the basis of di-chlorodiane 2,2-di-(n-carboxiphenyl) ethylene and diane (PAr), prepared by three different polycondensation methods: acceptor-catalytic (PAr-1), high-temperature (PAr-2) and low-temperature (PAr-3) ones. The comparison of theoretical curves $P_s(N)$ (Figs. 91–93) and experimental MWD curves (Fig. 97) allows to make several general conclusions. Firstly, MWD curves width for PAr-3 is much larger than for PAr-1 and PAr-2. By the analogy with Fig. 92 the conclusion can be made that for PAr-3 σ^2 is higher than for two remaining polyarylates. Secondly, for PAr-1 the greatest value N, corresponding to MWD (N_{max}) curve maximum is observed. The comparison with Fig. 91 supposes that macromolecular coils of PAr-1 have the greatest D_f value for all considered polyarylates. This is somewhat an unexpected result, since it is known, that high-temperature (equilibrium) polycondensation forms the macromolecular coils of PAr with the smallest exponents a_η in Mark-Kuhn-Houwink equation [5]. Since the values a_η and D_f are linked between themselves by the Eq. (4), then the greatest D_f values for PAr-2 should be expected. However, MWD curves (Fig. 97) analysis has shown that the strongest decay of the curve in large N domain is observed for PAr-1 and the comparison with the plots of Fig. 93 demonstrates, that for this polymer the greatest value b is expected. This means that the destructive process most clearly is expressed for PAr-1 and therefore for it D_f higher magnitude is expected [133].

The comparison of Figs. 91–93, on the one hand, and Fig. 97, on the other hand, shows that direct simulation of MWD curves by curves $P_s(N)$ for PAr, as and for PDMDAAC, is impossible to obtain, since MWD curves have too low N_{max} values. Nevertheless, the conformity of experiment and theory can be obtained for these curves as well with the aid of the described above renormalization, which correctness is confirmed by excellent conformity of experimental and theoretical curves for PAr shown in Fig. 97.

The estimation of macromolecular coils mobility in solution according to the Eqs. (166) and (167) has shown that the value ϑ increases in sequence PAr-1→PAr-2→PAr. This increase results in N_{max} reduction and random collisions probability enhancement, that is, in σ^2 increase that defines MWD curves shape. The supplementary factor is destruction high probability for acceptor-catalytic polycondensation that somewhat increases D_f value for PAr-1.

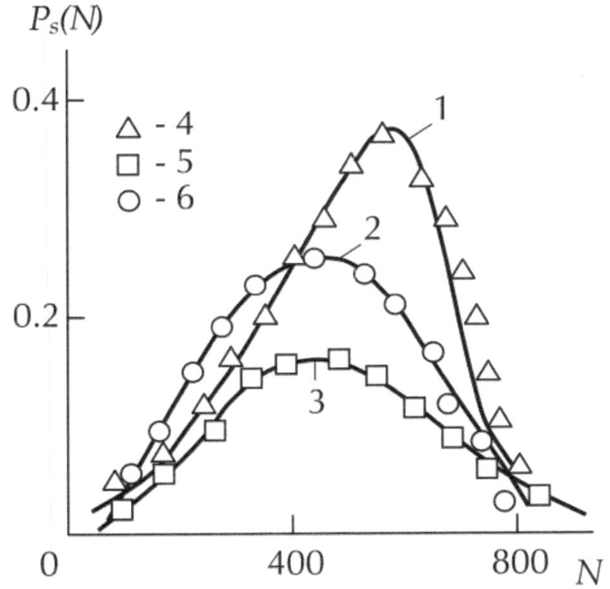

FIGURE 97 MWD curves for PAr-1 (1, 4), PAr-2 (2, 5) and PAr-3 (3, 6). 1–3—experimental results, 4–6—calculation according to the Eq. (164) after renormalization.

Hence, the stated above results have shown that molecular weight distribution of polymers, prepared by different modes of synthesis, can be simulated and predicted within the framework of irreversible aggregation cluster–cluster model. MWD curve shape and position are controlled by factors number, which are common for any synthesis method, namely, by the macromolecular coil structure, coil environment stochastic contribution in synthesis process intensity and coil destruction level in the indicated process. The coil mobility in reactionary medium exerts very strong influence on MWD curve shape.

In Refs. [208, 217–219] the dynamical scaling description of cluster sizes distribution in the cluster–cluster diffusion-limited aggregation (DLA) model was considered. As it was mentioned above, this model was applicable completely for PDMDAAC radical polymerization description by the following reason. The estimated according to the Eq. (4) value

D_f=1.65 for this polymer is typical fractal dimension for the aggregates, received in cluster–cluster DLA process [131]. Proceeding from this, the dynamical scaling function can be used for the description of PDMDAAC MWD curves, which looks like [208]:

$$n_s(t) \sim S^{-2} t^{z} \tag{168}$$

Let us consider the estimation methods of included in the relationship (168) parameters. ns(t) values for the gelation point were determined from the experimental curves $W_i(N)$ as follows. It was assumed N=S and then the polymer fraction, corresponding to S_i, is equal to total weight part of polymer fraction W_i. Dividing W_i by S_i and multiplying by maximum conversion degree Q, let us obtain ns(t_g) value or n_i for S_i (t_g is the time of gelation point attainment). In its turn, by the analogy with the Eq. (166) the diffusivity D_s value within the framework of irreversible aggregation models is accepted usually as follows [208, 220]:

$$D_s \sim S^{\beta} \tag{169}$$

D_s value can be calculated according to the known equation [221]:

$$D_s = \frac{kT}{6\pi\eta_0 R_g \alpha}, \tag{170}$$

where K is Boltzmann constant, T is testing temperature, η_0 is initial viscosity of reactionary medium, R_g is gyration radius of diffusible particle (macromolecular coil), α is numerical coefficient, defined by a boundary conditions on particle surface.

From the Eqs. (169) and (170) comparison it can be received

$$\beta \sim -\ln \eta_0 \tag{171}$$

And at last, the exponent z in the Eq. (168) is determined as follows [220]:

$$z = D_f(1-\beta) - (d-2) \tag{172}$$

As it has been shown in [208, 219], β value defines a distribution function shape. For β<–0.5 cupola-like distribution curves, having the maximum, were obtained. For β>–0.5 monotonously reducing distribution curves were obtained. For PDMDAAC MWD curves have a typical cupola-like shape. Proceeding from η_0 absolute values, the authors [222, 223] supposed in the Eq. (171) the equality sign and in this case β values are varied within the limits of −1/–3 (η_0 values are given in relative units) and this range corresponds completely to the experimental MWD curves shape. Besides, as it has been shown in [209], MWD curve maximum position N_{max} is defined by the Eq. (167). It is easy to see, that c_0 increase, leading to ${}_0$ growth, defines absolute value growth and N_{max} enhancement.

In Fig. 98 the dependences ln (S^2n_s) on St^z for four PDMDAAC polymers, synthesized at different c_0, are shown. As it follows from the plots of Fig. 98, all four MWD curves are described by the sole curve. This is the most important result, confirming correctness of irreversible aggregation models usage for the description of polymerization process [222, 223].

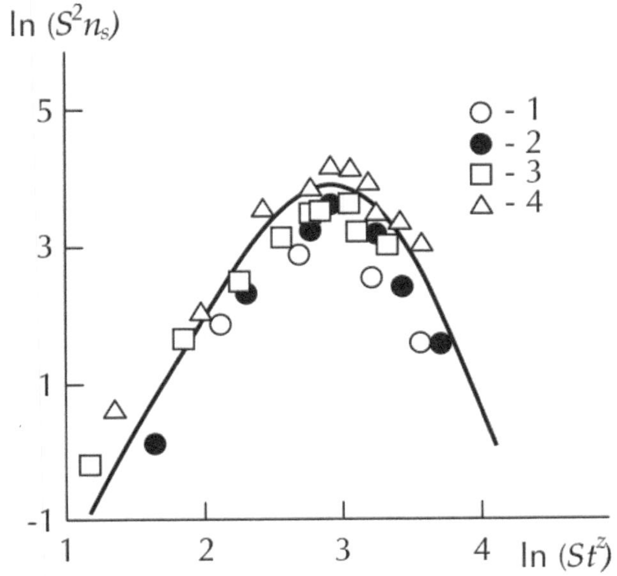

FIGURE 98 The generalized MWD curve in the shape of the dependence ln (S^2n_s) on St^z for PDMDAAC, synthesized at initial concentration of monomer c_0: 1.0 (1), 2.5 (2), 4.0 (3) and 5.0 (4) mol/l.

Using the shown in Fig. 98 generalized curve, MWD dynamics as a function of polymerization process duration t can be predicted theoretically. With this purpose at first the value S=N is given and St^z magnitude is determined at arbitrary t. Then from the plot of Fig. 98 the corresponding to it value S^2n_s is found and according to the described above procedure W_i is determined. Further the same calculation repeats for other S and so on. In Fig. 99 as an example MWD curves for c_0=4.0 mol/l at t=t_g=108 min and t=53 min comparison is adduced. As one can see from curves 1 and 2 in Fig. 99, t reduction (t<t_g) leads to MWD maximum displacement to lower values of molecular weight and MWD narrowing. Besides, in Fig. 99 MWD curve for c_0=1.0 mol/l at t=t_g (curve 3) is adduced. From the comparison of all three MWD curves of Fig. 99 it follows, that m and c_0 reduction gives the similar effect.

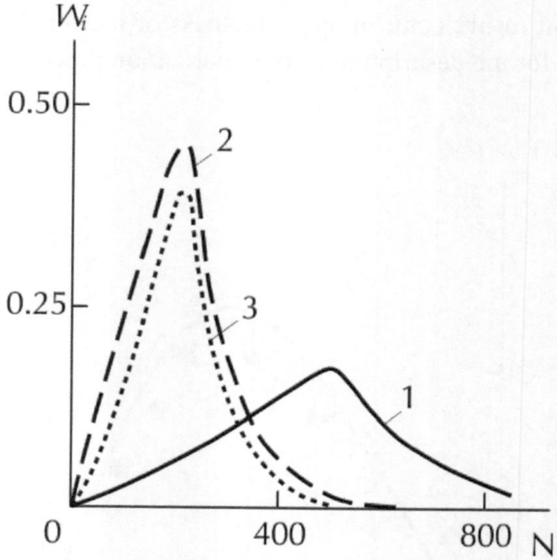

FIGURE 99 The experimental MWD curves in gelation point for PDMDAAC, synthesized at c_0: 4.0 (1) and 1.0 (3) mol/l. Theoretical MWD curve (2) at t=53 min for PDMDAAC, synthesized at c_0=4.0 mol/l.

Hence, the stated above results have shown correctness of the description of polymers MWD on the example of PDMDAAC within the framework of dynamic distribution function of irreversible aggregation cluster–

cluster model. This is confirmed by the generalized distribution curve construction, which allows one MWD change dynamics prediction as a function of the initial concentration of monomer c_0 and reaction duration t [223].

2.12 THE ANALYSIS OF A SOLUTION OF POLYMERS WITH COMPLEX MOLECULAR ARCHITECTURE

At present the polymer solutions behavior analysis is performed with the aid of the numerous equations, obtained within the framework of different approaches (including empirical ones) and using a large number of either approximations. Thus, the equations number in Budtov's monography makes up about 900. It is obvious, that such large equations number defines corresponding number of parameters, characterizing polymer solutions. This leads to the fact that each author chooses either set of characteristics, on the basis of which the subsequent model is constructed. The obvious principal deficiency of the described above approaches is the united structural basis absence—all the mentioned equations give a polymer solutions properties and macromolecules dimensional characteristics combination. Nevertheless, such structural basis can be obtained within the framework of fractal analysis—at present it is well-known through Refs. [10, 25], that a macromolecule in solution represents the fractal object, characterized by its fractal dimension D_f, which is the structural characteristic in strict physical significance of this term, since it describes macromolecule elements distribution in space [224]. It becomes obvious, that the most general characteristics of macromolecule are connected with dimension D_f by simple formulas. Thus, the exponent a_η in Mark-Kuhn-Houwink equation is determined according to the Eq. (4), bulk interactions parameter ε, characterizing interactions of a macromolecule in solution—according to the Eq. (39) and gyration radius of macromolecular coil R_g is calculated according to one from the main relationships of fractal analysis—the Eq. (8).

This enumeration can be continued. For the example, Huggins constant k_x is also connected with dimension D_f (see the Eq. (50)). This situation is defined by the objective key reasons: any property of an object depends on its structural state in one way or another [188]. Proceeding from the said above, the authors [225] offered the structural treatment of polymer solutions behavior with the fractal analysis methods participation. With this purpose as the experimental data the results of Ref. [226] were used, describing molecular and hydro dynamical characteristics of star-like polystirenes with fullerene C_{60} molecules as branching centers (PS-C_{60}).

The authors [226] constructed the dependence $[\eta](MM)$ in double logarithmic coordinates for PS-C_{60} solutions in two solvents—tetrahydrofuran and chloroform. These dependences proved to be linear, that allows to estimate the exponent a_η for the studied polymers, which proved to be equal to −0.06±0.07 for tetrahydrofuran and 0.2±0.2—for chloroform. However, this construction causes a questions number. Firstly, in the well-known Mark-Kuhn-Houwink equation (the formula (1)) coefficient K_η depends on both exponent a_η and macromolecular repeated link molecular weight m_0. One from the relationship between K_η, a_η and m_0 variants is the Eq. (2). The plots $[\eta](MM)$ in double logarithmic coordinates construction and subsequent the exponent a_η determination from their slope assumes the condition K_η=const, that is not far from obvious.

Secondly, a_η negative value for PS-C_{60} solutions in tetrahydrofuran according to the Eq. (4) means the condition D_f>d, where d is the dimension of Euclidean space, in which a fractal is considered (it is obvious, that in our case d=3, that is physically incorrect [86].

Thirdly, the authors [226] showed that Huggins constant k_x for polymers PS-C_{60} varied within the wide enough limits: 0.25–0.90. As it is known through Ref. [1], Huggins constant characterizes thermo-dynamical interaction polymer-solvent and hydro dynamical solution behavior and therefore the macromolecular coil structure invariability (a_η=const) at the indicated above k_x variation induces certain doubts. The constant k_x is linked with a macromolecule in solution characteristics as follows [1]:

$$k_x + a_\eta = 1.1 \pm 0.1, \tag{173}$$

and

$$k_x = 0.25 + \frac{0.30}{\alpha^6}, \tag{174}$$

where α is a coil swelling coefficient.

Let us note that the coefficient α depends also on macromolecular coil structure or dimension D_f according to the Eq. (78).

Fourthly, as it follows from the plots of Fig. 4 of Ref. [226], at MM=45×10^3 [η] values are close by absolute value for the initial polystirene and PS-C_{60} with beams number f=6. This assumes the similarity of the indicated polymers macromolecular coil structure.

Fifthly, the authors [226] used for PS-C_{60} with beams numbers f=12 and 22 two flexible junctions between branching centers (molecules of fullerene C_{60}): $-Si(CH_3)-$ and $-(CH_2)_4-$. It is not quite clear, how the flexible junction type influences the macromolecule structure at the condition a_η=const for each from the used solvents (and accounting for errors interval it can be assumed without large strained interpretation a_η=const for both solvents or a=0).

Sixthly, even from schematic picture of PS-C_{60} with different beams number, adduced in Ref. [226], it is intuitively clear, that f increasing should result in macromolecule structure densification or the condition D_f=variant with the corresponding condition a_η=variant according to the Eq. (4).

Let us consider the adduced above considerations within the framework of fractal analysis [56]. The value D_f evaluation can be performed by two methods according to the experimentally received k_x values: by the Eqs. (173) and (4) and also the Eqs. (174) and (78) usage in pairs. Such evaluations give D_f values within the range of 1.714–2.724 and 1.763–2.711, respectively, with the average discrepancy of approx. 3%. This D_f range corresponds to a_η variation approx. 0.10–0.75, that is far from postulated in Ref. [226] condition a_η=const≈-0.06–0.20. The indicated variation of dimension D_f assumes not only quantitative macromolecular coil structure change at f growth, but also qualitative one: if PS-C_{60} is assumed as a branched polymer, then at f=6 D_f value corresponds to the coil in good solvent, at f=12—in -solvent and at f growth higher than 22 PS-C_{60} has tendency to the compact globule state (see Fig. 100) [10, 13]. Let us also note qualitative influence of a flexible junction type on the structure of PS-C_{60} macromolecular coil in solution: if for $-Si(CH_3)-$ D_f value corresponds to the coil in good solvent with transition to θ-solvent, then for $-(CH_2)_4-$ D_f value is higher than the corresponding dimension of coil in θ-solvent. In other words, the junction $-(CH_2)_4-$ is more flexible than $-Si(CH_3)-$.

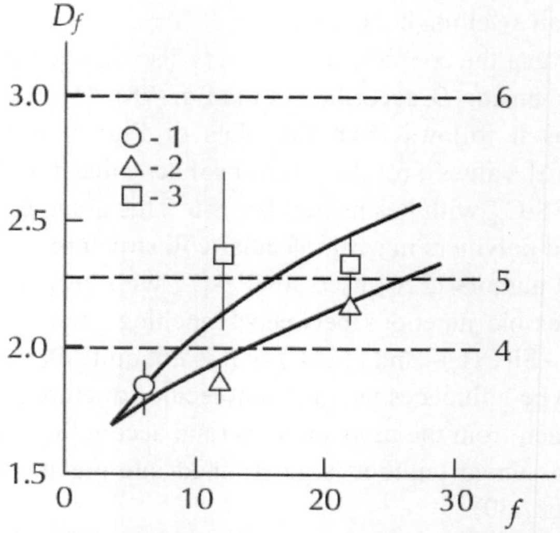

FIGURE 100 The dependences of macromolecular coil fractal dimension D_f on beams number f for PS-C_{60} without flexible junction (1) and with flexible junctions –Si(CH$_3$)– (2), –(CH$_2$)$_4$– (3). The horizontal stroked lines indicate D_f values for coil in good solvent (4), θ-solvent (5) and compact globule (6) in a branched polymers case.

D_f value for the initial PS can be evaluated according to the Eq. (18), which at a_D=0.55 in tetrachloroethane gives D_f=1.818, that is close to the values D_f=1.763–1.947 for PS-C_{60} solutions in tetrahydrofuran and chloroform. As it has been noted above, such result was supposed on the basis of [η] values closeness for PS and PS-C_{60} according to the data of Fig. 4 in Ref. [226].

Let us consider further theoretical treatment of experimentally received condition [η]=const for PS-C_{60} [226]. The authors [6–8] received Mark-Kuhn-Houwink equation fractal variant (the Eq. (10)), where $c(\alpha)$=2.91. For the Eq. (10) application m_0 value estimation is necessary, what can be made with the aid of the Eq. (8) in assumption m_0=MM/N and at using the obtained above D_f values, experimental R_g magnitudes [226] and the constant 0.349. As these estimations have shown the systematic m_0 reduction at D_f growth is observed. In Fig. 101 the dependence $D_f(m_0)$ is adduced, from which it follows, that the indicated m_0 reduction is a linear one. By its physical significance this dependence reflects D_f reduction at polymer chain rigidity enhancement. The indicated chain rigidity enhancement can be characterized by m_0 growth or Kuhn segment length A increasing. The dependence of D_f on A has the following form [56]:

$$D_f = K_1 - K_2 A, \tag{175}$$

where K_1 and K_2 are constants, dependent on polymer class (see the Eqs. (54) and (55)).

It is easy to see, that the dependence $D_f(m_0)$ corresponds completely to the corresponding correlation $D_f(A)$, described by the Eq. (175). The equation for analytic expression of the adduced in Fig. 101 dependence $D_f(m_0)$ looks like [225]:

$$D_f = 3 - 3.13 \times 10^{-3} m_0 \tag{176}$$

Let us note that the lowest m_0 value corresponds to molecular weight of monomer link for the initial PS. Further the bulk interactions parameter ε can be calculated according to the Eq. (39). In Fig. 102 the dependence $m_0(\varepsilon)$ is adduced for star-like PS, which turns out to be linear and is described analytically by the following equation [225]:

$$m_0 = 1.07 \times 10^3 \left(\varepsilon + 0.333\right) \tag{177}$$

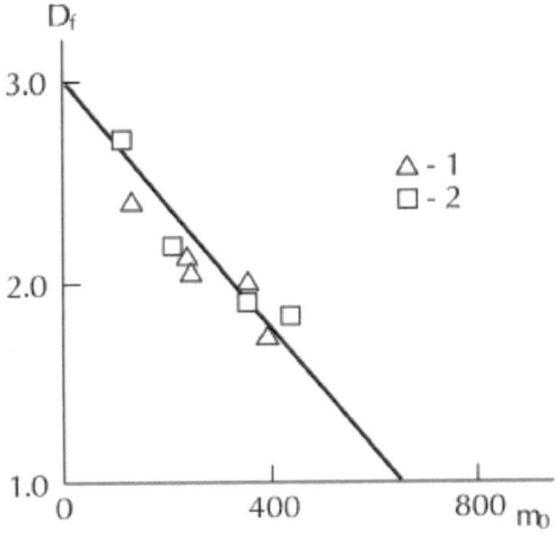

FIGURE 101 The dependence of macromolecular coil fractal dimension D_f on molecular weight of repeated link m_0 for PS-C_{60} solutions in tetrahydrofuran (1) and chloroform (2).

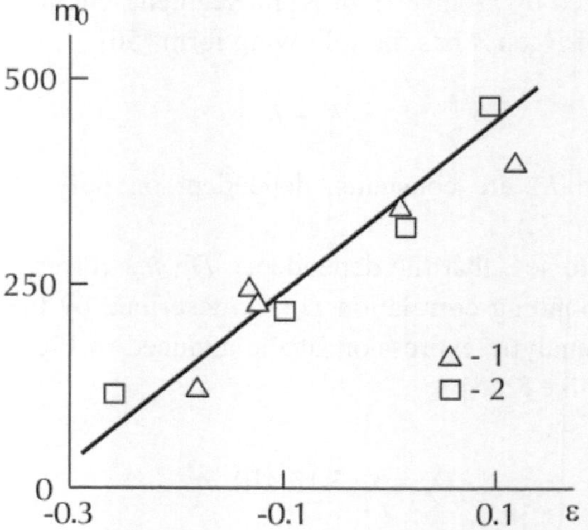

FIGURE 102 The dependence of molecular weight of repeated link m_0 on bulk interactions parameter ε for PS-C_{60} solutions in tetrahydrofuran (1) and chloroform (2).

The Eq. (177) physical significance is obvious: the smaller parameter ε or the stronger interactions of a macromolecular coil elements between themselves, is the lower mobility (freedom) of the indicated elements or the smaller their size m_0 is. At ε=-0.333, that corresponds to D_f=3.0 or compact globule, m_0=0. In other words, the dense package defines suppression of all fragments with arbitrary m_0 mobility [225].

The Eqs. (175)–(177) reflect the change dynamics of macromolecular coil structure in solution and nonaccounting for this factor serious enough errors in estimation can be introduced. The intrinsic viscosity $[\eta]^T$ values, calculated according to the Eq. (10), are adduced in Table 23, from which their both qualitative and quantitative correspondence to this parameter experimental values [η] follows (the average discrepancy of [η] and $[\eta]^T$ makes up 8%).

TABLE 23 The molecular weight and hydro dynamical characteristics of star-like PS.

Sample №	f	Solvent	k_x	D_f	$MM_w\times10^3$	R_g, nm	[η], cm³/g	$[\eta]^T$, cm³/g
1	6	Tetrahydrofuran	0.25	1.763	45±4	5.4	22±1	20.2
		Chloroform	0.50	1.947	—	4.4	12.3±0.5	11.0
2	12	Tetrahydrofuran	0.51	1.957	110±7	7.0	19±1	18.0
		Chloroform	0.36	1.819	—	7.4	24±1	21.4
3	12	Tetrahydrofuran	0.60	2.060	102±4	6.8	20±1	17.8
		Chloroform	0.90	2.710	—	6.3	15.4±0.4	14.3
4	22	Tetrahydrofuran	0.64	2.110	207±13	9.0	22.0±0.6	20.4
		Chloroform	0.70	2.220	—	8.3	17.2±0.5	16.0
5	22	Tetrahydrofuran	0.80	2.410	182±12	8.0	18±1	16.3

Macromolecule dynamics in solution appreciation means that it is impossible to consider as polymers the same (having the same characteristics), consisting of the same chemical groups. Thus, the authors [227] have shown that for modified by Dendron's PS of first-fourth generation Kuhn segment length A varies within the limits of 3.9–23.3 nm. It is obvious that each generation represents itself the individual polymer, differing essentially by chain rigidity.

The stated above results suppose the necessity of elaboration of a polymer solutions behavior model, in the basis of which the macromolecule structure is put, defining all its main characteristics. Such model can be received within the framework of fractal analysis. No less important the problem aspect is accounting for macromolecule in solution structure change dynamics at the action of either factors.

The exponent a_η value defines usually a macromolecular coil in solution conformation: for linear polymers its value makes up 0.7–0.8, for dendrimers it is practically equal to zero and superbranched polymers the value $a_\eta \approx 0.2$–0.4 [228]. a value for dendrimers allows to suppose compactization of dendrimers macromolecules shape in comparison with classical macromolecular coils. Proceeding from these considerations the authors [229] performed theoretical description of intrinsic viscosity [η] change as a function of generations number N_D for solutions of polystirene, modified by Dendron's (PSD) [227]. The necessary for these experimental characteristics of the indicated solutions are accepted according to the Ref. [227] data and listed in Table 24.

For calculation of the intrinsic viscosity $[\eta]$ of PSD solutions in tetrahydrofuran Mark-Kuhn-Houwink equation fractal variant (the Eq. (10)) was used, where the constant $c(\alpha)$ is accepted equal to 2.91. MM and m_0 values, received experimentally, are adduced in Table 24 and D_f value for PSD was calculated as follows. As it is known through Ref. [1], for flexible-chain polymers, having $\alpha \geq 1$, Huggins constant can be determined according to the Eq. (174), from which it follows that the condition $\alpha \geq 1$ is fulfilled for the values k_x0.55. The Table 24 data showed that the last condition was correct for PSD with the least generations number (PSD-1 and PSD-2) and therefore the Eq. (174) is applicable only for them. For the dimension D_f calculation in PSD-3 and PSD-4 case the relationship (8) was used with the proportionality coefficient β=0.349 [225], if R_g is given in nm. D_f values estimated by the indicated mode, are reduced from 1.783 up to 1.520 at the increase of generation number N_D from 1 up to 4 [229].

TABLE 24 The conformational characteristics of PSD macromolecules.

The polymer conventional sign	N_D	$MM\times10^{-6}$	m_0	k_x	R_g, nm	D_f	a_η	$[\eta]$, dl/g	$[\eta]^T$, dl/g
PSD-1	1	1.10	671	0.32	22.2	1.783	0.683	0.69	0.69
PSD-2	2	2.19	1459	0.43	28.0	1.724	0.740	0.57	0.45
PSD-3	3	2.70	2907	0.95	28.0	1.563	0.919	0.43	0.53
PSD-4	4	5.78	5803	1.62	32.8	1.520	0.974	0.43	0.41

The comparison of the received experimentally $[\eta]$ and calculated according to the Eq. (10) $[\eta]^T$ intrinsic viscosity values are adduced in Table 24. As one can see, a good correspondence of theory and experiment is obtained (the average discrepancy of $[\eta]$ and $[\eta]^T$ makes up approx. 12%).

The Eq. (10) allows to perform analysis of dependence $[\eta](MM)$ as a function of m_0 and D_f. In Fig. 103 theoretical dependences $[\eta]^T(MM)$ are adduced at m_0=const and D_f=const. From the adduced comparison it follows that experimentally observed [η] change with *MM* is due to simultaneous variation of m_0 and D_f. Besides, the plots of Fig. 103 allow to determine [η] limiting values for PSD, which correspond to this parameter value at plateau reaching: [η]=0.06–10.1 dl/g.

Hence, the Eq. (10) supposes the dependence of [η] on three parameters, namely, *MM*, m_0 and D_f, from which the last accounts for solvent influence: the higher solvent quality in respect to polymer, is the lesser D_f is. However, in case of the same solvent the indicated above parameters are mutually dependent. In Fig. 104 the dependence of D_f on polymerization degree $N=MM/m_0$ is adduced. As one can see, D_f linear growth with *N* increase is observed, described analytically by the following empirical equation [229]:

$$D_f = 1.16 + 3.93 \times 10^{-4} N \quad (178)$$

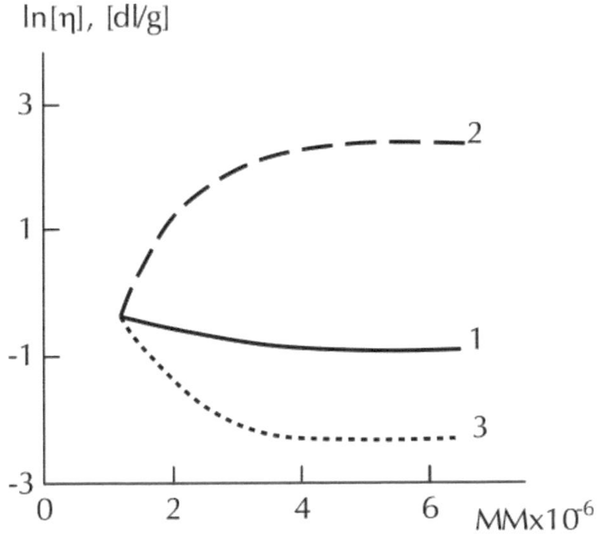

FIGURE 103 The dependences of intrinsic viscosity [η] on molecular weight *MM* in logarithmic coordinates for PSD. 1—experimental data; 2, 3—calculation according to the Eq. (10) at m_0=const=671 (2) and D_f=const=1.783 (3).

Hence, the minimum D_f value for PSD is equal to 1.16 in the case, if N=1 or m_0=MM. The maximum D_f value is reached at N=4680, that is, the considered polymers with larger N value are impossible to obtain. If to account, that N value is linearly decreasing function of generations number N_D, then the plot $N(N_D)$ has shown that the condition N=1 is reached at the eighth generation realization.

And in consummation it is necessary to make the following important conclusion. The affirmation [227, 228], that for dendrimers the exponent a_η value in Mark-Kuhn-Houwink equation is close to zero and even can be negative is incorrect (see Table 24). This affirmation is based on the plots $N(MM)$ construction in the supposition K_η=const. As a matter of fact, the constant K_η is a function of m_0 [1, 5]. a_η value can be changed within the limits of 0–2, that follows from the equation (4) at general variation of D_f=1–3 [56]. D_f value and, hence, a_η depends on concrete molecular architecture of polymers.

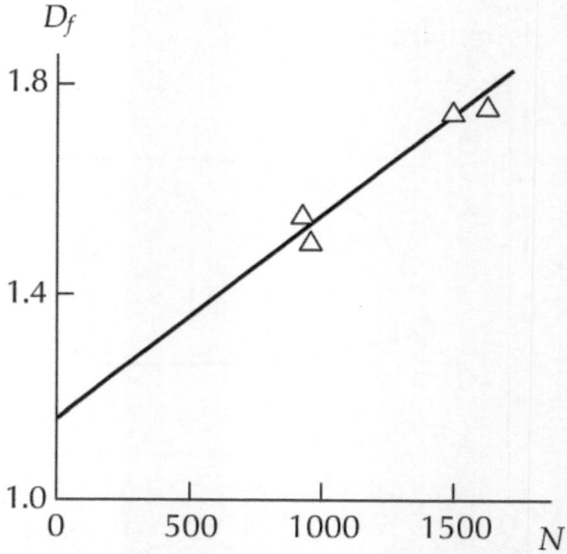

FIGURE 104 The dependence of macromolecule fractal dimension D_f on polymerization degree N for PSD.

Let us consider further solvent influence on the structure of a macromolecular coil in solution for the polymers with complex architecture.

The authors [228] supposed, that for dendrimers the exponent a_η value was equal to about zero, that corresponds to D_f=d=3.0, that is, to compact globule [25]. However, in Ref. [230] it has been shown that for dendrimers of the third generation a_η=0.24 or D_f=2.42 according to the Eq. (4). As it is known through Ref. [180], the fractal object density ρ is determined according to the equation (a more precise variant of the equation (139)):

$$\rho = \rho_0 \left(\frac{R_g}{a_m} \right)^{D_f - d}, \tag{179}$$

where ρ_0 is density of Euclidean object of the same material, R_g is object radius, a_m is lower linear scale of fractal behavior, d is the dimension of Euclidean space, in which a fractal is considered (it is obvious, that in our case d=3).

In case of typical for macromolecular coil gyration radius R_g=7 nm [226] and a_m=1 nm let us obtain $\rho=\rho_0$ for D_f=d=3.0 and $=0.323\rho_0$ for D_f=2.42. If a_m value will be reduced up to 0.3 nm (the dependence on scale, typical for fractal objects), then $\rho=0.161\rho_0$, that is, for fractal object approx. 68–84% of its volume is accessible for the solvent, that can change essentially its structure [230].

As it is known through Ref. [13], D_f value change in different solvents is due to interactions polymer-solvent variation and within the framework of fractal analysis this effect is described by the Eq. (18) of Chapter 1. It is easy to see, that in case of point zero-dimensional solvent molecules (δ_f=0) the Eq. (18) of Chapter 1 gives fractal dimension of swollen (i.e., with appreciation of the excluded volume effects) macromolecular coil, namely, the Eq. (12) of Chapter 1.

Hence, at any rate for linear polymers (for which the spectral dimension d_s=1.0 [45]) the problem of the dimension D_f prediction in different solvents comes to δ_f value prediction. Therefore, the authors [231] performed the study of solvent structure in solutions of star-like polystyrene (PS-C_{60}), possessing complex enough molecular architecture [226], within the framework of models, considered in detail in Section 2.4.

As it is known through Ref. [1], Huggins constant k_x, characterizing thermo-dynamical interaction polymer-solvent and solutions hydro dynamical behavior [226], is linked with the exponent a_η in

Mark-Kuhn-Houwink equation by a simple relationship (the Eq. (173)), that gives the possibility to determine D_f value according to the Eq. (4) by k_x known values.

As it has been noted above, the calculation of solvent fractal dimension δ_f according to the Eq. (18) of Chapter 1 shows, that the value $\delta_f=0$ (a separate molecule of low-molecular solvent is a point or zero-dimensional object [25]) is obtained for only $D_f=5/3$ or a macromolecular coil in good solvent [232]. Since such molecule always remains a zero-dimensional object, then it should be supposed, that the dimension δ_f characterizes the structure of low-molecular solvent molecules totality ("swarm"), which participates in polymers dissolution process. The bulk interactions parameter ε of a macromolecular coil is connected with dimension D_f by the Eq. (39), from which it follows, that $D_f=2$ (the coil in θ-solvent, attractive and repulsive interactions are balanced) $\varepsilon=0$, at $D_f>2$ $\varepsilon<0$ characterizes attractive interactions of coil elements between themselves and at $D_f<2$ $\varepsilon>0$—repulsive interactions.

The intersections (contacts) N_{it} of two fractal objects (in the considered case—macromolecular coil and "swarm" of low-molecular solvent molecules) can be determined according to the Eq. (13) of Chapter 1 with replacement of dimension D_{f2} by δ_f. As a matter of fact, in such form the Eq. (13) of Chapter 1 gives the evaluation of solvent molecules number in their "swarm." This relationship analysis in the very general terms demonstrates, that at $\delta_f<0$ the exponent of a macromolecular coil gyration radius is smaller than zero ($D_f \leq d$) and N_{int} value is very small (as a matter of fact $N_{int} \rightarrow 0$). In other words, in this case the solvent molecules "swarm" is not formed. In Fig. 105 the dependence $\varepsilon(N_{int})$ for star-like PS-C_{60} is adduced, from which it follows, that at $\varepsilon<0$ (attractive interactions) N_{int} fast growth or solvent molecules number increase in "swarm" is observed, which is required for the indicated interactions neutralization.

The authors [97] proposed the Eq. (82) for the dimension δ_f evaluation with using the solvent δ_s and polymer δ_p solubility parameters. The calculation according to this equation has shown that for star-like PS-C_{60} δ_p value changes from 9.78 $(cal/cm^3)^{1/2}$ for usual PS in tetrahydrofuran solution up to 30.7 $(cal/cm^3)^{1/2}$ for star-like PS-C_{60} with 22 beams in chloroform solution. It is obvious, that the effective δ_p value characterizes polymer dissolution complication at dimension D_f growth or macromolecular coils densification. In Fig. 106 the dependence $\delta_p(\varepsilon)$ is adduced, from which the expected δ_p growth follows at intensification of attractive interactions of macromolecular coil elements between themselves. δ_p value can be determined alternatively according to the Eq. (83), from which it follows, that δ_p growth means polymer cohesion intensification.

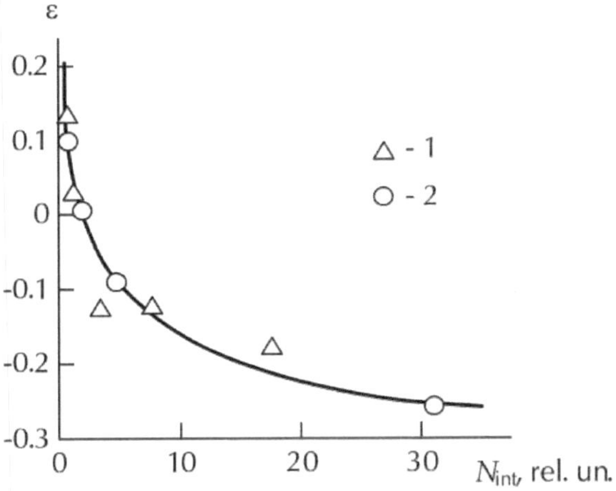

FIGURE 105 The dependence of bulk interactions parameter ε on intersections number N_{int} for PS-C_{60} solutions in tetrahydrofuran (1) and chloroform (2).

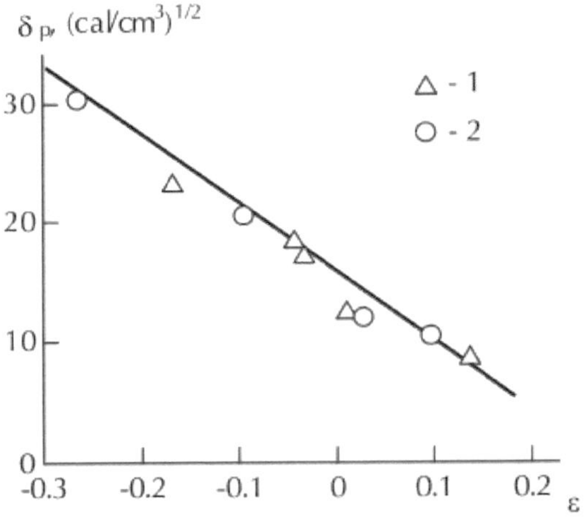

FIGURE 106 The dependence of polymer effective solubility parameter δ_p on bulk interactions parameter ε for star-like PS-C_{60} solutions in tetrahydrofuran (1) and chloroform (2).

As it is known through Ref. [1], solvent quality in respect to a polymer is often estimated by the difference $\Delta\delta=|\delta_{p-}\delta_s|$—the smaller it is, the better-indicated quality is. The plot $N_{int}(\Delta\delta)$ construction has shown, as it was expected, N_{int} increase at $\Delta\delta$ growth, but for each from the solvents (tetrahydrofuran and chloroform) the different linear dependences are obtained, that assumes the existence, at any rate, of one more factor, influencing on solvent quality. The macromolecular coil and solvent dimensions can be this factor: the average D_f value is equal to 2.06 for the solutions in tetrahydrofuran and D_f=2.17 for the solutions in chloroform, the mean value δ_f=1.27 for tetrahydrofuran and δ_f=1.32 for chloroform [231].

The authors [104] proposed to estimate solvent quality by the difference $|\delta_s^{1/2}\text{-}27/\delta_p|$ or by δ_f value (see the Eq. (82)). In Fig. 107 the dependence $N_{int}(\delta_f)$ is adduced, from which the common, although not linear, dependence $N_{int}(\delta_f)$ for both considered solvents follows. Let us note, that N_{int} sharp growth or star-like PS-C_{60} solubility change for the worse begins at $\delta_f\approx1.35$ or D_f2.15, that corresponds to the greatest dimension of a macromolecular coil for linear polymers [71].

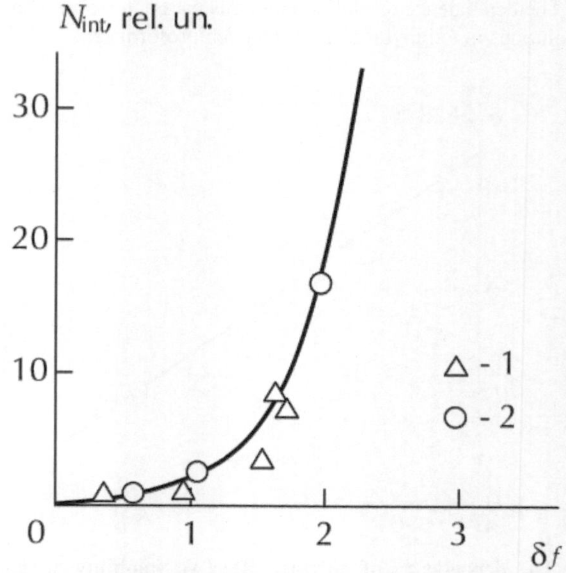

FIGURE 107 The dependence of intersections number N_{int} on dimension $_f$ of solvent molecules "swarm" for star-like PS-C_{60} solutions in tetrahydrofuran (1) and chloroform (2).

Within the framework of classical theory a macromolecular coil gyration radius R_g value can be determined according to the known equation [233]:

$$R_g = l_0 \left(\frac{C_\infty MM}{6M_0} \right)^{1/2}, \qquad (180)$$

where l_0 is C-C bond length, equal for polystirene (PS) to 0.154 nm, C_∞ is characteristic ratio, equal to 10 for PS [69], MM is polymer molecular weight, M_0 is molar mass per one bond of the main chain, equal to 52 for PS [233) and the exponent 1/2 presents itself Flory exponent for θ-solvent.

Modification of polystirene by Dendron's of first-fourth generations results in its molecular characteristics essential change: so, the monomer link molecular weight m_0 changes from 104 for PS up to 5803 for modified by Dendron's PS of the fourth generation (PSD-4) [227] (see also Table 24). It is obvious, that such molecular characteristics changes should result in the corresponding changes of R_g value. Actually, at PSD transition from the first generation up to the fourth one R_g increase from 22.2 up to 32.8 nm is observed (Table 24) at MM growth within the range of $(1.10–5.78)\times10^6$ [227]. The authors [234] performed the analysis of the indicated above PSD characteristics influence on the macromolecular coil gyration radius value within the framework of both classical and fractal models.

In Fig. 108 the comparison of calculated according to Stox formula R_g and the Eq. (180) R_g^T values of gyration radius of the macromolecular coil for PS, modified by Dendron's (PSD) at the indicated above l_0, C_∞ and M_0 values is adduced. As it follows from the data of this figure, calculation according to the Eq. (180) gives the overstated R_g^T values, which can exceed R_g twice for the greatest molecular weight MM from the used ones. This discrepancy should be attributed to a large change of both molecular characteristics and macromolecule structure at its modification by Dendron's. Therefore, it is necessary to replace M_0 value in the Eq. (180) by molecular weight of monomer link m_0 that reflects PSD chain rigidity enhancement at generation number increasing from 1 up to 4. Besides, the exponent 1/2 in the indicated equation should be replaced by $1/D_f$ value, where D_f is the macromolecular coil fractal dimension, which is the structural characteristic in strict physical significance of this term, since it describes macromolecule elements distribution in space [224].

Such replacement follows from the well-known Eq. (1) of Chapter 1. In such variant the Eq. (180) acquires the following form [234]:

$$R_g = l_0 \left(\frac{C_\infty MM}{6m_0} \right)^{1/D_f}, \tag{181}$$

and accounting for the relationship $MM/m_0=N$, where N is polymerization degree, let us receive finally [234]:

$$R_g = l_0 \left(\frac{C_\infty N}{6} \right)^{1/D_f}, \tag{182}$$

where D_f values for PSD were accepted according to the data of Table 24 [229].

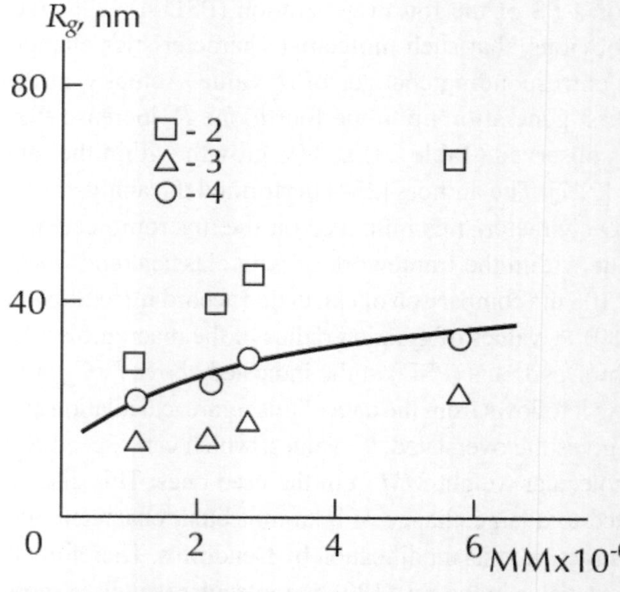

FIGURE 108 The dependences of macromolecular coil gyration radius R_g on molecular weight MM for PSD. 1—calculation according to Stox formula [227]; 2—calculation according to the Eq. (180); 3, 4—calculation according to the Eq. (182) at l_0=0.154 nm (3) and l_0=0.270 nm (4).

The comparison of calculated according to Stox formula and the equation (182) the macromolecular coil gyration radius values for PSD is also adduced in Fig. 108, from which it follows, that the Eq. (182) gives understated R_g^T values. This discrepancy should be attributed to l_0 value change at PS modification by Dendron's. As it is known through Ref. [69], l_0 values for a large number of flexible-chain polymers with simple enough chemical structure are varied within very narrow limits (0.147–0.154 nm), but at polymer chain chemical structure complication l_0 values are increased and are changed within the limits of 0.209–0.788. Proceeding from this general tendency the value l_0=0.270 nm was chosen for PSD on the basis of the best conformity of theory and experiment. As it follows from the data of Fig. 108, in this case the Eq. (182) gives a good correspondence to experiment (the mean discrepancy of R_g and R_g^T makes up 4.6%).

The Eq. (182) at the condition l_0=const=0.270 nm and C_∞=const=10 is reduced to a purely fractal form, that is, to the Eq. (8) with B=0.349. Let us note essential distinctions of the Eqs. (180) and (8). Firstly, if the first from the indicated equations takes into account object mass *MM* only, then the second one uses elements number N of macromolecule, that is, takes into account dynamics of molecular structure change. Secondly, the Eq. (8) takes into account real structural state of macromolecule with the aid of its fractal dimension D_f. The indicated above factors appreciation defines correct description by the equation (8) the dependence of macromolecular coil gyration radius R_g on molecular weight *MM* of polymer [235].

Thus, the fractal model for the description of a macromolecular coil gyration radius change at polymer molecular weight variation is offered. This model takes into account both macromolecule molecular structure change dynamics at chemical composition variation and its structural state, defined, for example, by solvent choice [71]. These factors appreciation allows the concrete theoretical description of the indicated dependence.

The authors [228, 236] synthesized polymers of the type of so-called molecular brushes-poly(sodiumoxi) methylsylseskvioxanes (PSMSO). Molecular brushes one names comb-like polymers with high density of grafted side chains. They found anomalous low values of solutions viscosity of molecular brushes PSMSO and on this basis supposed, that at side branch length growth the qualitative changes of polymer macromolecule in solution and block occurred [228, 236, 237], which consisted of transition from conformation of a loose macromolecular coil, characteristic to basic linear polymer, to molecular globular particle with high density of the monomer links in space. These conclusions are based on intrinsic viscosity [η] reduction from 0.27 dl/g for the initial polydimethylsiloxane up to 0.026 dl/g for comb-like polymers on its basis. The authors [238] performed quantitative estimation of the indicated above supposed transition within the framework of fractal analysis [235] on the example of comb-like PSMSO. The necessary data for further theoretical estimations, namely, polymer molecular weight *MM*, branching factor g', intrinsic viscosity [η] for PSMSO were accepted according to the results of Ref. [236].

As it is known through Ref. [1], two variants of branching factor determination exist: the factor g is determined as the ratio of gyration radii of branched R_θ and linear $R_{l,\theta}$ analogs (the Eq. (104)) and factor g'—as the ratio of corresponding intrinsic viscosities $[\eta]_\theta$ and $[\eta]_{l,\theta}$ (the Eq. (105)) for polymers with the same molecular weight *MM*. Between the indicated branching factors the following relationship exists [1]:

$$g \sim (g')^2 \tag{183}$$

Assuming to the first approximation $g \approx (g')^2$, the branching centers number per polymer macromolecule can be calculated with the aid of the Eq. (129). As the estimations according to the indicated equation have shown, m value reduces systematically at molecular weight *MM* decrease for PSMSO (Table 25). In Fig. 109 the dependence $m(MM)$ is adduced for the considered polymers, which proves to be linear and passing through coordinates origin. This circumstance allows one to determine the molecular weight of a chain part between branching centers MM_b as MM/m. From the plot of Fig. 109 the condition MM_b=const≈440 follows.

TABLE 25 The molecular and hydro dynamical characteristics of PSMSO.

Fraction №	$MM\times10^{-3}$	g'	m	m_0	[η], dl/g	$[\eta]^T$, dl/g
1	25.0	0.54	60	417	0.059	0.060
2	19.0	0.59	40	475	0.053	0.051
3	14.0	0.60	37	378	0.043	0.045
4	10.5	0.69	22	477	0.040	0.036
5	9.0	0.75	20	450	0.039	0.036
6	8.0	0.73	18	445	0.035	0.034

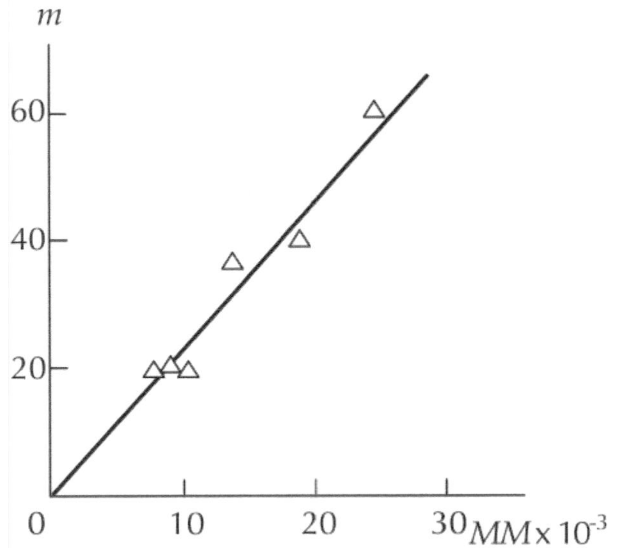

FIGURE 109 The dependence of branching centers number per one macromolecule *m* on molecular weight *MM* for PSMSO.

As it is known through Ref. [45], the macromolecule spectral (fraction) dimension d_s value, characterizing its connectivity degree, varies from 1.0 for linear chains up to 1.33 for very branched (cross-linked) macromolecules. Hence, this dimension also characterizes macromolecules branching degree. The parameters d_s and m are linked between themselves by the Eq. (130). As it follows from the data of Table 25, for all considered polymers $m>4$ and, hence, for them $d_s=1.33$, that speaks about high (limiting) PSMSO macromolecules branching degree. As it is known through Ref. [25], the fractal dimension D_f value of a swollen macromolecular coil is determined according to the Eq. (12) of Chapter 1. Calculation according to the indicated equation gives $D_f=2.0$, that corresponds to the branched polymer dimension in good solvent (for the linear polymer the same dimension is equal to 5/3). The indicated result was obtained within the framework of both fractal analysis [25] and mean field theory [232].

The exponent a_η value in Mark-Kuhn-Houwink equation depends on D_f according to the Eq. (4). Hence, the transition from linear chain to a branched macromolecule results in a_η reduction from 0.70 up to 0.5 without changing macromolecule conformational state. Correctness of D_f value estimation verification can be performed as follows. The dimension D_f can be calculated alternatively according to the Eq. (18) of Chapter 1, where $_f$ value is determined according to the Eq. (82). Calculation according to the Eqs. (18) of Chapter 1 and (82) gives $D_f \approx 2.05$ that corresponds well to D_f determination according to the Eq. (12) of Chapter 1. For correctness verification of the obtained above molecular and structural characteristics of PSMSO in toluene [236] let us perform the intrinsic viscosity [η] estimation according to Mark-Kuhn-Houwink equation fractal variant—the Eq. (10). The comparison of the received experimentally [η] [236] and calculated according to the Eq. (10) $[\eta]^T$ intrinsic viscosity values for PSMSO in toluene (at D_f=const=2.0) have shown their good correspondence (the mean discrepancy of [η] and $[\eta]^T$ makes up 6.2%). Let us note the principally important aspect of this calculation: the using in the Eq. (10) any value D_f2.0 will give the function [](*MM*) change tendency, differing from an experimental one, since m_0=const.

The value [η] for PSMSO can be changed by a solvent variation. So, using solvent, poor for PSMSO (for example, nitromethane with δ_s=12.9 (cal/cm^3)$^{1/2}$ [14]) will give the value δ_f=1.20 according to the Eq. (82), D_f=2.27 according to the Eq. (18) of Chapter 1 and []=0.029 dl/g according to the Eq. (10) for PSMSO with MM=25×10^3. Let us note that the indicated D_f value corresponds to the branched polymer macromolecular coil conformation in θ-solvent [13, 232], that is, in the given case the conformation qualitative change is reached at invariable chemical structure of a macromolecule.

Thus, the fractal analysis methods were used above for treatment of comb-like poly(sodiumoxi) methylsylseskvioxanes behavior in solution. It has been shown that the intrinsic viscosity reduction at transition from a linear analog to a branched one is due to the sole factor, namely, to a macromolecule connectivity degree enhancement, characterized by spectral dimension. This conclusion is confirmed by a good correspondence of the experimental and calculated according to Mark-Kuhn-Houwink equation fractal variant intrinsic viscosity values. It has been shown that qualitative transition of the structure of branched polymer macromolecular coil from a good solvent to θ-solvent can be reached by a solvent change.

The time of polymer solutions free relaxation $\langle\tau_d\rangle$, determined experimentally, is the parameter, sensitive to macromolecules sizes and conformation change. It characterizes the rate, with which the optically anisotropic molecules primary orientation disappears, having established under the action of external electric field, applied to solution [227]. This supposes, that $\langle\tau_d\rangle$ value is controlled by the macromolecule structure, which is characterized by the fractal dimension D_f within the framework of fractal analysis. The authors [239] performed this supposition verification on the example of polymers with complex molecular architecture—dendrimers. With this purpose the experimental data for polystirene, modified by Dendron's (PSD) of the first-fourth generations [227] and cylindrical dendrimer on the basis of L-aspagine acid (LPA-3) of the third generation [230]. For the fractal dimension D_f of macromolecule in solution calculation the Eq. (4) was used.

In Fig. 110 the dependence $\langle\tau_d\rangle(D_f)$ is adduced for the considered dendrimers is adduced, from which the fast decay of relaxation time at macromolecule fractal dimension growth follows. Such shape of the dependence $\langle\tau_d\rangle(D_f)$ can be explained by two factors influence. The macromolecular coil gyration radius R_g is linked with dimension D_f and polymerization degree N by the Eq. (8). As one can see, the indicated equation supposes very strong (power) dependence of R_g on D_f. The estimation according to the Eq. (4) has shown that for the considered dendrimers D_f value varies within the limits of 1.52–2.42. Using reasonable values N=1000 [227] and B=0.35 [234], let us obtain R_g variation within the limits of approx. 6–33 nm, that is, in 5.5 times, for the considered dendrimers, that corresponds well to the experimental data [227, 230]. It is obvious, that reorientation of small macromolecules in solution occurs much easier than large ones and requires much less duration, which is, $\langle\tau_d\rangle$.

Sharp increase of a macromolecular coil density ρ at D_f growth can be the second factor. The value of a coil as a fractal object can be estimated according to the Eq. (179). Accepting as the lower linear scale of fractal behavior a_m dimensional boundary between low- and high-molecular substances (a_m=1 nm) and R_g=32 nm, let us obtain according to the Eq. (179) ρ increase from 0.0059 up to 0.134 relative units within the range of D_f=1.52–2.42, that is, in approx. 23 times, for the considered dendrimers. It is obvious, that such macromolecular coil density growth will hamper its fragments orientation in external electric field, which facilitates their relaxation at this field removal, that is, decreases $\langle\tau_d\rangle$.

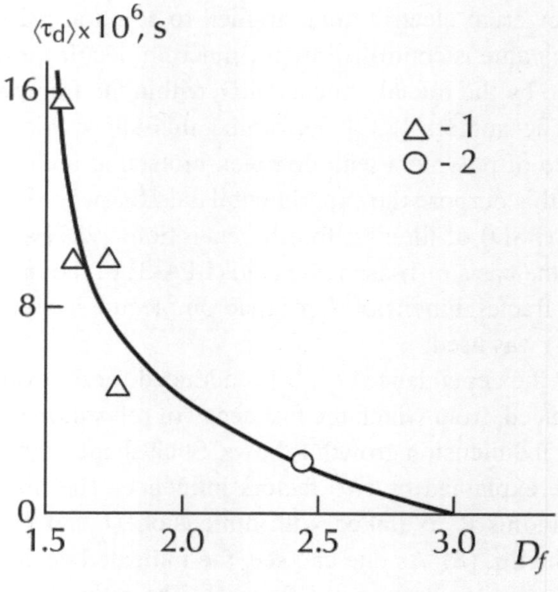

FIGURE 110 The dependence of free relaxation time $\langle\tau_d\rangle$ on macromolecule fractal dimension D_f for solutions of dendrimers PSD in tetrahydrofuran (1) and LPA-3 in chloroform (2).

For the made above supposition verification in Fig. 111 the dependence $\langle\tau_d\rangle(R_g)$ is adduced for the considered dendrimers. As it was to be expected, $\langle\tau_d\rangle$ linear reduction at R_g decreasing is observed, described analytically by the following empirical equation [239]:

$$\langle\tau_d\rangle = 0.43\times10^{-6} R_g \text{, s,} \tag{184}$$

where R_g is given in nm.

The dependence $\tau_d\rangle(R_g)$ passes through coordinates origin for the obvious reason: at $R_g \to 0$ all molecule orientations are the same, since it is a zero-dimensional (point) object and thus its relaxation from any position does not have physical significance.

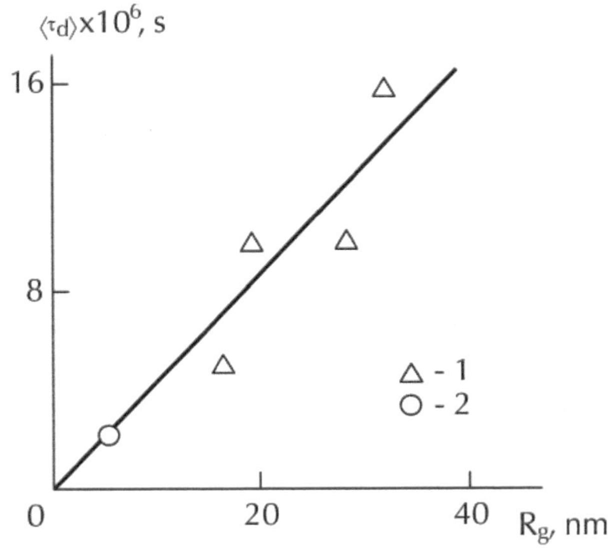

FIGURE 111 The dependence of free relaxation time $\langle \tau_d \rangle$ on macromolecule gyration radius R_g for solutions of dendrimers PSD in tetrahydrofuran (1) and LPA-3 in chloroform (2).

Hence, the free relaxation time of dendrimers macromolecule in solution is a function of its structure that was to be expected by virtue of the general postulate: any property of an object is controlled by its structure [240].

The new regularly branched polymers on the basis of PS were synthesized recently [227]. Each repeated link of these polymers macromolecules contains the Dendron's of Frechet type as a side radical. The study of this modified by Dendron's PS solutions has shown, that the generation number increase from 1 up to 4 results in hydro dynamical sizes and macromolecules equilibrium rigidity enhancement in comparison with the initial PS (see Table 24). The similar results were obtained for other polymers as well, modified by Dendron's. These studies were performed within the framework of classical experimental methods, which are distinguished by both high complexity and united theoretical basis absence, taking into account macromolecule structure in solution. Therefore, the authors [240] performed the theoretical description within the framework of fractal analysis of PS, modified by Dendron's, solutions study results [227], whose chemical structure is adduced in Fig. 112.

As it has been shown above, a polymer macromolecule, having either conformation (statistical coil, rod-like form and so on), is the main element of polymer solutions. The structure of macromolecule in solution can be characterized with the aid of the fractal dimension D_f, which is the structural characteristic in strict physical significance of this term, since it describes macromolecule elements distribution in space [224]. D_f value for the considered polymers (PSD) can be evaluated as follows. As it is known through Ref. [1], for flexible-chain polymers, having the swelling coefficient $\alpha \geq 1$, Huggins constant can be determined according to the Eq. (174), from which it follows, that the condition $\alpha \geq 1$ is fulfilled for k_x values ≤ 0.55.

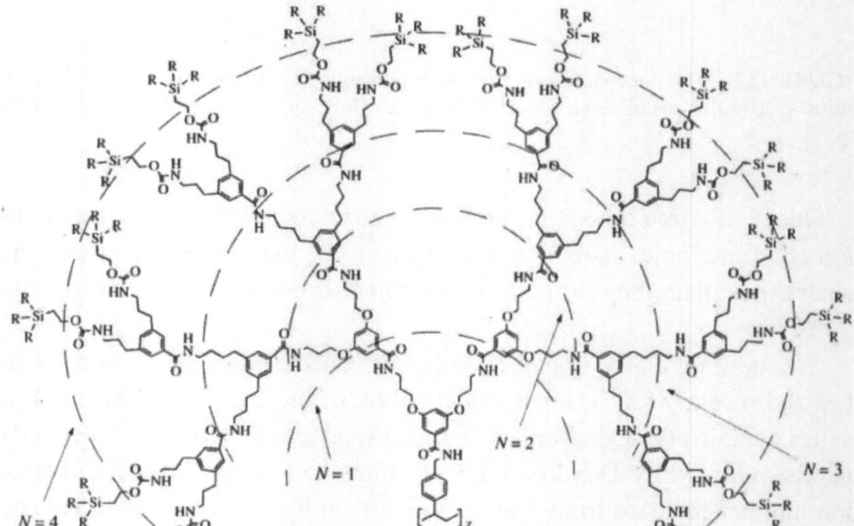

FIGURE 112 The chemical structure of PSD macromolecule repeated link (R=CH_3, N is polymerization degree, N_D is Dendron generation number).

The data of Ref. [227] have shown that the last condition is correct for PSD with the least generation number (PSD-1 and PSD-2) and therefore the Eq. (174) is applicable only for them. At the same time the general fractal Eq. (8) exists between a macromolecular coil gyration radius R_g and polymerization degree N. For D_f value quantitative estimation according to the last relationship it is necessary to determine for it the proportionality coefficient B, since the experimental R_g and N values for the considered PSD are adduced in Ref. [227]. For this α value for PSD-1 was calculated according to the Eq. (174), which corresponds to the condition α≥1 (k_x=0.32 [227], =1.275) and then the dimension D_f was determined according to the Eq. (78). Further B value in the Eq. (8) can be calculated, which turns out to be equal to 0.349, if R_g is given in nm. As R_g this parameter values, calculated according to Stox formula, were used, since they are close to the similar values, determined with the aid of translational diffusivity [227]. Therefore, any from the indicated gyration radius values using does not give essential discrepancy at D_f evaluation. The dimensions D_f, estimated by the indicated mode, are reduced from 1.783 up to 1.520 at Dendron's generation number N_D increase from 1 up to 4. Hence, N_D increasing leads to PSD macromolecules curdling degree—if at N_D=1 D_f value is a typical for the linear polymer in good solvent, then at N_D=4 D_f value is close to the permeable coil dimension [56]. This conclusion corresponds completely to the results, obtained in Ref. [227] because of quite other reasons.

As it has been shown in Ref. [227], N_D increase from 1 up to 4 leads to essential growth of PSD macromolecules equilibrium rigidity: Kuhn segment length A increases from 3.9 up to 23.3 nm. The authors [56] received the relationships (54) and (55) between structural D_f and molecular A macromolecule parameters for flexible- and rigid-chain polymers, respectively. In Fig. 113 the similar dependence $D_f(A)$ for the considered polymers PSD is adduced, which, as earlier, has shown D_f reduction at A growth and is described analytically by the following empirical correlation [240]:

$$D_f = 1.82 - 1.45\times10^{-3} A, \tag{185}$$

where A value is given again in Angströms (Å).

In Fig. 113 the dependences, constructed according to the Eqs. (54) and (55), are also adduced. As one can see from the adduced comparison, PSD behavior in solution is much closer to rigid-chain polymers behavior, then to a flexible-chain one, that was to be expected, proceeding from PSD macromolecule equilibrium rigidity fast growth, characterized by Kuhn segment length enhancement. The Eq. (185) allows to define the condition of attainment by PSD macromolecule of a rod-like (stick-like) shape, which is characterized by the criterion D_f=1.0 [25]. The last criterion is realized at A=566 Å, that is, at A value, typical for rigid-chain polymers. The constant value 1.82 in the Eq. (185) is determined for the initial polystyrene because of the following reasons. The translational diffusivity D_0^{PS} for PS can be calculated according to the following equation [227]:

$$D_0^{PS} = 1.2 \times 10^{-4} MM^{-0.55}, \qquad (186)$$

where MM is polymer molecular weight.

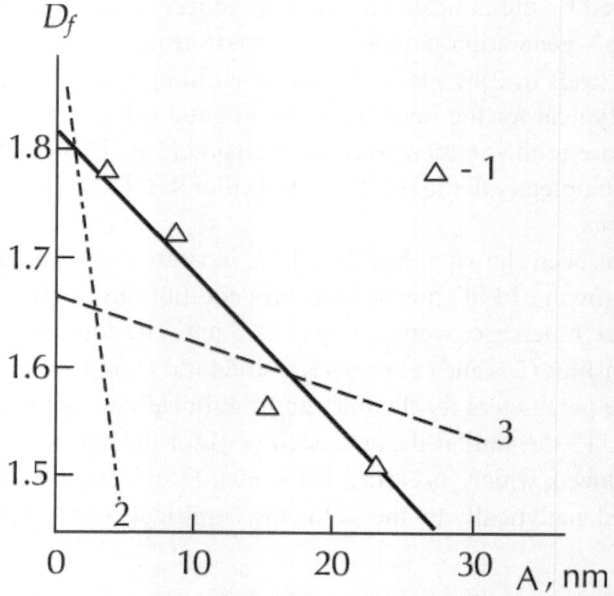

FIGURE 113 The dependences of macromolecular coil in solution fractal dimension D_f on Kuhn segment length A. 1—the data for PSD; 2, 3—the dependences, constructed according to the Eqs. (54) (2) and (55) (3).

The exponent a_D for molecular weight in the Eq. (186) is linked to the dimension D_f by the Eq. (18), that for the initial PS gives D_f≈1.82. Thus, the indicated coefficients in the Eq. (185) represents dimension D_f for the initial PS. The data of Fig. 113 demonstrate that PS modification by Dendron's increases progressively macromolecule rigidity, characterized by Kuhn segment length A, and reduces chain curdling degree, characterized by the dimension D_f.

In Fig. 114 the dependence of Kuhn segment length A on generation number N_D is adduced for the considered PSD. As one can see, A value and, hence, PSD macromolecule rigidity grows at generations number N_D increasing. The plot $A(N_D)$ extrapolation to $N_D>4$ demonstrates, that a macromolecule rod-like shape (D_f=1.0 or A=566 Å) attainment is realized at six generations availability. Let us note an important feature of PSD macromolecules. The condition D_f=1.0 for them is reached at A=566 Å, whereas for rigid-chain polymers according to the Eq. (55)—at $A\approx$1500 Å. The authors [241] noted that the typical rigid-chain polymer polybutylisozianate (A=1000 Å) had macromolecules with conformation of weakly bended thin rod, corresponding to D_f=1.271 according to the Eq. (55), that although is close to D_f=1.0, but is somewhat larger than this limiting dimension. The same conformation PSD will have at A=379 Å, that is, at N_D=5, as it follows from the data of Fig. 114. It is obvious, that this distinction should be attributed to Dendron's availability in PSD macromolecules.

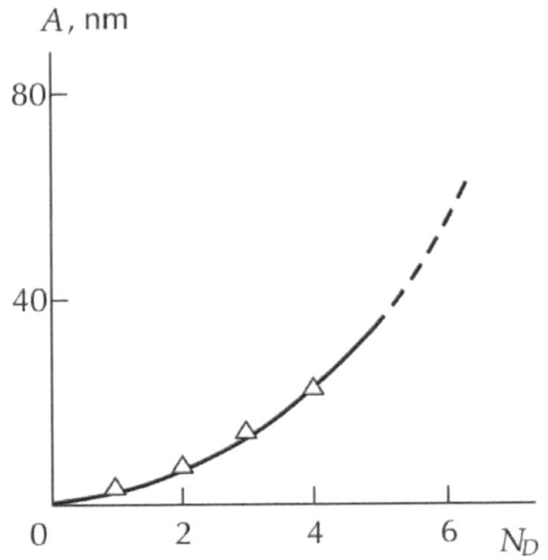

FIGURE 114 The dependence of Kuhn segment length A on Dendron generation number N_D for PSD.

Hence, the performed within the framework of fractal approach analysis of behavior of polystirene, modified by Dendron's, in diluted solutions gave the same conclusions, as the analysis within the framework of classical approaches. The main distinction of the indicated approaches is the fact, that the structural model, allowing to describe quantitatively macromolecules structural state and conformation, was placed in the fractal approach base. Other characteristics (gyration radius, Kuhn segment length and so on) are the function of the indicated structural state of a macromolecule. The fractal analysis methods, used for the description of linear flexible-chain polymers behavior, can be applied successfully also in case of polymers with more complex macromolecular architecture.

KEYWORDS

- **Aarony-Stauffer rule**
- **elementary link**
- **Flory-Fox equation**
- **leaking coil**
- **Mark-Kuhn-Houwink equation**
- **nanogel**
- **nanoreactor**
- **Ptitsyn-Eizner approximation**
- **rigid-chain polymers**
- **Rimman-Liouville fractional derivative**

REFERENCES

1. Budtov, V. P.; *Physical Chemistry of Polymer Solutions*. Sankt-Peterburg, Chemistry, **1992,** 384 p.
2. Frenkel, S. Ya. *Introduction in Statistical Theory of Polymerization*. Moscow-Leningrad, Science, **1965,** 268 p.
3. Karmanov, A. P.; Monakov, Yu. B.; The fractal structure of lulk- and end-wise-dehydropolymers. High-Molecular Compounds. B, **1995,** *37(2),* 328–331.

4. Kozlov, G. V.; Shustov, G. B.; Zaikov, G. E.; The fractal and scaling analysis of chemical reactions. *J. Appl. Polymer Sci.* **2004,** *93(5),* 2343–2347.
5. Askadskii, A. A.; Physics-Chemistry of Polyarylates. Moscow, Chemistry, **1968,** 214 p.
6. Kozlov, G. V.; Dolbin, I. V.; Fractal variant of Mark-Kuhn-Houwink equation. *High-Molecular Compounds, B*, **2002,** *44(1),* 115–118.
7. Kozlov, G. V.; Dolbin, I. V.; Zaikov, G. E. A fractal form of Mark-Kuhn-Houwink equation. In book: New Perspectives in Chemistry and Biochemistry. Ed. Zaikov, G.; New York, Nova Science Publishers, Inc.; **2002,** 41–47.
8. Kozlov, G. V.; Dolbin, I. V.; Zaikov, G. E.; Fractal form of Mark-Kuhn-Houwink equation. J.; Balkan Tribologic Association, **2003,** *9(2),* p. 240–245.
9. Shustov, G. B.; Kozlov, G. V.; The Techniques of Fractal Dimensions Determination for Polymeric Materials. (Preprint). Nal'chik, Publishers KBSU, **2006,** 53 p.
10. Baranov, V. G.; Frenkel, S. Ya.; Brestkin, Yu. V.; The dimensionality of linear macromolecule different states. Reports of Academy of Sciences of SSSR, **1986,** *290(2),* 369–372.
11. Kozlov, G. V.; Temiraev, K. B.; Ovcharenko, E. N.; Lipatov, Yu.S.; The description of low-temperature polycondensation process within the framework of irreversible aggregation models. Reports of National Academy of Sciences of Ukraine, **1999,** *12,* 136–140.
12. Temiraev, K. B.; The evaluation of aromatic copolyethersulfoneformals molecular weight by the fractal analysis methods. Manuscript deposited to VINITI RAS, Moscow, 20.07.1998, *22*, V91-V98.
13. Family, F.; Fractal dimension and grand universality of critical phenomena. J.; Stat. Phys.; **1984,** *36(5/6)*, 881–896.
14. Wieche, I. A.; Polygon mapping with two-dimensional solubility parameter. Ind. Engng. Chem. Res.; **1995,** *34(2),* 661–673.
15. Shogenov, V. N.; Beloshenko, V. A.; Kozlov, G. V.; Varyukhin, V. N.; The prediction of mechanical properties of heterochain polyethers. Physics and Engineering of High Pressures, **1999,** *9(3),* 30–36.
16. Askadskii, A. A.; *Structure and Properties of Thermostable Polymers*. Moscow, Chemistry, **1981,** 320 p.
17. Pavlov, G. M.; Korneeva, E. V.; Mikhailova, N. A.; Anan'eva, E. P.; Hydrodynamic and molecular characteristics of fractions of mannane, formed by yeasts phodotorula rubra. Biophysics, **1992,** *37(6),* 1035–1040.
18. Pavlov, G. M.; Korneeva Polymaltotrioza, E. V.; Molecular characteristics and equilibrium chains rigidity. Biophysics, **1995,** *40(6),* 1227–1233.
19. Kozlov, G. V.; Dolbin, I. V.; Express method of fractal dimension estimation of biopolymers macromolecular coils in solution. Biophysics, **2001,** *46(2),* 216–219.
20. Kozlov, G. V.; Dolbin, I. V.; Zaikov, G. E.; Rapid method of estimating the fractal dimension of macromolecular coils of biopolymers in solution. In book: Perspectives on Chemical and Biochemical Physics. Ed. Zaikov, G.; New York, Nova Science Publishers, Inc.; **2002,** 217–223.
21. Kozlov, G. V.; Dolbin, I. V.; Zaikov, G. E.; Rapid method of estimating the fractal dimension of macromolecular coils of biopolymers in solution. In book: Polymer Year-

book 18. Ed. Pethrich, R.; Zaikov, G.; Shawbury, Rapra Technology Limited, **2003,** 393–400.
22. Tager, A. A.; Kolmakova, L. K.; Solubility parameter, its evaluation methods, connection with polymers solubility. High-Molecular Compounds. A, **1980,** *22(3),* 483–496.
23. Lee, H. -R.; Lee, Y. -D.; Solubility behavior of an organic soluble polyimide. *J. Appl. Polymer Sci.* **1990,** *40(1*1–12, 2087–2099.
24. Kozlov, G. V.; Dolbin, I. V.; Mashukov, N. I.; Burmistr, M. V.; Korenyako, V. A.; Prediction of the macromolecular coils fractal dimension in diluted solution on the basis of two-dimensional solubility parameter model. Problems of Chemistry and Chemical Technology, **2001,** *6,* 71–77.
25. Vilgis, T. A.; Flory theory of polymeric fractals—intersection, saturation and condensation. Physica, A.; **1988,** *153(2),* 341–354.
26. Vasilova, O. I.; Zaitseva, V. V.; Kucher, R. V.; The solubility parameter of three-component copolymers of styrene with acrylonitrile and methylmethacrylate. High-Molecular Compounds. B, **1987,** *29(12),* p. 912–916.
27. Kozlov, G. V.; Shustov, G. B.; Zaikov, G. E.; The fractal physics of the polycondensation processes. J.; Balkan Tribological Association, **2003,** *9(4),* 467–514.
28. Kozlov, G. V.; Shustov, G. B.; The fractal physics of the polycondensation processes. In book: Achievements in the Polymers Physics-Chemistry Field. Ed. Zaikov, G. a.a. Moscow, Chemistry, **2004,** 341–411.
29. Encyclopaedia of Polymers. V. 1. Ed. Kargin, V.; Moscow, Soviet Encyclopaedia, **1972, 1223,** p.
30. Andrianov, K. A.; Pavlova, S. A.; Tverdokhlebova, I. I.; Pertsova, N. V.; Temnikovskii V. A.; Hydrodynamical properties of poly(methyl phenylsiloxane). High-Molecular Compounds. A, **1972,** *14(8),* p. 1816–1821.
31. Liu, M.; Peng, S.; Gao, G.; Wu J.; The diluted solution properties and molecular chain sizes of atactic polystirene. China Synth. Rubber Ind.; **1991,** v. 1*4(3),* 206–210.
32. Kireev, V. V.; High-Molecular Compounds. Moscow, Higher Schools, **1992,** 378 p.
33. Kozlov, G. V.; Dolbin, I. V.; Shustov, G. B.; The evaluation of exponent in Mark-Kuhn-Houwink equation for linear polymers within the framework of fractal analysis. Proceedings of XVII Mendeleev's Congress by General and Applied Chemistry "Achievements and Perspectives of Chemical Science." Kazan, 21–26 October **2003,** 279.
34. Topchiev, D. A.; Malkanduev Yu.A.; Cationic Polyelectrolytes of, N.; N-dialkyl-N, N-diallyl-ammonium Halogenides Set: Formation Processes Features, Properties and Application. Nal'chik, Publishers KBSU, **1997,** 181 p.
35. Gribel, T.; Kulicke, W. -M.; Hanzemzadeh, A.; Characterization of water-soluble polyelectrolytes as examinated by poly(acrylamid-cotrimethylammonium-thacrylate chloride) and establishment of structureproperty relationships. Colloid Polymer Sci.; **1991,** *269(1),* 113–120.
36. Kozlov, G. V.; Malkanduev Yu.A.; Ulybina, A. S.; The analysis of hydrodynamic properties of copolymers of acrylamide with trimethylammonium methylmethacrylatechloride solutions within the framework of fractal analysis. Bulletin of Kabardino-Balkarian State University, Chemical Sciences, **2003,** *5,* 175–181.
37. Kozlov, G. V.; Sanditov, D. S.; Anharmonic Effects and Physical-Mechanical Properties of polymers. Novosibirsk, Science, **1994,** 261 p.

38. Ptitsyn, O. B.; Eizner, Yu. E.; Hydrodynamics of polymer solutions. II.; Hydrodynamical properties of macromolecules in good solvents. Journal of Technical Physics, **1959,** *29(9),* 1117–1134.
39. Kozlov, G. V.; Afaunov, V. V.; Temiraev, K. B.; Fractal dimension of biopolymers macromolecular coil in solution as a bulk interactions measure. Manuscript deposited to VINITI RAS, Moscow, 08.01.1998*(9*-V98.
40. Havlin, S.; Ben-Avraham, D.; Theoretical and numerical study of fractal dimensionality in self-avoiding walks. Phys. Rev. A, **1982,** *26(3),* p. 1728–1734.
41. Kozlov, G. V.; Temiraev, K. B.; Shustov, G. B.; The fractal analysis of polysaccharides behavior in solutions. Proceedings of Higher Educational Institutions, North-Caucasus region, natural sciences, **1998,** *3,* 82–85.
42. Timofejeva, G. J.; Pavlova, S. A.; Wandrey Ch.; Jaeger, W.; Hahn, M.; Linow K. -J.; Görnitz, E. On the determination of the molecular weight distribution of poly(dimethyl diallyl ammonium chloride) by ultracentrifugation. Acta Polymerica, **1990,** *41(9),* 479–484.
43. Pavlov, G. M.; Evlampieva, N. P.; Dynamical two-beam refraction of polymaltotrioze solutions. Biophysics, **1995,** *40(6),* 1220–1226.
44. Rammal, R.; Toulouse, G.; Random walks on fractal structures and percolation clusters. *J. Phys. Lett.* (Paris), **1983,** *44(1),* p. L13-L22.
45. Alexander, S.; Orbach, R.; Density of states on fractals: "fractons." *J. Phys. Lett.* (Paris), **1982,** *43(17),* p. L625-L631.
46. Kozlov, G. V.; Temiraev, K. B.; Afaunov, V. V.; Hydrodynamical and molecular characteristics of aromatic polyethersulfoneformals: the fractal analysis. Manusript deposited to VINITI RAS, Moscow, 08.01.1998, 10, V98.
47. Temiraev, K. B.; Shustov, G. B.; Mikitaev, A. K.; Synthesis and properties of copolyethersulfoneformals. High-Molecular Compounds. B, **1988,** *30(6),* 412–415.
48. Vilenchik, L. Z.; Sklizkova, V. P.; Tennikova, T. B.; Bel'nikevich, N. G.; Nesterov, V. V.; Kudryavtsev, V. V.; Belen'kii, B. G.; Frenkel, S. Ya.; Koton M. M.; Chromatographic study of poly *(4),* 4′-oxydiphenylene) pirromellitimide acid solutions. High-Molecular Compounds. A, **1985,** v. 2*7(5),* 927–930.
49. Berstein, V. A.; Egorov, V. M.; Differential Scanning Calorimetry in Physics-Chemistry of Polymers. Leningrad, Chemistry, **1990,** 256 p.
50. Privalko, V. P.; Lipatov, Yu.S.; The influence of macromolecular chain flexibility on glass transition temperature of linear polymers. High-Molecular Compounds. A, **1971,** *13(12),* 2733–2737.
51. Korshak, V. V.; Pavlova, S.-S. A.; Timofeeva, G. I.; Kroyan, S. A.; Krongauz, Ye. S.; Travnikova, A. P.; Paubach Ch.; Schultz, G.; Gnauk, R.; Hydrodynamic properties of randomly branched polyphenylquinoxalines with low degrees of branching. High-Molecular Compounds. A, **1984,** *26(9),* 1868–1876.
52. Kozlov, G. V.; Temiraev, K. B.; Sozaev, V. A.; The evaluation of fractal dimension of macromolecular coil in diluted solution by the characteristics of viscosity. Journal of Physical Chemistry, **1999,** *73(4),* 766–768.
53. Samarin, A. F.; Starkman, B. P.; Viscosimetric express-method of determination of polymers molecular weight. High-Molecular Compounds. A, **1985,** *27(5),* 1101–1103.
54. Büller, K. U.; Heat- and Thermostable Polymers. Moscow, Chemistry, **1984, 1056,** p.

55. Brown, D.; Sherdron, G.; Kern, V.; The practical handbook by synthesis and study of the polymers properties. Ed. Zubkov, V.; Moscow, Chemistry, **1976,** 384 p.
56. Kozlov, G. V.; Dolbin, I. V.; Zaikov, G. E.; Fractal *Physical Chemistry of Polymer Solutions*. In book: Chemical and Biological Kinetics. New Horizons. V. 1. Chemical Kinetics. Ed. Burlakova, E.; Shilov, A.; Varfolomeev, S.; Zaikov, G.; Leiden, Boston, Brill Academic Publishers, **2005,** 448–483.
57. Dubrovina, L. V.; Pavlova, S. -S. A.; Bragina, T. P.; Properties of a solutions of polyarylates with side aliphatic group. High-Molecular Compounds. A, **1988,** *30(5),* 995–1000.
58. Papkov, S. P.; The evaluation of Kuhn segment value of rigid-chain polymers according to the viscosimetric properties of diluted solutions. High-Molecular Compounds. B, **1982,** *24(11),* 869–873.
59. Kozlov, G. V.; Temiraev, K. B.; Mikitaev, A. K.; The correlations coil gyration radius—Kuhn segment size for rigid-chain polymers: the fractal analysis. Proceedings of Higher Educational Institutions. Chemistry and Chemical Technology, **2002,** *45(2),* 30–33.
60. Grosberg Yu.A.; Khokhlov, A. R.; Physics in Polymers World. Moscow, Science, **1989,** 208 p.
61. Khamrakulov, G. P.; Musaev Kh.N.; Budtov, V. P.; Kuleznev, V. N. On properties of diluted solutions of diacetate cellulose in different solvents. High-Molecular Compounds. A, **1993,** *35(6),* 705–709.
62. Novikov, V. U.; Kozlov, G. V. A polymers structure and properties within the framework of fractal approach. Successes of Chemistry, **2000,** *69(6),* 572–599.
63. Novikov, V. U.; Kozlov, G. V.; The foundations of fractal approach to polymers structure. Non-Euclidean physics of polymers. In: Applied Synergetics, Fractals and Structures Computer Simulation. Ed. Oksogoev, A.; Tomsk, TSU, **2002,** 268–302.
64. Dolbin, I. V.; Kozlov, G. V.; The intercommunication of polymer chain rigidity and macromolecular coil structure in diluted solutions. Proceedings of All-Russian Sci. Conf. "Perspective-2003." V. 4. Nal'chik, KBSU, **2003,** 108–111.
65. Dondos, A. A new relation between the intrinsic viscosity and the molecular mass of polymers derives from the blob model: determination of the statistical segment length of flexible polymers. Polymer, **2001,** *42(4),* 897–901.
66. Gans, C.; Schnee, J.; Scherf, U.; Staikos, G.; Pierri, E.; Dondos, A.; Viscosimetric determination of the statistical segment length of wormlike polymers. Polymer, **1998,** *39(17),* 4155–4158.
67. Kozlov, G. V.; Shustov, G. B.; Dolbin, I. V. A fractal dimension and interactions of a macromolecular coil in a solution. Proceedings of I-th International Sci. Conf. "Modern Problems of Organic Chemistry, Ecology and Biotechnology." V. 1. Organic Chemistry. Luga, June **2001,** 17–18.
68. Kozlov, G. V.; Burya, A. I.; Shustov, G. B.; Temiraev, K. B.; Shogenov, V. N.; The influence of main characteristics of polymer and solvent on fractal dimension of a macromolecular coils in solution. Bulletin of University of Dnepropetrovsk. Chemistry, **2000,** *5,* 96–102.
69. Aharoni, S. M. On entanglements of flexible and rodlike polymers. Macromolecules, **1983,** *16(9),* 1722–1728.

70. Wu S.; Chain structure and entanglement. J.; Polymer Sci.: Part B: Polymer Phys.; **1989,** *27(4),* 723–741.
71. Kozlov, G. V.; Temiraev, K. B.; Kaloev, N. I.; Influence of solvent nature on structure and formation mechanism of polyarylate in conditions of low-temperature polycondensation. Reports of Academy of Sciences, **1998,** *362(4),* 489–492.
72. Nigmatullin, R. R.; Fractional integral and its physical interpretation. Theoretical and Mathematical Physics, **1992,** *90(3),* 354–367.
73. Bolotov, V. N.; The positron annihilation in fractal mediums. Letters to Journal of Technical Physics, **1995,** *21(10),* 82–84.
74. Meilanov, R. P.; Sveshnikova, D. A.; Shabanov, O. M.; Sorption kinetics in systems with fractal structure. Proceedings of Higher Educational Institutions, North-Caucasus region, natural sciences, **2001**, *1,* 63–66.
75. Kekharsaeva, E. R.; Mikitaev, A. K.; Aleroev, T. S.; The model of deformation—strength characteristics of chloro-containing polyethers on the basis of derivatives of fractional order. Plastics, **2001,** *3,* 35.
76. Kozlov, G. V.; Shustov, G. B.; Zaikov, G. E.; The role of polymer melt structure in the heterochain polyethers thermooxidative degradation process. Journal of Applied Chemistry, **2002,** *75(3),* 485–487.
77. Kozlov, G. V.; Batyrova, H. M.; Zaikov, G. E.; The structural treatment of a number of effective centers of polymer chain in the process of thermooxidative degradation. *J. Appl. Polymer Sci.* **2003,** *89(7),* p. 1764–1767.
78. Kozlov, G. V.; Dolbin, I. V.; Shogenov, V. Kh.; Zaikov, G. E.; Estimation of end-to-end distances for a polycarbonate chain within the framework of the fractional derivatives theory. In book: Fractal Analysis of Polymers: From Synthesis to Composites. Ed. Kozlov, G.; Zaikov, G.; Novikov, V.; New York, Nova Science Publishers, Inc.; **2003,** 123–130.
79. Kozlov, G. V.; Dolbin, I. V.; Shogenov, V. Kh.; Zaikov, G. E.; Estimation of end-to-end distance for a polycarbonate chain within the framework of the fractional derivatives theory. J.; Balkan Tribological Association, **2003,** v. *9(3),* 428–434.
80. Dolbin, I. V.; Sargaeva, T. N.; Kozlov, G. V.; Shogenov, V. Kh. The end-to-end distance estimation for polycarbonate chain within the framework of the theory of fractaional derivatives. Proceedings of All-Russian Sci. Conf. "Perspective-2004." Nal'chik, KBSU, **2004,** *3,*178–181.
81. Kozlov, G. V.; Dolbin, I. V.; Shogenov, V. Kh.; Zaikov, G. E.; Application of the fractional derivative theory to the estimation of the end-to-end distance of polycarbonate chains. *J. Appl. Polymer Sci.* **2004,** *91(6),* 3765–3768.
82. Schnell, G.; Chemistry and Physics of Polycarbonates. Moscow, Chemistry, **1967,** 229 p.
83. Shogenov, V. Kh.; Shkhanykov-Lafishev, M. Kh.; Beshtoev Kh.M.; The Fractional Derivatives: Interpretation and some Applications in Physics. (Preprint). Dubna, OIYaI, **1997,** 14 p.
84. Temiraev, K. B.; Kozlov, G. V.; Sozaev, V. A.; The prediction of macromolecular coil fractal dimension by polymer molecular characteristics. Bulletin of Kabardino-Balkarian State University, series physical sciences, **1998,** *3,* 24–28.

85. Aharoni, S. M.; Correlations between chain parameters and failure characteristics of polymers below their glass transition temperature. Macromolecules, **1985,** *18(12),* 2624–2630.
86. Feder, E.; Fractals. New. York, Plenum Press, **1990,** 249 p.
87. Novikov, V. U.; Kozlov, G. V. A macromolecules fractal analysis. Successes of Chemistry, **2000,** *69(4),* 378–399.
88. Balankin, A. S.; Synergetics of Deformable Body. Moscow, Publishers of Ministry Defence of SSSR, **1991,** 404 p.
89. Okatova, O. V.; Lavrenko, P. N.; The temperature influence on intrinsic viscosity and curling degree of para- and metha-aromatic polyamides in sulfur acid. High-Molecular Compounds. B, **1992,** *34(7),* 9–18.
90. Pfeifer, P.; Avnir, D.; Farin, D.; Scaling behavior of surface irregularity in the molecular domain: from adsorption studies to fractal catalysts. J.; Stat. Phys.; **1984,** *36(5/6,* 699–716.
91. Malkanduev Yu.A.; Kozlov, G. V.; The description of flocculated ability of, N.; N-dimethyl-N, N-diallyl ammoniumchloride within the framework of fractal analysis. Proceedings of Sci.-Pract. Conf. "Actual Problems of Chemistry, Biology and Ecology." Nal'chik, KBSU, **1997,** 36–37.
92. Kozlov, G. V.; Temiraev, K. B.; Lipatov, Yu.S.; The temperature dependence of the fractal dimension of macromolecular coils in diluted solutions. In book: Fractals and Local Order in Polymeric Materials. Ed. Kozlov, G.; Zaikov, G.; New York, Nova Science Publishers, Inc.; **2001,** 1–9.
93. Tsvetkov, V. N.; Rigid-Chain Polymer Molecules. Leningrad, Chemistry, **1986,** 328 p.
94. Gel'mont, M. M.; Braverman, L. P.; Smirnova, V. N.; Kulichikhin, V. G.; Efros L. S.; The influence of different-linking connected with asymmetry of one from monomers, on polyamidobenzimidazoles properties. High-Molecular Compounds. A, **1987,** *29(3),* 537–543.
95. Ryskina, I. I.; Zhiganova, I. Yu. The temperature influence on conformation of triacetate cellulose macromolecules in benzile spirit. High-Molecular Compounds. B, **1994,** *36(3),* 503–506.
96. Korshak, V. V.; Vinogradova, S. V.; Nonequilibrium Polycondensation. Moscow, Science, **1972,** 695 p.
97. Dolbin, I. V.; Kozlov, G. V.; The evaluation of fractal dimension of organic solvent molecules. Proceedings of All-Russian Sci. Conf. "Perspective-2002." Nal'chik, KBSU, **2002,** 84–89.
98. Meakin, P.; Diffusion-controlled flocculation: the effect of attractive and repulsive interactions. J.; Chem. Phys.; **1983,** *79(5),* 2426–2429.
99. Nekhaenko, E. A.; Rogovina, L. Z.; Slonimskii, G. A.; Genin Ya.V. On possibility of properties regulation for polymer films, formed from solution. High-Molecular Compounds. B, **1979,** *19(3),* 279–281.
100. Manenok, G. S.; Artamonov, V. A.; Shingel, I. A.; Prokopchuk, N. R.; The study of properties of films, prepared from copolyamide solutions depending on solvent nature. High-Molecular Compounds. B, **1986,** *28(6),* 448–451.
101. Dolbin, I. V.; Kozlov, G. V.; Zaikov, G. E.; Structural Stabilization of Polymers: The Fractal Models. Moscow, Publishers "Academy of Natural Sciences", **2007,** 328 p.

102. Mashukov, N. I.; Temiraev, K. B.; Shustov, G. B.; Kozlov, G. V.; Modelling of solid state polymer properties at the stage of synthesis: fractal analysis. Chapters of 6th International Chaptershop of Polymer Reaction Engng. Berlin, 5–7 October, **1998,** *134,* 429–438.
103. Dolbin, I. V.; Kozlov, G. V.; Ovcharenko, E. N.; The physical nature of solvent structure in diluted polymer solutions. Proceedings of 3th International Conf. "Actual Problems of Modern Science." Natural Sciences, parts 4–6. Samara, SSU, **2002,** 47.
104. Kozlov, G. V.; Dolbin, I. V.; The physical significance and evaluation methods of low-molecular solvents structural parameters in diluted polymer solutions. Proceedings of Higher Educational Institutions, North-Caucasus region, natural sciences, **2004,** *3,* 69–71.
105. Kozlov, G. V.; Dolbin, I. V.; Zaikov, G. E.; The description of polymer dissolution as the transition nonequilibrium-quasiequilibrium system. Proceedings of V-th All-Russian Seminar "Simulation of Nonequilibrium Systems." Krasnoyarsk, KSU, **2002,** 96–97.
106. Kozlov, G. V.; Novikov, V. U.; The cluster model of polymers amorphous state. Achievements of Physical Sciences, **2001,** *171(7),* 717–764.
107. Kozlov, G. V.; Ovcharenko, E. N.; Mikitaev, A. K.; Structure of the Polymers Amorphous State. Moscow, Publishers of the Mendeleev RCTU, **2009,** 392 p.
108. Nolynskaya, A. V.; Godovskii Yu.K.; Papkov, V. S.; Dissolution heats of amorphous polymers. High-Molecular Compounds. A, **1979,** *21(5),* 1059–1063.
109. Lipatov, Yu.S.; Privalko, V. P.; Demchenko, S. S.; Besklubenko Yu.D.; Titov G. V.; Shumskii V. F. On structural "memory" of amorphous polystirene. High-Molecular Compounds. A, **1986,** 2*8(3),* 573–579.
110. Beloshenko, V. A.; Kozlov, G. V.; Lipatov, Yu.S.; Glass transition mechanism of cross-linked polymers. Physics of Solid Body, **1994,** *36(10),* 2903–2906.
111. Kozlov, G. V.; Gazaev, M. A.; Novikov, V. U.; Mikitaev, A. K.; The simulation of amorphous polymers structure as percolation cluster. Letters to Journal of Technical Physics, **1996,** *22(16),* 31–38.
112. DiBenedetto, A. T.; Trachte, K. L.; The brittle fracture of amorphous thermoplastic polymers. *J. Appl. Polymer Sci.* **1970,** *14(11),* 2249–2262.
113. Kozlov, G. V.; Temiraev, K. B.; Shustov, G. B.; The intercommunication of structure of macromolecular coil in solution with structure and properties of linear polyarylates condensed state. Proceedings of Higher Educational Institutions, North-Caucasus region, natural sciences, **1999,** *3,* 77–81.
114. Kozlov, G. V.; Temiraev, K. B.; Shustov, G. B.; Mashukov, N. I.; Modelling of solid state polymer properties at the stage of synthesis: fractal analysis. *J. Appl. Polymer Sci.* **2002,** *85(6),* 1137–1140.
115. Dubrovina, L. V.; Pavlova, S. -S. A.; Ponomareva, M. A.; Hydrodynamical and thermo-dynamical properties of polyblock copolymers solutions. High-Molecular Compounds. A, **1985,** *27(4),* 780–785.
116. Kargin, V. A.; The Selected Chapters: A Polymers Structure and Mechanical Properties. Moscow, Chemistry, **1979,** 354 p.
117. Kozlov, G. V.; Temiraev, K. B.; Malamatov, A. Kh. The genetic intercommunication of structures reaction products, condensed state of polymers and their properties. Chemical Industry, **1998,** *4,* 48–50.

118. Dubrovina, L. V.; Ponomareva, M. A.; Shirokova, L. V.; Storozhuk, I. P.; Valetskii, R. M.; The study of poly(arylatearylenesulfonoxide) block-copolymers synthesis mode influence on their some properties. High-Molecular Compounds. A, **1981,** *23(5),* 384–388.
119. Dubrovina, L. V.; Pavlova, S. -S. A.; Ponomareva, M. A.; Properties of poly(arylatearylenesulfonoxide) block-copolymers solutions. High-Molecular Compounds. A, **1983,** *25(7),* 1536–1543.
120. Shogenov, V. N.; Kozlov, G. V.; Mikitaev, A. K.; The prediction of mechanical behavior, structure and properties of film polymer samples at quasistatic tension. In book: Collection of selected chapters "Polycondensation Reactions and Polymers." Nal'chik, Publishers KBSU, **2007,** 252–270.
121. Kozlov, G. V.; Zaikov, G. E.; The generalized description of local order in polymers. In book: Fractals and Local Order in Polymeric Materials. Ed. Kozlov, G.; Zaikov, G.; New York, Nova Science Publishers, Inc.; **2001,** 55–63.
122. Hoy, K. L.; New values of the solubility parameters from vapor pressure data. J.; Paint Technol.; **1970,** *42(1),* 76–118.
123. Graessley, W. W.; Edwards, S. F.; Entanglement interactions in polymers and the chain contour concentration. Polymer, **1981,** *22(10),* 1329–1334.
124. Lin, Y. -H.; Number of entanglement strands per cubed tube diameter, a fundamental aspect of topological universality in polymer viscoelasticity. Macromolecules, **1985,** *18(12),* 3080–3083.
125. Shaboldin, V. A.; Sukhomudrenko, A. G.; Krasheninnikov, A. I.; Morozov, V. A.; The study of diluted solutions of low-molecular rubber SKN-18. High-Molecular Compounds. A, **1972,** *14(7),* 1462–1466.
126. Martin, L.; Lavin, M.; Backe, S. T.; Size execlusion chromatography of poly(ethylene terephthalate) using 0-chlorophenol. J.; Liquid Chromatogr.; **1992,** *15(11),* 1817–1830.
127. Kozlov, G. V.; Burya, A. I.; Shustov, G. B.; The influence of solvent nature on low-temperature polycondensation: the fractal analysis. Bulletin of Academy of Sciences of Kazakhstan, Engineering Sciences, **2007,** *(5–6)*, 23–32.
128. Korshak, V. V.; Vinogradova, S. V.; Vasnev, V. A.; The study of solvent nature influence on low-temperature polycondensation. High-Molecular Compounds. A, **1968,** *10(6),* 1329–1335.
129. Meakin, P.; Formation of fractal clusters and networks by irreversible diffusion-limited aggregation. *Phys. Rev. Lett.* **1983,** *51(13),* 1119–1122.
130. Kolb, M.; Botet, R.; Jullien, R.; Scaling of kinetically growing clusters. *Phys. Rev. Lett.* **1983,** *51(13),* 1123–1126.
131. Kokorevich, A. G.; Gravitis Ya.A.; Ozol-Kalnin, V. G.; The development of scaling approach at study of lignin supramolecular structure. Chemistry of Wood, **1989,** *1,* 3–24.
132. Jullien, R.; Kolb, M.; Hierarchical method for chemically limited cluster–cluster aggregation. J.; Phys. A, **1984,** *17(12),* p. L639-L643.
133. Botet, R.; Jullien, R.; Diffusion-limited aggregation with disaggregation. *Phys. Rev. Lett.* **1985,** *55(19),* 1943–1946.
134. Alexandrowicz, Z.; Kinetics of formation and mean shape of branched polymers. *Phys. Rev. Lett.* **1985,** *54(13),* 1420–1423.

135. Shogenov, V. N.; Kozlov, G. V.; Fractal Clusters in Physics-Chemistry of Polymers. Nal'chik, Polygraphservice and, T.; **2002,** 268 p.
136. Klymko, P. W.; Kopelman, R.; Fractal reaction kinetics: exciton fusion on clusters. J.; Phys. Chem.; **1983,** *87(23),* 4565–4567.
137. Kozlov, G. V.; Zaikov, G. E.; The physical significance of reaction rate constant in Euclidean and fractal spaces at polymers thermooxidative degradation consideration. Theoretical Principles of Chemical Technology, **2003,** *37(5),* 555–557.
138. Dolbin, I. V.; Kozlov, G. V.; The physical significance of reactionary medium heterogeneity for polymer solutions and melts. Reports of Adygskoi (Cherkesskoi) International Academy of Sciences, **2004,** *7(1),* 134–137.
139. Meakin, P.; Stanley, H. E.; Coniglio, A.; Witten Surfaces, T. A.; Interfaces and screening of fractal structures. Phys. Rev. A, **1985,** *32(4),* 2364–2369.
140. Kozlov, G. V.; Shustov, G. B.; Dolbin, I. V.; Structural memory of the polyarylates macromolecular coil: fractal analysis. Proceedings of XVII Mendeleev's Congress by General and Applied Chemistry "Achievements and Perspectives of Chemical Science." Kazan, 21–26 October **2003,** p. 442.
141. Kozlov, G. V.; Shustov, G. B.; Fractal analysis of the structural memory of macromolecular coil of polyarylates. Reports of Adygskoi (Cherkesskoi) International Academy of Sciences, **2007,** *9(2),* 138–141.
142. Kozlov, G. V.; Dolbin, I. V.; Zaikov, G. E.; Fractal *Physical Chemistry of Polymer Solutions*. J.; Balkan Tribological Association, **2005,** *11(3),* p. 335–373.
143. Kozlov, G. V.; Serdyuk V. D.; Dolbin, I. V.; Fractal geometry of chain and deformability of amorphous glassy polymers. Materials Sciences, **2000,** *12,* 2–5.
144. Afaunova, Z. I.; Synrgesis and Study of Polyurethanarylates. Mater's Thesis of Chemical Sciences. Moscow, MCTI, **1973,** 181 p.
145. Afaunova, Z. I.; Kozlov, G. V.; Bazheva, R. Ch. The prediction of glass transition temperature of polyurethanarylated, prepared by different modes of polycondensation. Electronic Journal "Studied in Russia." *73,* 809–813, **2001,** http: //zhurnal.ape.relarn.ru/articles/2001/073.pdf.
146. Kozlov, G. V.; Dolbin, I. V.; Zaikov, G. E.; Fractal *Physical Chemistry of Polymer Solutions*. In book: Focus on Natural and Synthesis Polymer Science. Ed. Zaikov, G.; New York, Nova Science Publishers, Inc.; **2006,** 131–175.
147. Vasnev, V. A.; Vinogradova, S. V.; Markova, G. D.; Voitekunas V. Yu. Macromolecular design in nonequilibrium polycondensation. High-Molecular Compounds. A, **1997,** *39(3),* 412–421.
148. Kozlov, G. V.; Shustov, G. B.; Zaikov, G. E.; The macromolecular coil structure influence on functional groups activity at copolycondensation. Chemical Physics and Mesoscopy, **2008,** *10(3),* 332–335.
149. Kozlov, G. V.; Shustov, G. B.; Zaikov, G. E.; The influence of macromolecular coil structure on activity of functional groups at copolycondensation. Encyclopaedia of Engineer-Chemist, **2011,** *6,* 9–12.
150. Kozlov, G. V.; Bejev, A. A.; Zaikov, G. E.; The physical reasons for the homogeneous and nonhomogeneous reactions of haloid-containing epoxy polymer curing. *J. Appl. Polymer Sci.* **2003,** *90(5),* 1202–1205.
151. Bashorov, M. T.; Kozlov, G. V.; Mikitaev, A. K.; The nanodimensional effects in epoxy polymers curing process. Chemical Technology, **2010,** *11(2),* 83–87.

152. Kozlov, G. V.; Bashorov, M. T.; Mikitaev, A. K.; Zaikov, G. E.; Transition nanoreactor-nanoparticle in epoxy polymers curing process. Chemistry and Chemical Technology, **2008,** *2(4),* 281–284.
153. Kozlov, G. V.; Bashorov, M. T.; Mikitaev, A. K.; Zaikov, G. E.; The transition nanoreactor-nanoparticle in curing process of epoxy polymers. Polymers Research, J.; **2009,** *3(1),* 95–102.
154. Kozlov, G. V.; Bashorov, M. T.; Mikitaev, A. K.; Zaikov, G. E.; The transition nanoreactor-nanoparticle in curing process of epoxy polymers. In book: Chemistry and Biochemistry. From Pure to Applied Science. New Horisons. Ed. Pearce, E.; Zaikov, G.; Kirshenbaum, G.; New York, Nova Science Publishers, Inc.; **2009,** 345–352.
155. Kozlov, G. V.; Bashorov, M. T.; Mikitaev, A. K.; Zaikov, G. E.; The nanodimensional effects in curing process of epoxy polymers in the fractal space. In book: Trends in Polymer Research. Ed. Zaikov, G.; Jimenez, A.; Monakov Yu. New York, Nova Science Publishers, Inc.; **2009,** 87–94.
156. Kozlov, G. V.; Bejev, A. A.; Lipatov, Yu.S.; The fractal analysis of curing process of epoxy resins. *J. Appl. Polymer Sci.* **2004,** *92(4),* 2558–2568.
157. Kozlov, G. V.; Bejev, A. A.; Dolbin, I. V.; Change of microgel structures at curing epoxy polymers in fractal space. J.; Balkan Tribological Association, **2004,** *10(1),* 31–35.
158. Kozlov, G. V.; Shustov, G. B.; Zaikov, G. E.; The fractal and scaling analysis of chemical reactions. *J. Appl. Polymer Sci.* **2004,** *93(5),* 2343–2347.
159. Kozlov, G. V.; Shustov, G. B.; Zaikov, G. E.; The fractal physics of the polycondensation processes. J.; Balkan Tribological Association, **2003,** *9(4),* 467–514.
160. Smirnov, B. M.; The Physics of Fractal Clusters. Moscow, Science, **1991,** 136 p.
161. Sergeev Nanochemistry, Moscow, G. B.; Book House "University", **2006,** 336 p.
162. Buchachenko, A. L.; Nanochemistry—direct way to high technologies of new sentury. Successes of Chemistry, **2003,** *72(5),* 419–437.
163. Sahimi, M.; McKarnin, M.; Nordahl, T.; Tirrell, M.; Transport and reaction on diffusion-limited aggregates. Phys. Rev. A, **1985,** *32(1),* 590–595.
164. Kozlov, G. V.; Malkanduev Yu.A.; Zaikov, G. E.; Gelation in the radical polymerization of dimethyl diallyl ammonium chloride. *J. Appl. Polymer Sci.* **2004,** *93(3),* 1394–1396.
165. Ivanova VV. S.; Kuzeev, I. R.; Zakirnichnaya, M. M.; Synergetics and Fractals. Universality of Mechanical Behavior of Materials. Ufa, Publishers USSTU, **1998,** 366 p.
166. Kozlov, G. V.; Malkanduev Yu.A.; Novikov, V. U.; Synergetics of radical polymerization of dimethyl diallyl ammonium chloride. Proceedings of International Interdisciplinary Symposium "Fractals and Applied Synergetics, FaAS-03", Moscow, Publishers MSOU, **2003,** 114–117.
167. Burya, A. I.; Kozlov, G. V.; Temiraev, K. B.; Malamatov, A. Kh. The branching influence on fractal dimension of macromolecular coil in solutions. Problems of Chemistry and Chemical Technology, **1999,** *3,* 26–28.
168. Shustov, G. B.; Temiraev, K. B.; Afaunova, Z. I.; Kozlov, G. V.; The factors, influencing on fractal dimension of branched polymers macromolecular coils in diluted solutions. Bulletion of KBSC RAS, **2000,** *2,* 92–94.
169. Temiraev, K. B.; Shustov, G. B.; Mikitaev, A. K.; Bromine—containing aromatic copolyethersulfones. High-Molecular Compounds. B, **1993,** *35(12),* 2057–2059.

170. Kozlov, G. V.; Dolbin, I. V.; Spectral dimension of macromolecular oil in solution: the theoretical evaluation. Proceedings of All-Russian Sci.-Pract. Conf. "Chemistry in Technology and Medicine." Makhachkala, DSU, **2002,** 45–48.
171. Kuchanov, S. I.; Kinetical Methods of Calculations in Polymers Chemistry. Moscow, Chemistry. Moscow, Chemistry, **1978,** 368 p.
172. Kozlov, G. V.; Dolbin, I. V.; Zaikov, G. E.; Description of the structure of branched polyphenyquinoxaline macromolecular coil within the framework of fractal analysis. *J. Appl. Polymer Sci.* **2006,** *99(6),* 3574–3577.
173. Kozlov, G. V.; Dolbin, I. V.; Zaikov, G. E.; The description of the structure of branched polyphenyquinoxaline macromolecular coil within the framework of the fractal theory. In book: Molecular and High Molecular Chemistry: Theory and Practice. Ed. Monakov Yu.; Zaikov, G.; New York, Nova Science Publishers, Inc.; **2006,** 111–118.
174. Kozlov, G. V.; Dolbin, I. V.; Zaikov, G. E.; Fractal Physical Chemistry of polymer solution. Polymer Research, J.; **2007,** *1,* (½, 167–210.
175. Kozlov, G. V.; Shustov, G. B.; Mikitaev, A. K.; The intercommunication of fractal dimension and branching factor of macromolecular coils. Bulletin of KBSC RAS, **2009,** *3,* 130–134.
176. Kozlov, G. V.; Burya, A. I.; Shustov, G. B.; The dependence of the chain branching degree on molecular weight: fractal analysis. Chemical Industry and Chemical Engineering Quarterly, **2008,** *14(8),* 181–184.
177. Magomedov, G. M.; Kozlov, G. V.; Zaikov, G. E.; Structure and Properties of Cross-Linked Polymers. Shawbury, A Smithers Group Company, **2011,** 492 p.
178. Kozlov, G. V.; Zaikov, G. E.; Mikitaev, A. K. Th Fractal Analysis of Gas Transport in Polymers. The Theory and Practical Aplications. New York, Nova Science Publishers, Inc.; **2009,** 238 p.
179. Kozlov, G. V.; Malkanduev Yu.A.; Zaikov, G. E.; The usage of poly, N.; N-dimethyl-N, N-diallyl ammonium chloride for solution of ecological problems. In book: Homolytic and Heterolytic Reactions: Problems and Solutions. Ed. Zaikov, G.; Monakov Yu.; Jimenez, A.; New York, Nova Science Publishers, Inc.; **2004,** 91–119.
180. Brady, L. M.; Ball, R. C.; Fractal growth of copper eletrodeposites. Nature, **1984,** *309(***5, 965,** 225–229.
181. Esmurziev, A. M.; Malkanduev Yu.A.; Kozlov, G. V.; The fractal analysis of dimethyl diallyl ammonium chloride flocculating ability. Bulletin of Kabardino-Balkarian State University. Chemical Sciences, **2001,** *4,* p. 101–104.
182. Kozlov, G. V.; Novikov, V. U.; The fracton conception of polymers failure. Material Science, **1997,** *8–9,* 3–6.
183. Shnaider, M. A.; Kolganova, I. V.; Ter-Minasyan, R. I.; Topchiev, D. A.; The soy-bean oil purification by polymeric flocculators. Oil-Fat Industry, **1990,** *10,* 17–21.
184. Kozlov, G. V.; Malkanduev Yu.A.; Mirzoeva, A. A.; The fractal model of flocculation process of low-molecular admixtures by polymeric flocculator. Proceedings of IV International Sci.-Pract. Conf. "The Economics of Usage and Protection of Nature." Penza, PSU, **2001,** 153–156.
185. Botet, R.; Jullien, R.; Kolb, M.; Hierarchicl model for irreversible kinetic cluster formation. J.; Phys. A, **1984,** *17(2),* p. L75-L79.
186. Witten, T. A.; Meakin, P.; Diffusion-limited aggregation at multiple growth sites. Phys. Rev. A, **1983,** *28(10),* 5632–5642.

187. Hentschel, H. G. E.; Deutch, J. M.; Meakin, P.; Dynamic scaling and the growth of diffusion-limited aggregates. J.; Chem. Phys.; **1984,** *81(5),* 2496–2502.
188. Bartenev, G. M.; Frenkel, S. Ya. Physics of Polymers. Leningrad, Chemistry, **1990,** 432 p.
189. Kurbanaliev, M. K.; Dustov, I. K.; Malkin, A. Ya. A preceding history influence of films formation from solutions on their longevity. High-Molecular Compounds. A, **1982,** *24(11),* 2291–2297.
190. Shogenov, V. N.; Belousov, V. N.; Potapov, V. V.; Kozlov, G. V.; Prut, E. V.; The description of stress-strain curves of glassy polyarylatesulfone within the framework of high-elasticity conceptions. High-Molecular Compounds. A, **1991,** *33(1),* 155–160.
191. Dolbin, I. V.; Kozlov, G. V.; The polymer films structure formation: Witten-Sander model. Bulletin of KBSC RAS, **2004,** *2,* 40–44.
192. Kozlov, G. V.; Shogenov, V. N.; Kharaev, A. M.; Mikitaev, A. K.; The temperature dependence of parameters, characterizing inelastic deformation of polymers, in the conditions of impact loading. High-Molecular Compounds. B, **1987,** *29(4),* 311–314.
193. Botet, R.; Jullien, R.; Kolb, M.; Gelation in kinetic growth models. Phys. Rev. A, **1984,** *30(4),* 2150–2152.
194. Witten, T. A.; Sander, L. M.; Diffusion-limited aggregation a kinetical critical phenomena. *Phys. Rev. Lett.* **1981,** *47(19),* 1400–1403.
195. Kozlov, G. V.; Beloshenko, V. A.; Varyukhin, V. N.; Simulation of cross-linked polymers structure as diffusion-limited aggregate. Ukrainian Physical Journal, **1998,** *43(3),* 322–323.
196. Kozlov, G. V.; Shogenov, V. N.; Mikitaev, A. K.; Local order in polymer—the description within the framework of irreversible colloidal aggregation model. Engineering-Physical Journal, **1998,** *71(6),* 1012–1015.
197. Kozlov, G. V.; Yanovskii Yu.G.; Kubica, S.; Zaikov, G. E. A nanofiller particles aggregation in elastomeric nanocomposites: the irreversible aggregation model. Przetworstwo Tworzyw, **2011,** *5,* 413–416.
198. Mandelkern, L.; The relation between structure and properties of crystalline polymers. Polymer, J.; **1985,** *17(1),* 337–350.
199. Narisawa, I.; The Strenght of Polymeric Materials. Moscow, Chemistry, **1987,** 400 p.
200. Mikitaev, A. K.; Kozlov, G. V.; Fractal Mechanics of Polymeric Materials. Nal'chik, Publishers KBSU, **2008,** 312 p.
201. Shogenov, V. N.; Kozlov, G. V.; Mikitaev, A. K.; The prediction of failure process parameters of rigid-chain polymers. High-Molecular Compounds. B, **1989,** *31(11),* 809–811.
202. Novikov, V. U.; Kozlov, G. V.; Boronin, D. V.; The fractal approach to numerical study of cross-linked polymers: computer program elaboration. Material Science, **1999,** *3,* 25–28.
203. Kozlov, G. V.; Polymer phase behavior in nanocomposites. Polymer Research, J.; **2011,** *4(2/*3, 113–159.
204. Belousov, V. N.; Kozlov, G. V.; Mashukov, N. I.; Lipatov, Yu.S. A dislocation analogies application for the description of yield process in crystallizing polymers. Reports of Academy of Sciences, **1993,** *328(6),* 706–708.

205. Kozlov, G. V.; Sanditov, D. S.; Serdyuk V. D. On suprasegmental formations type in polymers amorphous state. High-Molecular Compounds. B, **1993,** *35(12),* 2067–2069.
206. Kozlov, G. V.; Beloshenko, V. A.; Stroganov, I. V.; Lipatov, Yu.S.; The intercommunication between glass transition temperature change and cross-linked polymers structure at heat aging. Reports of National Academy of Sciences of Ukraine, **1995,** *10,* 117–118.
207. Rafikov, S. R.; Pavlova, S. A.; Tverdokhlebova, I. I.; Methods of High-Molecular Compounds Molecular Weights and Polydispersity Determination. Moscow, Publishers Academy of Sciences of SSSR, **1963,** 368 p.
208. Meakin, P.; Vicsek, F.; Dynamic cluster-size distribution in cluster–cluster aggregation: effects of cluster diffusivity. Phys. Rev. B, **1984,** *31(1),* 564–569.
209. Botet, R.; Jullien, R.; Size distribution of clusters in irreversible kinetic aggregation. J.; Phys. A, **1984,** *17(12),* 2517–2530.
210. Shiyan, A. A.; Stationary distributions of macromolecular fractals masses in diluted polymer solutions. High-Molecular Compounds. B, **1995,** *37(9),* 1578–1580.
211. Kozlov, G. V.; Malkanduev Yu.A.; Zaikov, G. E.; Molecular weight distribution of poly(dimethyl diallyl ammonium chloride): analysis within the framework of irreversible aggregation model. In book: Fractal Analysis of Polymers: From Synthesis to Composites. Ed. Kozlov, G.; Zaikov, G.; Novikov, V.; New York, Nova Science Publishers, Inc.; **2003,** 131–139.
212. Kozlov, G. V.; Malkanduev Yu.A.; Zaikov, G. E.; Molecular weight distribution of poly(dimethyl diallyl ammonium chloride): analysis within the framework of irreversible aggregation models. J.; Balkan Tribological Association, **2003,** *9(3),* 442–448.
213. Kozlov, G. V.; Malkanduev Yu.A.; Zaikov, G. E.; Molecular weight distribution of poly(dimethyl diallyl ammonium chloride): analysis within the framework of irreversible aggregation models. *J. Appl. Polymer Sci.* **2004,** *91(5),* 3140–3143.
214. Bityurin, N. M.; Genkin, V. N.; Zubov, V. P.; Lachinov, M. B. On Mechanism of gel-effect at radical polymerization. High-Molecular Compounds. A, **1981,** *23(8),* 1702–1709.
215. Topchiev, D. A.; Malkanduev Yu.A.; Korshak Yu.V.; Mikitaev, A. K.; Kabanov, V. A.; Kinetics of, N.; N-dimethyl-N, N-diallyl ammonium chloride radical polymerization in concentrated water solutions. Acta Polymerica, **1985,** *36(7),* 372–374.
216. Kozlov, G. V.; Batyrova Kh.M.; Shustov, G. B.; Mikitaev, A. K.; Molecular weight distribution of polyarylates: the fractal analysis. Proceedings of International Sci.-Pract. Conf. "New Polymer Composite Materials." Moscow, **2000,** 44.
217. Kolb, M.; Aggregation phenomena and fractal structures. Physica, A.; **1986,** *140(1–2,* 416–420.
218. Kolb, M.; Herrmann, H. J.; The sol-gel transition modeled by irreversible aggregation of clusters. J.; Phys. A, **1985,** *18(4),* p. L435-L441.
219. Vicsek, T.; Family, F.; Dynamic scaling for aggregation of clusters. *Phys. Rev. Lett.* **1984,** *52(14),* 1669–1672.
220. Kolb, M.; Unified description of static and dynamic scaling for kineti cluster formation. *Phys. Rev. Lett.* **1984,** *53(17),* 1653–1656.

221. Shamurina, M. V.; Poldugin, V. I.; Pryamova, T. D.; Vysotskii V. V.; Influence of particles modification on aggregates structure and conductivity of metal-filled film composites. Colloidal Journal, **1994,** *57(4),* 580–584.
222. Kozlov, G. V.; Malkanduev Yu.A.; Zaikov, G. E.; Fractal analysis of polymers molecular mass distribution. Dynamic scaling. J.; Balkan Tribological Association, **2003,** *9(2),* 252–256.
223. Kozlov, G. V.; Malkanduev Yu.A.; Zaikov, G. E.; Fractal analysis of polymers molecular mass distribution: dynamic scaling. *J. Appl. Polymer Sci.* **2003,** *89(9),* 2382–2384.
224. Ebeling V.; The formation of structures at irreversible processes. Moscow, Wored, **1979,** 275 p.
225. Kozlov, G. V.; Dolbin, I. V.; Zaikov, G. E.; Fractal models for polymer solutions behavior analysis. Applied Analytical Chemistry, **2012,** *3(2),* p. 18–22.
226. Filippov, A. P.; Romanova, O. A.; Vinogradova, L. V.; Molecular and hydrodynamical characteristics of star-like polystirenes with one or two molecules of fullerene C_{60} as branching center. High-Molecular Compounds. A, **2010,** *52(3),* 371–377.
227. Lezov, A. V.; Mel'nikov, A. B, Filippov, S. K.; Polushina, G. E.; Antonov, E. A.; Mikhailova, M. E.; Ryumtsev, E. I.; Hydrodynamical and conformational properties of polystirene, modified by dendrons. High-Molecular Compounds. A, **2006,** *48(3),* 508–515.
228. Muzafarov, A. M.; Vasilenko, N. G.; Tatarinova, E. A.; Ignat'eva, G. M.; Myakushev, V. M.; Obrezkova, M. A.; Meshkov, I. B.; Voronina, N. V.; Novozhilov O. V.; Macromolecular nanoobjects—perspective direction of polymers chemistry. High-Molecular Compounds. C, **2011,** *53(7),* p. 1217–1230.
229. Kozlov, G. V.; Dolbin, I. V.; Mil'man, L. D.; The Mark-Kuhn-Houwink equation fractal variant for polystirene, modified by dendrons. Proceedings of VIII International Sci.-Pract. Conf. "New Polymer Composite Materials." Nal'chik, Publishers "Print-Center", **2012,** 124–127.
230. Tsvetkov, N. V.; Andreeva, L. N.; Filippov, S. K.; Bushin, S. V.; Bezrukova M. A.; Marchenko, I. A.; Stremina, I. A.; Alyab'eva, V. P.; Girbasova N. V.; Bilibin, A. Yu. Hydrodynamical, optical and electrooptical properties of cylindrical dendrimers of third generation in chloroform and dichlorobite acid. High-Molecular Compounds. A, **2010,** *52(1),* 11–22.
231. Kozlov, G. V.; Dolbin, I. V.; Malamatov, A. Kh. A solvent structure in star-like polystirene solutions. Proceedings of VIII International Sci.-Pract. Conf. "New Polymer Composite Materials." Nal'chik, Publishers "Print-Center", **2012,** 120–123.
232. Isaacson, J.; Lubensky, T. C.; Flory exponents for generalized polymer problems. *J. Phys. Lett.* (Paris), **1980,** *41(19),* p. L469-L471.
233. Schnell, R.; Stamm, M.; Creton, C.; Direct correlation between interfacial width and adhesion in glassy polymers. Macromolecules, **1998,** *31(7),* 2284–2292.
234. Kozlov, G. V.; Dolbin, I. V.; The theoretical estimation of macromolecule gyration radius in solution: the fractal model. Proceedings of III-th International Sci.-Pract. Conf. "Young Scientifics in Solution of Science Actual Problems." Vladikavkaz, 18–20 May **2012,** 9–11.

235. Kozlov, G. V.; Dolbin, I. V.; Shogenov, V. N.; Dynamics of change of macromolecule structure in solution. Proceeding of III-th International Sci.-Pract. Conf. "New Polymer Composites Materials." Nal'chik, Publishers "Print-Center", **2012,** 128–131.
236. Obrezkova, M. A.; Vasilenko, N. G.; Myakushev, V. D.; Muzafarov, A. M.; Hydrolytic polycondensation of sodiumoxi (methyl) (dialcoxi) silans as the method of preparation of linear poly(sodiumoxi) methylsylsescvioxane. High-Molecular Compounds. B, **2009,** *51(12),* 2164–2171.
237. Chernikova, E. A.; Vasilenko, N. G.; Myakushev, V. D.; Muzafarov, A. M.; Synthesis of functional polymacromonomers with siloxane main chain. High-Molecular Compounds. A, **2004,** *46(4),* 682–691.
238. Kozlov, G. V.; Dolbin, I. V.; Shogenov, V. N.; Molecular and hydrodynamical characteristics of comb-like poly(sodiumoxi) methylsylsescvioxane. The New in Polymers and Polymer Composites, **2012,** *2,* 74–78.
239. Kozlov, G. V.; Dolbin, I. V.; Tlenkopachev, M. A.; The intercommunication of free relaxation time and macromolecule structure for dendrimers solutions. Proceedings of VI-th International Sci.-Techn. Conf. "Mathematical and Computer Simulation of Natural Scientific an Social Problems." Penza, PSU, **2012,** 133–136.
240. Kozlov, G. V.; Dolbin, I. V.; The fractal analysis of solutions of polystirene, modified by dendrons. The New in Polymers and Polymer Composites, **2012,** *3,* 40–44.
241. Filippov, A. P.; Belyaeva, E. V.; Tarabukina, E. B.; Amirova, A. I.; Properties of superbranched polymers in solutions. High-Molecular Compounds. C, **2011,** *53(7),* 1281–1292.

CHAPTER 3

THE FRACTAL ANALYSIS OF POLYMER MELTS

CONTENTS

3.1 Fractal Characterization of Polymer Melts 256
3.2 The Viscosity of Polymer Nanocomposites Melt 278
3.3 The Structural Model of Polymers Synthesis Reactions in Melt 297
Keywords 302
References 302

3.1 FRACTAL CHARACTERIZATION OF POLYMER MELTS

The polymer melts viscosity represents polymers important characteristic, restricting often their application in industrial conditions. This circumstance predetermines that large attention, which is paid to this problem study [1]. The given aspect is important particularly for hetero-chain semi-rigid-chain and rigid-chain polymers, which are characterized traditionally as having high melts viscosity, making difficult, and excluding at the same time their processing on standard industrial equipment [2].

At present a parameters number exists, with the aid of which polymer melts viscosity is estimated [3]. In the present chapter with this purpose the most simple and easily measured from them—melt flow index (MFI) will be used. Excepting the merits, noted above, it is connected with a polymer key characteristics number, namely, with molecular weight and macromolecular entanglements network density [4].

The indicated above importance of polymer melts viscosity causes the appearance of a considerable number of theoretical treatments, describing this property on the basis of either representations, mainly from the point of view of free volume [3]. In the present chapter polymer melts viscosity is treated within the framework of fractal analysis [5]. This is due to the fact, that the macromolecular coil in polymer solutions and melts is a fractal [6] that creates prerequisites for the polymer melt viscosity prediction quite at the synthesis stage. The authors of Ref. [7] demonstrated the possibility of polymer melts viscosity description and prediction within the framework of fractal analysis on the example of two polymers of different classes—aromatic polyethersulfonoformals (APESF) and high-density polyethylene (HDPE). MFI values were determined on the automatic capillary viscometer IIRT-A at temperatures and loads, listed in Table 1. The fractal dimension D_f of a macromolecular coil in solution was determined according to the Eq. (11).

TABLE 1 MFI values for APESF.

The formal blocks contents, mol. %	Load, N	Temperature, K	MFI, g/10 min
0	30	623	7.3
10	30	573	4.0
30	30	523	1.0
50	30	523	1.2
70	30	523	5.0
100	1.7	458	3.5–4.0

As it is well-known through Ref. [1], both temperature and pressure (or load), at which MFI is measured, influenced strongly on its value. Therefore, in Ref. [7] all MFI data reduction to the same temperatures and loads was performed. As standard temperature T_g+100 K for APESF and T_m+100 K for HDPE (T_g and T_m are glass transition and melting temperatures for APESF and HDPE, respectively) were accepted and as the standard load—load of 30 N. The correction was performed by multiplication by the ratio of corresponding parameters natural logarithms [1].

In Ref. [8] the polymer melt viscosity for modified HDPE at zero shear η_0 was determined as reciprocal of MFI:

$$\eta_0 \sim (\mathrm{MFI})^{-1} \tag{1}$$

Besides, in Ref. [8] within the framework of the scaling approach [9] it was demonstrated, that η_0 was a nonlinear function of the ratio (MM^3/M_e^2), where M_e was the molecular weight of macromolecule fragment between entanglements nodes. In Ref. [7] the possibility of M_e value estimation within the framework of fractal analysis was considered. As it is known through Ref. [6], the intersections number of two fractals N_{int} (in the considered case—of two macromolecular coils) with dimensions D_{f_1} and D_{f_2} can be determined according to the Eq. (13) of Chapter 1. Since for the macromolecular coils in solution $D_{f_1}=D_{f_2}=D_f$ regardless of R_g and MM, then in case d=3 the Eq. (13) of Chapter 1 is reduced to the following form [7]:

$$N_{\mathrm{int}} \sim R_g^{2D_f-3} \tag{2}$$

Since R_g and MM values are linked between themselves by the Eq. (2.8), then the Eq. (2) and Eq. (8) of Chapter 2 combination allows one to determine N_{int} value as a function of the known MM and D_f magnitudes [7]:

$$N_{\mathrm{int}} \sim MM^{(2D_f-3)/D_f} \tag{3}$$

Further it is supposed that the macromolecular entanglements (traditional macromolecular "binary hookings" [10]) network is formed by just such contacts of fractals (macromolecular coils), therefore the indicated network density ν_e will be proportional to N_{int}. The proportionality coefficient between ν_e and N_{int} values can be determined with the aid of the literary data for HDPE, ν_e value for which is equal to 0.32×10^{27} m^{-3} [11, 12]. In its turn, MM_e value can be determined according to the Eq. (149) of Chapter 2.

In Fig. 1, the dependence of ν_e, determined by the indicated mode, on formal blocks contents c_{form} for APESF is adduced. As one can see, this dependence is nonadditive and similar to the corresponding dependence $[\eta](c_{form})$ (see Fig. 48 of Chapter 2). Since the value of intrinsic viscosity $[\eta]$ is defined by MM and macromolecular coil shape, that is, by D_f [13], then such similarity was to be expected, since MM_e (or ν_e) depends on the same parameters (the Eq. (3)). In its turn, the indicated similarity assumes a certain correlation between $[\eta]$ and ν_e. This supposition is confirmed by the plot $[\eta]$ (ν_e), shown in Fig. 2.

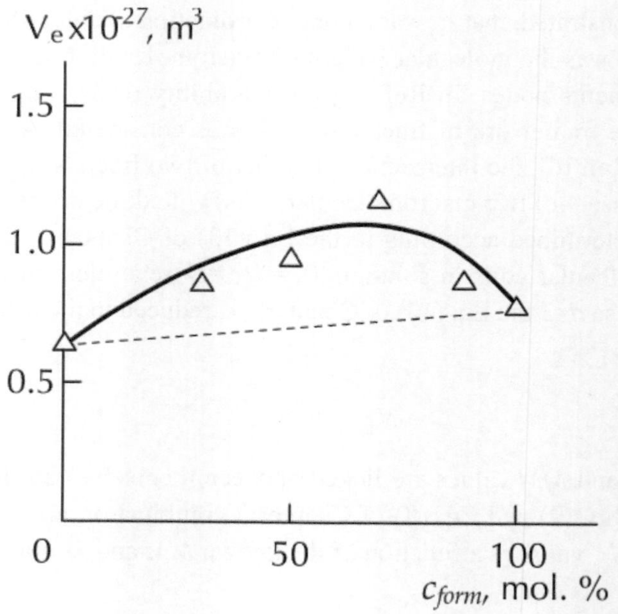

FIGURE 1 The dependence of macromolecular entanglements network density ν_e on formal blocks contents c_{form} for APESF.

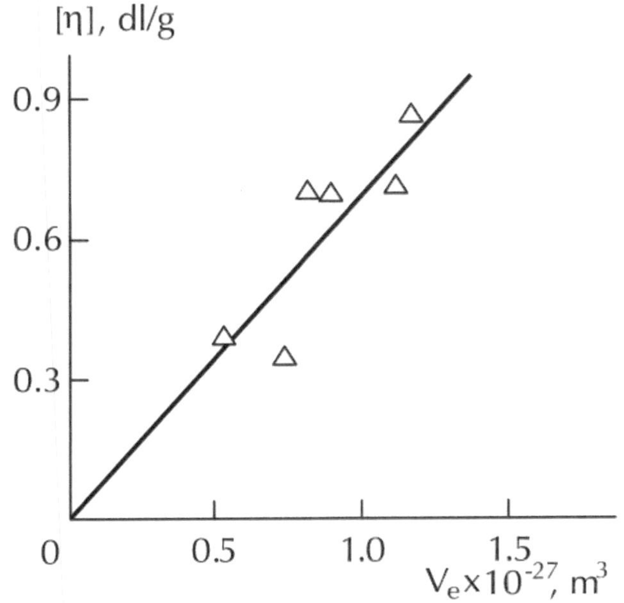

FIGURE 2 The dependence of intrinsic viscosity [η] on macromolecular entanglements network density ν_e for APESF.

ν_e increasing results in [η] growth, in addition the extrapolation to ν_e=0 gives zero value [η]. This means that at any value [η] macromolecular coils contacts are observed, which do not result to chemical interactions. In other words, the coils entanglements are formed in even diluted solutions at polymer concentration c_p=0.5%.

In Fig. 3 the dependence of η_0 on the ratio (MM^{β}/M_e^2) value for APESF and HDPE, modified by the high-disperse mixture Fe/FeO(Z) is adduced. As one can see, this dependence for the two so different polymers is approximated by one linear correlation, passing through coordinates origin. This means that the proposed in Ref. [9] approach is universal enough and can be used for polymers melt viscosity η_0 or MFI prediction according to the Eq. (1). The dependence $\eta_0(MM^{\beta}/M_e^2)$, shown in Fig. 3, can be written analytically as follows [7]:

$$\eta_0 = 3.04\left(MM^3 / MM_e^2\right), \text{ 10 min/g.} \tag{4}$$

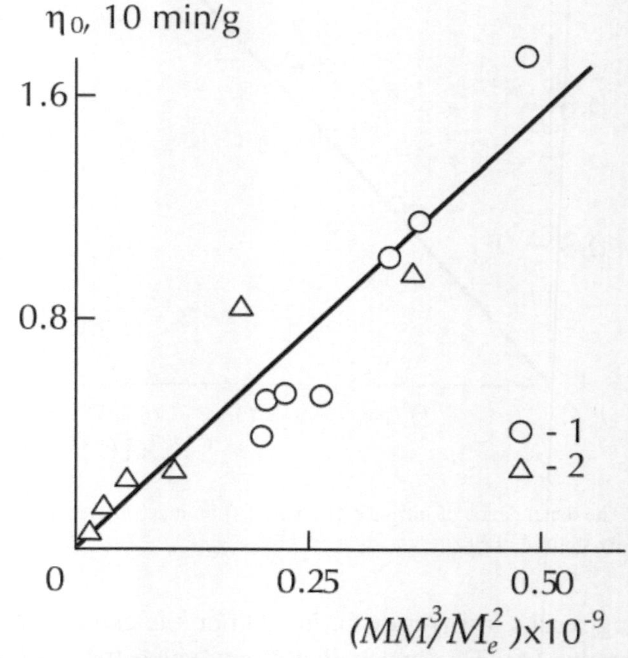

FIGURE 3 The dependence of melt viscosity η_0 on parameter (MM^3/M_e^2) value for HDPE+Z (1) and APESF (2).

The comparison of experimental values MFI and calculated according to the Eqs. (1) and (4) MFI^T magnitudes of melt flow index for APESF and HDPE+Z is adduced in Table 2. As it follows from the adduced comparison, the theory and experiment good correspondence is obtained in both qualitative and quantitative aspects. Hence, *MM* and MM_e values prediction techniques availability allows one to predict MFI (or η_0) value and the mentioned above corrections usage allows to estimate these parameters at different temperatures and loads.

TABLE 2 The comparison of experimental MFI and theoretical MFI^T melt flcw index values of APESF and modified HDPE.

Polymer	The formal blocks contents, c_{form}, mol. %	The modifier Z contents, mass. %	MFI, g/10 min	MFI^T, g/10 min
APESF	0	—	7.3	7.70
	10	—	4.0	3.10
	30	—	1.0	0.95
	50	—	1.2	1.92
	70	—	5.0	6.15
	100	—	3.5–4.0	6.40
HDPE+Z	—	0	2.50	1.90
	—	0.01	2.50	2.06
	—	0.03	1.25	1.34
	—	0.05	0.71	0.92
	—	0.07	1.11	1.19
	—	0.20	2.50	2.15
	—	1.0	3.33	2.26

Let us make in conclusion one comment. Quite many conceptions exist, defining polymer melt viscosity only as a function of *MM*. For example, in Ref. [14] the relationship, received within the framework of the thermofluctuation concept of macromolecular entanglements network, was proposed:

$$\eta_0 \sim MM^{3.46 \pm 0.05} \tag{5}$$

However, there are several methods (for example, described in Ref. [8]) of MM_e variation without *MM* change. This results in essential η_0 growth at constant molecular weight of polymer [8]. If to account, that ν_e weakly (approximately to the 0.25th power) increases with *MM* growth, then as a matter of fact the Eq. (4) gives the same exponent in the scaling dependence of η_0 on *MM*, as in the Eq. (5). However, if M_e value changes independently on *MM*, then the Eq. (5) will give an incorrect result.

Thus, the results stated above have shown the possibility of polymer melts viscosity prediction within the framework of scaling approach [9]. Let us note that the parameters necessary for this are determined just at synthesis stage. Therefore, the properties of both polymer melt and polymer in solid-phase state can be predicted and simulated, proceeding from the separate macromolecular coil structure.

As it was noted above, at present it becomes clear, that polymers in all their states and on different structural levels are fractals [16, 17]. This fundamental notion in principle changed the views on kinetics of processes, proceeding in polymers. In case of fractal reactions, that is, fractal objects reactions or reactions in fractal spaces, their rate ϑ_r with time t reduction is observed, that is expressed analytically by the Eq. (106) of Chapter 2. In its turn, the heterogeneity exponent h in the Eq. (106) of Chapter 2 is linked to the effective spectral dimension d_s' according to the following simple equation [18]:

$$d_s' = 2(1-h) \tag{6}$$

The experimental studies of reactions in polymers confirmed the correctness of the approach [18] as a whole and the Eq. (106) of Chapter 2 and Eq. (6) in particular [19]. The authors [20, 21] elaborated the dimension d_s' determination technique and verified its correctness on the example of two polymers melt: polyarylate (PAr) and block-copolymer polyarylatearylenesulfoxide (PAASO). These polymers polycondensation mode and their main characteristics (glass transition temperature T_g, mean weight molecular weight $\overline{M}_w$ and thermooxidative degradation rate k_r) are adduced in Table 3.

TABLE 3 The main characteristics of PAr and PAASO.

Polymer	Polycondensation mode	Conditional sign	T_g, K	$\overline{M}_w \times 10^{-3}$	$k_r \times 10^{-4}$*, s^{-1}
PAr	Interfacial	Par	462	69	0.17
PAASO	Low-temperature	PAASO-1	472	76	0.19
	High-temperature	PAASO-2	474	64	0.24

Footnote: * at temperature 623 K.

As it has been shown in Ref. [22], at chemical reactions on fractal objects study corrections on small clusters in system availability are necessary. Just such corrections require using in theoretical estimations not generally accepted spectral dimension d_s [23], but its effective value $d_s^{'}$ application. For percolation systems two cases are possible [24]:

(a) random walk is placed on the greatest percolation cluster of system, characterized by the dimension d_s;
(b) random walk can lit any cluster, including cluster of a small size. Such clusters set is characterized by the dimension $d_s^{'}$.

Let us consider, which of the two indicated cases is applicable to polymer melt. As it has been shown in Ref. [25], an amorphous polymer structure in glassy state represents percolation cluster, percolation threshold of which on temperature scale coincides with polymer glass transition temperature T_g. Such percolation cluster formation is defined by the local order domains availability in polymers at temperatures $T<T_g$. At $T>T_g$ such domains system is disintegrated [26] and a polymer structure is not a percolation system any more. The question arises about parameters, characterizing polymer melt structure. PAr and PAASO thermooxidative degradation process study was performed within the range of temperatures T=573–723°K and the range of glass transition temperatures of these polymers makes up 462–474 K (Table 3). As it is known through Ref. [27], the temperature of the so-called transition "liquid 1–liquid 2" T_{ll} can be estimated as follows:

$$T_{ll} \approx (1.20 \pm 0.05) T_g \tag{7}$$

From the adduced above T_g values and the Eq. (7) the condition $T>T_{ll}$ follows. At T_{ll} the polymer melt transition from "liquid with fixed structure" (where residual structural ordering is observed [27]) to truly liquid state or "structureless liquid" is observed [28]. Nevertheless, "structure absence" of melt at $T>T_{ll}$ is related to supramolecular structure absence, but macromolecular coil structure in melt remains an important structural factor (as a matter of fact, the only one at $T>T_{ll}$). Thus, a polymer melt structure can be considered as a set of separate clusters (macromolecular coils) large number and an oxidant (for example, oxygen) molecule (atom) trajectory in thermooxidative degradation process on such structure is simulated by random walk. In this case for d_s' evaluation the equation, received in Ref. [24], is applicable:

$$d_s' = d_s\left(2 - \frac{d}{d - \beta/\nu}\right), \tag{8}$$

where d is dimension of Euclidean space, in which a fractal is considered, β and ν are static percolation indices.

As it is known through Ref. [29], the fractal (Hausdorff) dimension Δ_f can be determined as follows:

$$\Delta_f = d - \frac{\beta}{\nu} \tag{9}$$

Besides, in Ref. [6] it has been shown that for polymer melts Δ_f value is determined as follows:

$$\Delta_f = \frac{d_s(d+2)}{2} \tag{10}$$

The Eqs. (8)–(10) combination allows to obtain the final formula, supposing d_s' calculation according to Δ_f known values [21]:

$$d_s' = \frac{2(2\Delta_f - d)}{d+2} \tag{11}$$

Let us consider some characteristic Δ_f values at d=3 and corresponding to them d_s' magnitudes. At transition from fractal objects to Euclidean ones (Δ_f=d) the value d_s'=1.2. As it is known through Ref. [30], for the last d_s=2, that is, jump-like d_s' change is observed at the indicated transition. For linear polymers the minimum value d_s=1.0 [23] and from the Eq. (10) let us obtain Δ_f=2.5. In this case d_s'=0.8. And at last, at Δ_f=1.5 (permeable macromolecular coils [31]) d_s'=0 [20].

For d_s' values calculation according to the Eq. (11) macromolecular coils in melt fractal dimension Δ_f preliminary evaluation is necessary. Such evaluation was performed according to the technique [20]. The formal kinetics of chemical reactions can be described by the following equation:

$$\frac{dQ}{dt} = k_r(1-Q), \tag{12}$$

where Q is reaction completion degree, t is its duration, k_r is reaction rate constant.

For the description of the chemical reactions kinetics the general fractal Eq. (79) of Chapter 2 is also used. Differentiating the Eq. (79) of Chapter 2 by time t and equating the derivative dQ/dt with the similar derivative in the equation (3.12), let us obtain the Eq. (112) of Chapter 2, in which the constant c_1 can be determined from boundary conditions. In thermo-oxidative degradation conditions Q value is determined as the ratio $N_{O_2}/N_{O_2}^{\infty}$, where N_{O_2} is consumed oxygen amount at an arbitrary time moment t, $N_{O_2}^{\infty}$ is oxygen amount, necessary for complete oxidation of aliphatic and aromatic groups of the studied polymers ($N_{O_2}^{\infty} \approx 25$ moles O_2/basic moles [20]).

The experimental values of heterogeneity exponent h were calculated according to the kinetic curves $N_{O_2}(t)$. The value k_{r_i}, corresponding to time moment t_i, was determined according to the equation [21]:

$$k_{r_i} = \frac{N_{O_2}^{i+1} - N_{O_2}^{i}}{t_{i+1} - t_i} \tag{13}$$

Then by the dependences $k_{r_i}(t_i)$ in double logarithmic coordinates construction the value h can be determined from their slope and the experimental values of effective spectral dimension d_s' can be calculated according to the Eq. (6). In Fig. 4 the comparison of theoretical values d_s' ($(d_s')^T$), calculated according to the Eq. (11), and evaluated according to the described method d_s' magnitudes is adduced. As one can see, a theory and experiment good correspondence is obtained (the mean discrepancy of $(d_s')^T$ and d_s' values makes up ~ 4.5% and maximum one ~ 9.3%).

Thus, theoretical technique of effective spectral dimension d_s' estimation was elaborated for polymer melts. The experimental and theoretical values of this parameter comparison has shown their good correspondence. d_s' theoretical evaluation is the first step for elaboration of computer technique for chemical reactions in polymer melts prediction.

As it has been shown above, proceeding of reactions in heterogeneous mediums, to which the polymers condensed state structure should be attributed [16, 17], does not obey to classical laws of the chemical reactions description [18, 22]. This distinction is due to structural features of the indicated polymers state [19]. So, within the simplest treatment thermooxidative degradation can be presented by a biomolecular reaction [22]:

$$A + B \rightarrow \text{oxidation product}, \tag{14}$$

where A is a polymer macromolecule, B is an oxygen molecule.

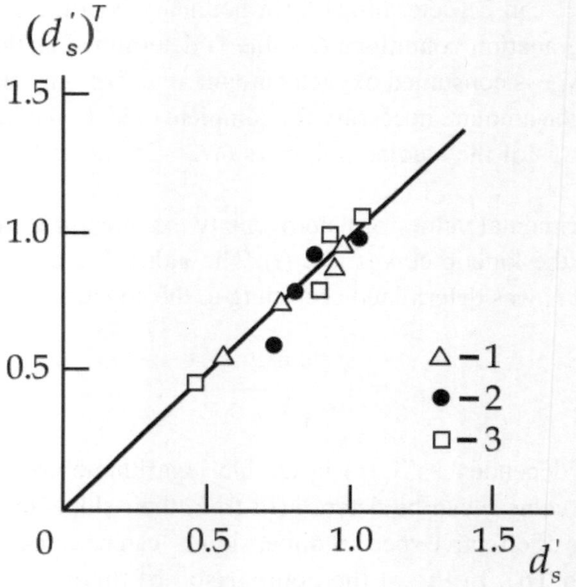

FIGURE 4 The relationship between experimental d_s' and calculated according to the Eq. (11) $(d_s')^T$ effective spectral dimension values for PAASO-1 (1), PAASO-2 (2) and PAr (3). The straight line shows the ratio 1:1.

Such reaction rate is described by the Eq. (106) of Chapter 2 and the parameters h and d_s' intercommunication gives the Eq. (6). The effective spectral dimension d_s' is linked to spectral (fracton) dimension d_s according to the subordination theorem as follows [22]:

$$d_s' = \beta_j d_s, \tag{15}$$

where β_j is the parameter, characterizing jumps times distribution (for example, oxidant molecules).

From the said above it follows, that for prediction of the rate of chemical reactions at all and thermooxidative degradation in particular it is necessary to be able to predict h value as a function of polymer melt characteristics and for this problem solution in the first place the parameter h physical significance should be elucidated. The authors [32] studied these two questions on the example of thermooxidative degradation of two heterochain polymers—PAr and PAASO.

As it has been shown above, d_s' value for polymers melt can be determined according to the Eq. (3.11) and from the Eqs. (6) and (11) let us obtain the combination [32]:

$$h = \frac{2(d+1-\Delta_f)}{d+2} \tag{16}$$

For three-dimensional Euclidean space the Eq. (16) is simplified up to [32]:

$$h = 0.4(4-\Delta_f) \tag{17}$$

For h values calculation according to the Eqs. (16) and (17) the preliminary calculation of macromolecular coil in melt fractal dimension Δ_f is necessary, that can be performed with the aid of the Eq. (112) of Chapter 2.

Thus, from the said above it follows, that thermooxidative degradation process of PAr and PAASO melts proceeds in the fractal space with dimension Δ_f. In such space degradation process can be presented schematically as "devil's staircase" [33]. Its horizontal sections correspond to temporal intervals, where the reaction does not proceed. In this case the degradation process is described with fractal time t using, which belongs to Cantor's set points [34]. If the reaction is considered in Euclidean space, then time belongs to real numbers sets.

For the evolutionary processes with fractal time description the mathematical calculus of fractional differentiation and integration is used [34]. As it has been shown in Ref. [35], in this case the fractional exponent

ν_{fr} coincides with fractal dimension of Cantor's set and indicates system states fractions, surviving during entire evolution. Let us remind, that Cantor's set is considered in one-dimensional Euclidean space (d=1) and therefore its fractal dimension d_f<1 by virtue of fractal definition [33]. For fractal objects in Euclidean spaces with higher dimensions (d>1) as ν_{fr} d_f (in the considered case—Δ_f) fractional part should be accepted according to the Eq. (74) of Chapter 2. Then ν_{fr} value characterizes a fractal (macromolecular coil) fraction, surviving in the degradation process.

In Fig. 5 the dependence of h on fractional exponent ν_{fr} for PAr and PAASO is shown, from which the linear correlation between these parameters follows. Hence, the heterogeneity parameter h physical significance can be defined as follows: the macromolecular coil structure heterogeneity is defined by this coil fraction, surviving in chemical reaction process. This means that the fractal object degradation process includes memory effects [36]. From the plot of Fig. 3.5 the analytical relationship between h and ν_{fr} can be obtained [32]:

$$h = 0.2 + 0.4\nu_{fr} \tag{18}$$

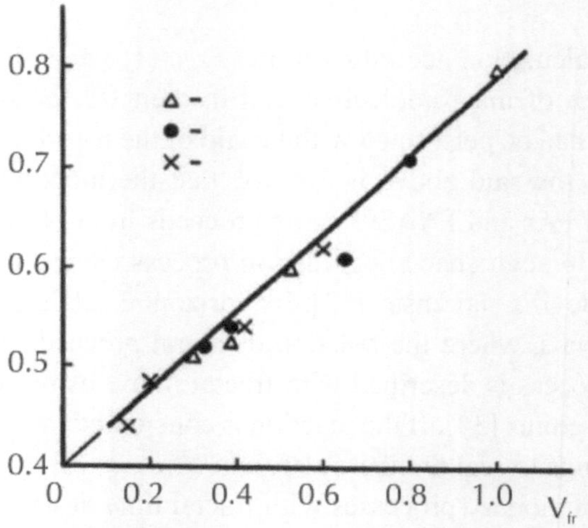

FIGURE 5 The dependence of heterogeneity parameter h on fractional exponent ν_{fr} for PAr (1), PAASO-1 (2) and PAASO-2 (3).

Let us consider the limiting cases for h. At h=0 ν_{fr}=-0.5 or, as it follows from the Eq. (74), Δ_f=1.5. The last Δ_f value defines the so-called "permeable" macromolecular coil, completely open for diffusion of various kinds of diffusates, including similar macromolecular coils [31]. As it has been noted above, this case corresponds to the classical behavior (h=0, k_r=const) [22]. At h=1.0 Δ_f=4, that is the upper critical dimension for separate macromolecular coil. In the last case the coil is Gaussian one (phantom one), that is, for it the excluded volume effects are absent [6].

The Eq. (16) allows to calculate Δ_f limiting values, that is, Δ_f for h=0 and Δ_f for h=1.0 for the case of Euclidean spaces with arbitrary dimension d. In table 3.4 Δ_f limiting values for d=1, 2, 3, 4, 5, 6 and 12 are adduced. As it follows from the data of this table, Δ_f and Δ_f values are a function of d and these dependences can be written analytically as follows [32]:

$$\Delta^0_f = d + 1, \tag{19}$$

$$\Delta^1_f = \frac{d}{2} \tag{20}$$

TABLE 4 The values of limiting fractal dimensions Δ^0_f and Δ^1_f of a macromolecular coil as a function of dimension d of surrounding Euclidean space.

d	Δ^0_f	Δ^1_f
1	2	0.5
2	3	1.0
3	4	1.5
4	5	2.0
5	6	2.5
6	7	3.0
12	13	6.0

Thus, the results stated above allow elucidating the heterogeneity parameter h physical significance: it is a linear function of a macromolecular coil fraction, surviving in chemical reaction process, realized in polymer melt [37]. The relationship (the Eq. (16)) was obtained, allowing *h* value theoretical calculation and showing its dependence on two dimensions: a macromolecular coil fractal dimension and dimension of surrounding Euclidean space. The conditions of the heterogeneity parameter limiting values (h=0 and h=1.0) realization were considered [32, 37].

As it is known through Ref. [38, 39], two main types of polymers oxidation kinetic curves exist: the curves of autoaccelerated type with clearly expressed induction period and the curves, exponentially reducing with time (of autodecelerated type). It is also supposed [40], that the sigmoid kinetic curves by their physical nature are two first types combination. Emanuel [40] assumed that the kinetic curve type was defined by polymer chemical composition and testing temperature. Thus, the autoaccelerated type curves are typical for polymers, containing C-H bonds at temperatures lower than ~ 473°K and the curves of autodecelerated type—for different carbo- and heterochain polymer oxidation at temperatures above 473°K. However, the authors [41], defining perfectly the structure role in a solid-phase polymers stabilization, underestimated this role for polymer melts case. Proceeding from the said above, the authors [42–44] pursued three aims. Firstly, it has been shown that the transition of kinetic curves at polymers thermooxidative degradation from autoaccelerated oxidation regime to an autodecelerated one is limited by the polymer physical structure. Secondly, the analytical condition of this transition was obtained within the framework of fractal analysis. And at last, it was demonstrated, that the kinetic curve type was defined by the space type (fractal or Euclidean ones), in which a thermooxidative degradation process proceeds. Let us consider these questions in more detail.

In Figs. 6 and 7 the temporal dependences of consumed oxygen amount $N_{O_2}(t)$ and a ruptures number per the initial macromolecule s(t), respectively, characterizing the thermooxidative degradation kinetics for PAA-SO (see Table 5) and polycarbonate (PC) (Table 6) are adduced. As one can see, in both cases the transition from autodecelerated (or sigmoid one) regime to autoaccelerated one is realized at the same heat aging temperature and invariable polymer chemical composition. The data comparison of Figs. 6 and 7, on the one hand, and Tables 5 and 6, on the other hand, shows that this transition is associated with the increase of the fractal dimension Δ_f of macromolecular coil in melt that supposes its purely structural nature [42].

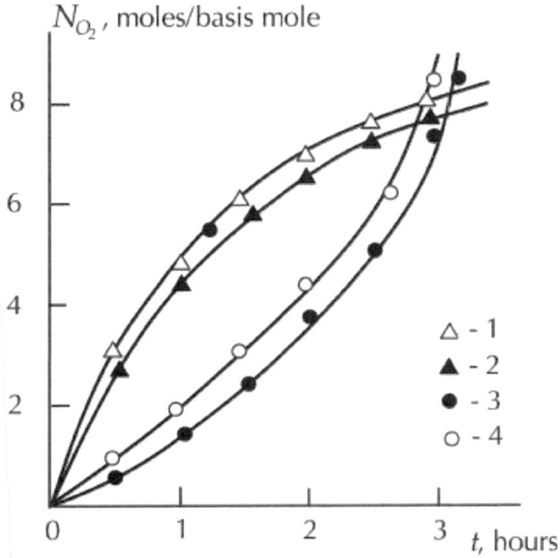

FIGURE 6 The dependences of consumed oxygen amount N_{O_2} on time t at T=623 K for PAASO-1 (1), PAASO-2 (2), PAASO-3 (3) and PAASO-4 (4).

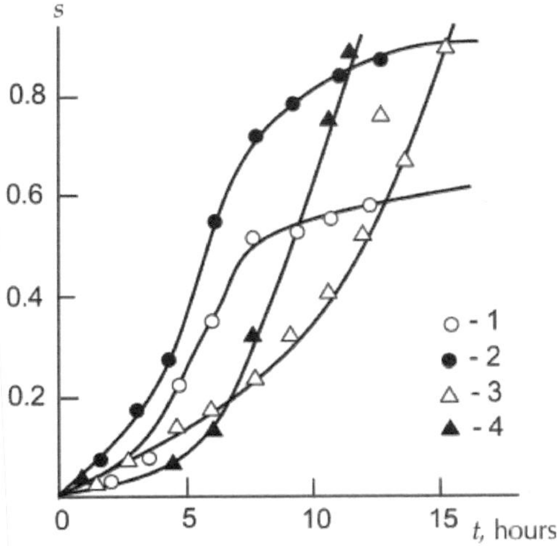

FIGURE 7 The dependences of ruptures number's per one macromolecule on thermooxidative degradation duration t at T=513 K for PC films, prepared from solutions in methylene chloride (1), chloroform (2), 1,4-dioxane (3) and tetrahydrofuran (4).

Δ_f growth means a macromolecular coil compactness enhancement [6] and, hence, oxygen diffusion in its internal regions weakening [38]. As the certain Δ_f value (Δ_f=2.54–2.69 for PC and 2.72–2.84 for PAASO) the kinetic curve type change occurs and, hence, polymer oxidation chemical mechanism [45, 46]. The authors [47] performed the oxidation curves scaling analysis and showed that the initial parts of oxidation kinetic curves of autoaccelerated and sigmoid types described oxidation proceeding in the identical conditions—in fractal space with the dimension, equal to the dimension of macromolecular coil surface. The further course of kinetic curve is defined by the type of space, in which oxidation proceeds: transition in Euclidean space gives sharp (practically linear) oxidation acceleration and transition in the fractal space with dimension, equal to macromolecular coil dimension,—sharp oxidation deceleration. Proceeding from this analysis, it can be supposed, that the observed kinetic curve type change occurs in the case, when the macromolecule reactive sites number on coil surface N_{surf} and its volume V_{vol} becomes equal [43]:

$$N_{surf} = N_{vol} \tag{21}$$

It has been shown above, that the fractional exponent ν_{fr}, determined by the Eq. (74) of Chapter 2, characterizes a fractal (macromolecular coil) states fraction, surviving in degradation process. Then the macromolecular coil fraction β_d, disintegrating in degradation process, is determined as follows [48]:

$$\beta_d = 1 - \nu_{fr} = 1 - \left[\Delta_f - (d-1)\right] = d - \Delta_f, \tag{22}$$

or, since in the considered case d=3 [48]:

$$\beta_d = 3 - \Delta_f \tag{23}$$

TABLE 5 The polycondensation mode and macromolecular coil fractal dimensions for PAASO.

Conditional sign	Polycondensation mode	Δ_f
PAASO-1	Low-temperature	2.60
PAASO-2	High-temperature	2.64
PAASO-3	Interfacial	2.84
PAASO-4	Emulsive	2.78

TABLE 6 The fractal dimension of macromolecular coil for PC samples, prepared from different solvents.

Solvent	Δ_f
Methylene chloride	2.45
Chloroform	2.54
Tetrahydrofuran	2.60
1,4-dioxane	2.74

Hence, the real value of reactive sites N_{vol} per volume of macromolecular coil can be written as follows [43]:

$$N_{vol} = N_m(3-\Delta_f) - N_{surf}, \tag{24}$$

where N_m is a reactive sites number per one macromolecule at the condition of their complete accessibility. It is obvious, that N_m value is defined by polymer chemical composition.

The parameters N_{vol} and N_{surf} relation is controlled by the fractal object volume V_{fr} and surface S_{fr} relation, which is equal to [49]:

$$\frac{V_{fr}}{S_{fr}} = \frac{R}{\Delta_f}, \tag{25}$$

where R is the fractal object radius, in the considered case equal to the macromolecular coil gyration radius R_g.

In its turn, R_g value for the considered polymers can be determined according to the following equations: for PAASO in chloroform [44]:

$$\langle R_g^2 \rangle = 0.211\times10^{-8}\overline{M}_w^{0.73}, \tag{26}$$

and for PC in chloroform [50]:

$$\langle R_g^2 \rangle = 0.988\times10^{-8}M_\eta^{0.50}, \tag{27}$$

where $\overline{M}_w$ and M_η are mean weight and mean viscosimetric molecular weight, respectively. In the equations (3.26) and (3.27) R_g values are obtained in nm.

The equations (3.21), (3.24) and (3.25) combination allows to obtain structural criterion of oxidation kinetic curves transition from autodecelerated (sigmoid) regime to autoaccelerated one [42–44]:

$$\frac{\Delta_f^{cr}}{d-\Delta_f^{cr}}=\frac{R_g}{2}, \quad (28)$$

where Δ_f^{cr} is Δ_f critical value at the indicated transition, d is dimension of Euclidean space, in which a fractal is considered and R_g is given also in nm.

Let us note two important features of the Eq. (28). Firstly, in this equation the parameter N_m is absent, that is, Δ_f^{cr} value is independent on polymer chemical composition. Secondly, R_g increase results in Δ_f^{cr} enhancement. This corresponds completely to assumption, served as the basis for the Eq. (28) derivation: N_{vol} value grows proportionally to R_g cube and N_{surf}—R_g quadrate, that is, R_g increasing results in the ration V_{fr}/S_{fr} enhancement.

Let us consider the boundary conditions for the Eq. (28). For Δ_f=0, that is, for a point object, R_g=0 and the criterion of Eq. (28) is valid for a zero-dimensional object. For Δ_f=d=3 $R_g \to \infty$ or, as it was to be expected, for Euclidean object measurement scale is of no importance [5].

The values Δ_f^{cr}, calculated according to the Eq. (28) with the Eqs. (26) and (27) using, proved to be equal to 2.76 for PC and 2.78—for PAASO that corresponds excellently to the indicated above Δ_f intervals for transition [42].

The simplest and most reliable method of confirmation of the fact, that the macromolecular coil fractal dimension Δ_f is unequivocal structural characteristics of polymer melt, is proposed in Ref. [48]. It has been shown earlier [50], that the dependences N_{O_2}/(1-K) (where N_{O_2} is the consumed oxygen amount, K is crystallinity degree) on thermooxidative degradation duration t for polypropylene are transformed in straight lines. The parameter (1–K) defines that structure part of semicrystalline polypropylene, in which the thermooxidative degradation processes are realized (amorphous phase). For PAr and PAASO melt the structure is characterized by a macromolecular coil fractal dimension Δ_f. As it has been shown above, the macromolecular coil fraction β_a, disintegrated in degradation process, is determined according to the Eq. (23). Thus, it should be assumed, that just parameter β_d should be used for N_{O_2} value normalization by analogy with the parameter (1–*K*) in case of solid-phase semicrystalline polymers. In Figs. 8 and 9, the dependences $N_{O_2}/(3-\Delta_f)$ on *t* for PC and PAASO, respectively, are adduced. The polyarylates on the basis of dichloranhydride 1,1-dichloro-2,2-di(n-carboxyphenyl) ethylene and diane, obtained by low-temperature (PAr-1), high-temperature (PAr-2) and interfacial (PAr-3) polycondensation. The adduced in Figs. 8 and 9 dependences can be linearized only at using as abscissa of $t^{1/2}$ value. This means that the chain breakage reaction order for PAr and PAASO melts is equal to 1/2. Different slopes of linear plots in Fig. 8 are due to different reaction rate constants [50]. The most important consequence of the linear plots, adduced in Figs. 8 and 9, is the fact that their linearization is reached by the fractal parameter $(3-\Delta_f)$ using. This means that polymer melt structure is described with the aid of the macromolecular coil fractal dimension Δ_f as unequivocally as semicrystalline solid-phase polymer structure with the aid of its crystallinity degree *K* [50].

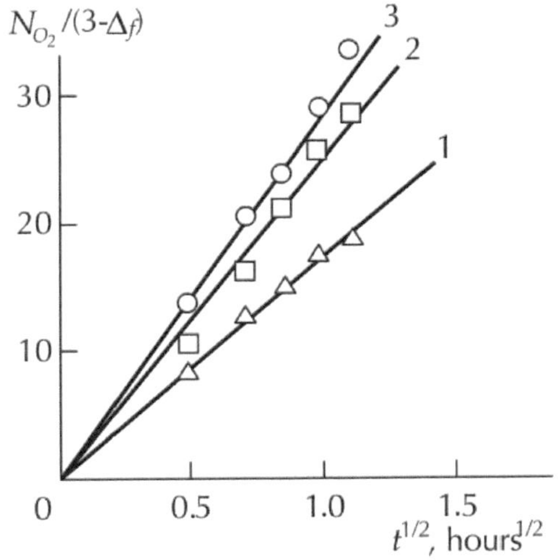

FIGURE 8 The dependences of normalized consumed oxygen amount $N_{O_2}/(3-\Delta_f)$ on thermooxidative degradation duration *t* at temperature 623°K for PAr-1 (1), 673°K for PAr-2 (2) and 723°K for PAr-3 (3).

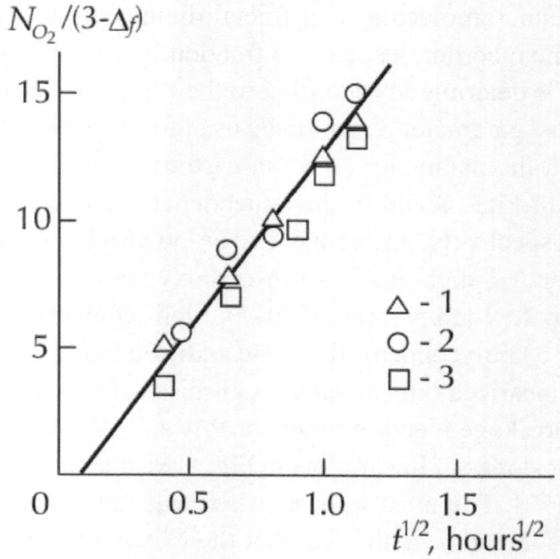

FIGURE 9 The dependences of normalized consumed oxygen amount $N_{O_2}/(3-\Delta_f)$ on thermooxidative degradation duration t at temperature 623°K for PAASO-1 (1), 673°K for PAASO-2 (2) and 723°K for PAASO-3 (3).

Between the plots of Figs. 8 and 9 the important distinction exists. The straight lines for PAr are passed through coordinates origin, whereas the straight line for PAASO is displaced by the abscissas axis. The similar effect was observed in Ref. [50] for atactic and isotactic polypropylene. The straight line for the last one was displaced by some time gap t_0. This displacement was explained by chain linear breakage on crystalline phase elements, playing the noticeable role in the very reaction beginning, when free radicals, participating in oxidation reaction, are captured by crystalline formations, in which oxygen is absent completely or its concentration is small. Too small badly formed crystallites can not capture free radicals, however at sizes and ordering enhancement of crystallites their role as free radicals traps grows, as a result of that the oxygen consummation rate decreases and the polymer oxidation induction period increases [50].

For the considered polymers (PAr and PAASO) this distinction can be explained by the following circumstance. As it has been shown in Ref. [51], the thermooxidative degradation activation energy at the same Δ_f is higher for PAASO than for PAr because of steric hindrances availability in PAASO chain, which can be substituents of various kinds, side branching, substituents in different positions (para-, orto-, metha-) and so on. In this case the indicated factor influence extent can be determined with the aid of Gammet's constant [52]. The indicated steric hindrances can be free radicals captures, just as in case of polypropylene, that decreases oxygen consummation rate and increases oxidation induction period of PAASO melts in comparison with PAr [48].

Thus, the results stated above have shown that the macromolecular coil fractal dimension is a parameter, unequivocally characterizing polymer melt structure. The steric hindrances in polymer chain can play the role of the traps for free radicals, creating the effect, similar to the effect of inhibitor presence.

The authors of Ref. [53] have shown, that frictional properties of fractal clusters can be different essentially for the usual results for compact (Euclidean) structures. It is known through Ref. [54], that the polymer melt structure can be presented as a macromolecular coils sets, which are fractal objects. Therefore, the authors [55] proposed general structural treatment of polymer melt viscosity within the framework of fractal analysis, using the model [53]. Within the framework of the indicated model the derivations for translational friction coefficient f(N) of clusters from N particles in three-dimensional Euclidean space were received, calculated according to Kirkwood-Riseman theory in the presence of hydrodynamical interaction between the cluster particles. The fundamental relationship of this theory is the following equation [53]:

$$\vec{F}_i + \zeta_0 \sum_{\substack{j=1 \\ (j\neq 1)}}^{N} \vec{T}_{ij} \cdot \vec{F}_j = \zeta_0 \vec{U}_i, \quad i=1, \ldots, N, \tag{29}$$

where $\vec{F}_i$ is the force exerted by *i*-th particle on the solvent, ζ_0 is the friction coefficient of each particle of cluster, $\vec{U}_i$ is the velocity of the solvent at *i*-th particle. In the Eq. (29) $\vec{T}_{ij}$ is the hydrodynamical interaction tensor.

In the nonfree draining limit where the hydrodynamical interaction will exclude solvent from the cluster interior one can proceed to the expression for polymer melt, which looks like [53]:

$$\ln\left[f(N)\right] = \ln c + \beta_f \ln N, \tag{30}$$

where N is a particles number in cluster.

Let us consider the Eq. (30) application for the description of melt viscosity of HDPE, modified by the high-disperse mixture Fe/FeO(Z). The compositions HDPE+Z melt viscosity was defined as reciprocal of MFI, that is, according to the Eq. (1). The coefficient c in the Eq. (30) was determined as follows [53]:

$$c = \left(a\rho_{fr}^{1/\Delta_f}\right)^{-1}, \tag{31}$$

where a is radius of the particle, forming cluster, ρ_{fr} is cluster density, Δ_f is the fractal dimension of macromolecular coil in melt.

Since $f(N)$ is calculated in relative units, than a=1=constant is accepted. Δ_f value is accepted equal to the fractal dimension d_f of polymer structure in solid-phase state, determined according to the Eqs. (151) and (152) of Chapter 2. The cluster density ρ_{fr} was calculated according to the Eq. (139) of Chapter 2, in which the macromolecular coil gyration radius R_g is linked to monomer links N_{pol} number in it (polymerization degree) by the Eq. (36) of Chapter 2.

The comparison of experimental (η_0=MFI^{-1}) and calculated according to the Eqs. (36), (151), (152) and (139) of Chapter 2, and Eqs. (30) and (31) as $f(N)$ values of compositions HDPE+Z melt viscosity has shown their good correspondence. Thus, the polymer melt viscosity η_0 grows at Δ_f reduction or macromolecular coil size R_g increase.

3.2 THE VISCOSITY OF POLYMER NANOCOMPOSITES MELT

It has been shown earlier [56] that HDPE modification by epoxy polymer ED-20 (EP) allows to obtain compositions, possessing an interesting properties number. So, at EP content c_{EP}=2.0–2.5 mass. % the elasticity modulus E maximum is observed at testing temperature T=293°K, moreover the increase makes up more than 30% in comparison with the initial HDPE. This stiffness maximum of solid-phase composition is accompanied by MFI extreme growth almost in three times in comparison with the initial HDPE, which corresponds to the similar melt viscosity reduction, that is, by the essential improvement of compositions HDPE-EP processing. The indicated compositions can be considered as specific polymer-polymer nanocomposites by virtue of two reasons: firstly, epoxy polymer particles sizes make up nanometer order and, secondly, the elasticity modulus of cross-linked EP is essentially higher than the corresponding parameter for HDPE devitrificated amorphous phase, where epoxy polymer is concentrated [56]. The authors of Ref. [57] elucidated the causes of MFI extreme enhancement or melt viscosity reduction for nanocomposites HDPE-EP.

As it has been noted above [7, 8], the authors of Ref. [9] proposed the scaling relationship for polymer melt viscosity η_0 determination at zero shear:

$$\eta_0 \sim \frac{\overline{M}_w^3}{MM_e^2}, \tag{32}$$

where $\overline{M}_w$ is mean weight molecular weight of polymer, MM_e is molecular weight of a chain part between its topological fixation points (chemical cross-linking nodes, nanoclusters, macromolecular "binary hookings," etc.).

The polymer melt structure can be characterized unequivocally by the macromolecular coil fractal dimension Δ_f. In this case the amount (density) of chain fixation points in melt is accepted equal to macromolecular coils intersections number N_{int}, which is determined by the Eq. (13) of Chapter 1. At $\Delta_{f_1}=\Delta_{f_2}=\Delta_f$ this relationship transforms in the Eq. (3). Δ_f value is accepted equal to the structure dimension d_f of solid-phase compositions HDPE-EP [58]. This dimension was determined according to the Eqs. (151) and (152) of Chapter 2. N_{int} value is scaled with MM_e according to the following relationship [15]:

$$MM_e = \frac{\rho_p N_A}{N_{int}}, \tag{33}$$

where ρ_p is polymer density, N_A is Avogadro number.

η_0 value calculation according to the Eqs. (151) of Chapter 2, (3), (32) and (33) (in relative units) has shown that it changes proportionally to reciprocal of MFI and the empirical correlation of these parameters looks like [57]:

$$\mathrm{MFI} = 7.0(\eta_0)^{-1}, \tag{34}$$

where MFI is given in dg/min.

In Fig. 10, the comparison of experimental and calculated according to the Eq. (34) melt flow index values as a function of epoxy polymer contents c_{EP} in the considered nanocomposites is adduced. As one can see, the proposed fractal model gives a good correspondence, both qualitative and quantitative, to the experimental data.

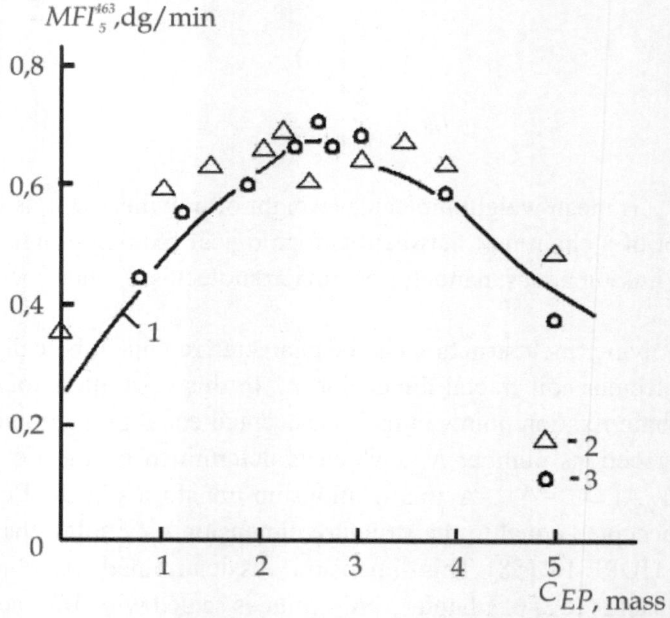

FIGURE 10 The dependences of melt flow index MFI on epoxy polymer contents c_{EP} for nanocomposites HDPE-EP. 1—the experimental data; 2—calculation according to the Eq. (34); 3—calculation according to the Eq. (37).

However, the indicated model does not explain the causes of nanocomposites HDPE-EP melt viscosity extreme reduction. Therefore, for this effect explanation the authors [57] used one more treatment. As it is known through Ref. [49], the extreme change of mixtures properties in case of their interaction (both chemical and physical ones) is realized at equimolar (stoichiometric) components content. Since for the considered nanocomposites extremum is reached at 2.0–3.0 mass. % of EP, then this means, that with epoxy polymer not the entire polyethylene matrix interacts, but only its part, consisting of 4–6 mass. % of HDPE. In this case for η_0 (further designated as η_0') estimation the relationship, applied for two components chemical reaction kinetics description can be used [59]:

$$\vartheta_{int} \sim [\text{HDPE}][\text{EP}]\,\eta_0' t^{(1-\Delta_f)/2}, \tag{35}$$

where ϑ_{int} is HDPE and EP interaction rate, [HDPE] and [EP] are concentrations of HDPE and EP interacting parts, respectively, accepted totally equal to 12 mass. %, t is interaction duration.

Assuming the values ϑ_{int}, t and Δ_f as constant (Δ_f variation makes up 2.606–2.686), η_0' value can be evaluated in relative units according to the relationship [57]:

$$\eta_0' \sim \frac{\text{const}}{[\text{HDPE}][\text{EP}]}, \tag{36}$$

where values [HDPE] and [EP] are given in mass. % and the constant in numerator of the relationship right-hand part can be determined by method of the best matching of theoretical and experimental results [57].

The estimations have shown that MFI and η_0' intercommunication could be described according to the following empirical equation [57]:

$$\text{MFI} = 0.078\left(\eta_0'\right)^{-1} = 0.078[\text{HDPE}][\text{EP}], \tag{37}$$

where MFI is given again in dg/min.

The comparison of experimental and calculated according to the Eq. (37) MFI values is also adduced in Fig. 10, from which a good correspondence of theory and experiment follows.

Hence, the stated above results have shown that melt viscosity extreme change of nanocomposites HDPE-EP could be described within the framework of the fractal model. The main structural parameter, controlling this effect, is the change of the fractal dimension of macromolecular coil in melt. The main physical cause, defining the mentioned effect, is a partial interaction of HDPE matrix and epoxy polymer particles. In this case the melt flow index extreme change is described correctly within the framework of chemical reactions fractal kinetics.

Inorganic nanofillers of various types usage for polymer nanocomposites production has been widely spread [56]. However, the mentioned nanomaterials melt properties are not studied completely enough. As a rule, when nanofiller application is considered, then compromise between mechanical properties in solid state improvement, melt viscosity at processing enhancement, nanofillers dispersion problem and process economic characteristics is achieved. Proceeding from this, the relation between nanofiller concentration and geometry and nanocomposites melt properties is an important aspect of polymer nanocomposites study. Therefore, the authors [60–62] performed an investigation and theoretical description of the dependence of nanocomposites polypropylene/calcium carbonate ($PP/CaCO_3$) melt viscosity on nanofiller contents.

For polymer microcomposites, that is, composites with filler of micron sizes, two simple relations between melt viscosity η, shear modulus G in solid-phase state and filling volume degree φ_f were obtained [63]. The relationship between η and G has the following form [63]:

$$\frac{\eta}{\eta_0} = \frac{G}{G_0}, \tag{38}$$

where η_0 and G_0 are melt viscosity and shear modulus of matrix polymer, respectively.

Besides, microcomposite melt viscosity increase can be estimated as follows (for $\varphi_f<0.40$) [63]:

$$\frac{\eta}{\eta_0} = 1 + \varphi_f \tag{39}$$

In Fig. 11, the dependences of ratios G_n/G_m and η_n/η_m, where G_n and η_n are shear modulus and melt viscosity of nanocomposite, G_m and η_m are the same characteristics for the initial matrix polymer, on $CaCO_3$ mass contents W_n for nanocomposites $PP/CaCO_3$ are adduced. Shear modulus G was calculated according to the general fractal Eq. (80) of Chapter 2. MFI reciprocal value was accepted as melt viscosity η measure. The data of Fig. 11 clearly demonstrate, that in case of the studied nanocomposites the

Eq. (38) is not fulfilled either qualitatively or quantitatively: the ratio η_n/η_m decay at W_n growth corresponds to G_n/G_m enhancement and η_n/η_m absolute values are much smaller than the corresponding G_n/G_m magnitudes.

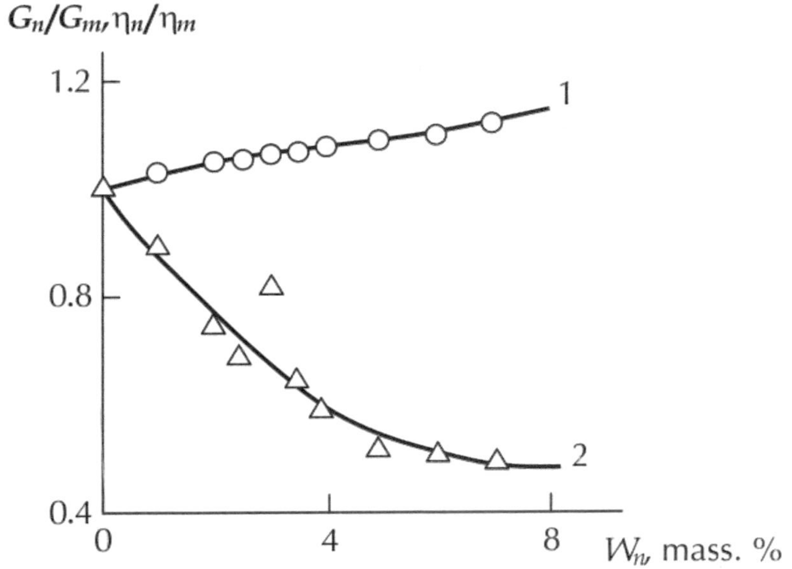

FIGURE 11 The dependences of shear moduli G_n/G_m (1) and melt viscosities η_n/η_m (2) ratios of nanocomposite G_n, η_n and matrix polymer G_m, η_m on nanofiller mass contents W_n for nanocomposites PP/$CaCO_3$.

In Fig. 12, the comparison of parameters η_n/η_m and $(1+\varphi_n)$ for nanocomposites PP/$CaCO_3$ is adduced. The discrepancy between the experimental data and the Eq. (39) is obtained again: absolute values η_n/η_m and $(1+\varphi_n)$ discrepancy is observed and $(1+\varphi_n)$ enhancement corresponds to melt relative viscosity reduction. At the plot of Fig. 12 construction the nominal φ_n value was used, which did not take into consideration nanofiller particles aggregation and estimated according to the equation [64]:

$$\varphi_n = \frac{W_n}{\rho_n}, \tag{40}$$

where ρ_n is nanofiller particles density, which was determined according to the formula [56]:

$$\rho_n = 188(D_p)^{1/3}, \text{ kg/m}^3, \tag{41}$$

where D_p is $CaCO_3$ initial particles diameter, which is given in nm.

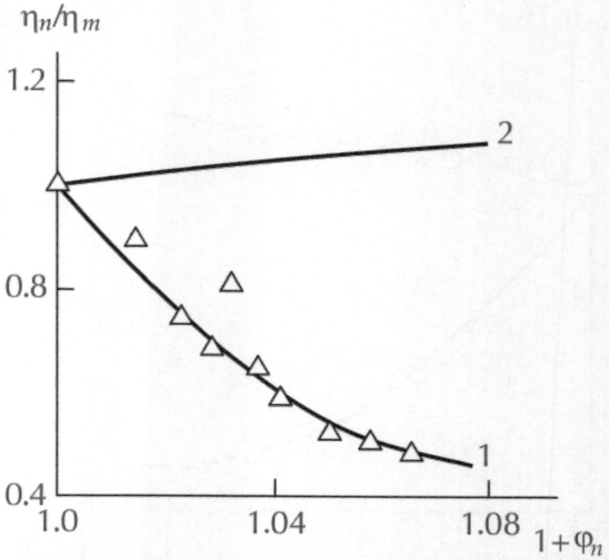

FIGURE 12 The dependence of nanocomposite and matrix polymer melt viscosities ratio η_n/η_m on nanofiller volume contents $(1+\varphi_n)$ (1) for nanocomposites PP/$CaCO_3$. The straight line 2 shows relation 1:1.

Hence, the data of Figs. 11 and 12 have shown, that the Eqs. (38) and (39) fulfilled in case of polymer microcomposites are incorrect for nanocomposites. In case of the Eq. (38) correctness and Kerner's equation application for G calculation the lower boundary of viscosity η_n can be obtained according to the equation [63]:

$$\frac{\eta_n}{\eta_m} = 1 + \frac{2.5\varphi_n}{1-\varphi_n} \tag{42}$$

Since η value is inversely proportional to MFI, then in such treatment the Eq. (42) can be rewritten as follows [62]:

$$\frac{\text{MFI}_m}{\text{MFI}_n} = 1 + \frac{2.5\varphi_n}{1-\varphi_n}, \tag{43}$$

where MFI_m and MFI_n are MFI values for matrix polymer and nanocomposite, respectively.

Three methods can be used for the value φ_n evaluation in the Eqs. (42) and (43). The first from them was described above, which gives nominal value φ_n. The second method is usually applied for microcomposites, when massive filler density is used as ρ_n, that is, ρ_n=const≈2000 kg/m^3 in case of $CaCO_3$. And at last, the third method also uses the Eqs. (40) and (41), but it takes into consideration nanofiller particles aggregation and in this case in the Eq. (41) the initial nanofiller particles diameter D_p is replaced by such particles aggregate diameter D_{ag}. To estimate $CaCO_3$ nanoparticles aggregation degree and, hence, D_{ag} value can be estimated within the framework of the strength dispersive theory [65], where yield stress at shear τ_n of nanocomposite is determined as follows:

$$\tau_n = \tau_m + \frac{Gb_B}{\lambda}, \tag{44}$$

where τ_m is yield stress at shear of polymer matrix, b_B is Burgers vector, λ is distance between nanofiller particles.

In case of nanofiller particles aggregation the Eq. (44) assumes the look [65]:

$$\tau_n = \tau_m + \frac{Gb_B}{k(\rho)\lambda}, \tag{45}$$

where $k(\rho)$ is an aggregation parameter.

The parameters included in the Eqs. (44) and (45) are determined as follows. The general relation between normal stress σ and shear stress τ assumes the look [66]:

$$\tau = \frac{\sigma}{\sqrt{3}} \tag{46}$$

Burgers vector value b_B for polymer materials is determined according to the Eq. (92). And at last, the distance λ between nonaggregated nanofiller particles is determined according to the following equation [65]:

$$\lambda = \left[\left(\frac{4\pi}{3\varphi_n}\right)^{1/3} - 2\right]\frac{D_p}{2} \tag{47}$$

From the Eqs. (45) and (47) $k(\rho)$ growth from 5.5 up to 11.8 within the range of W_n=1–10 mass. % for the considered nanocomposites follows. Let us consider, how such $k(\rho)$ growth is reflected on nanofiller particles aggregates diameter D_{ag}. The Eqs. (40), (41) and (47) combination gives the following expression [67]:

$$k(\rho)\lambda = \left[\left(\frac{0.251\pi D_{ag}^{1/3}}{W_n}\right)^{1/3} - 2\right]\frac{D_{ag}}{2}, \tag{48}$$

allowing at replacement D_p by D_{ag} in the indicated above equations to determine real, that is, with accounting for nanofiller particles aggregation, nanoparticles $CaCO_3$ aggregates diameter. Calculation according to the Eq. (48) shows D_{ag} increase (corresponding to $k(\rho)$ growth) from 80 up to 190 nm within the indicated W_n range (at value D_p=80 nm).

In Fig. 13, the electron micrographs of nanocomposites PP/$CaCO_3$ chips are adduced, which confirm D_{ag} estimations according to the Eq. (48). As one can see, for nanocomposite PP/$CaCO_3$ with $CaCO_3$ content of 1 mass. % the nanofiller separate particles with size of ~80 nm are observed, whereas W_n increase up to 3 mass. % results in $CaCO_3$ particles aggregates appearance, having sizes up to 300 nm that gives mean value D_{ag}≈125 nm for the indicated nanocomposite according to the Eq. (48) estimations.

Further the real value ρ_n for aggregated nanofiller can be calculated according to the Eq. (41) and real filling degree φ_n—according to the Eq. (40). In Fig. 14, the dependences $\text{MFI}_n^{-1}(W_n)$, obtained experimentally and calculated according to the Eq. (43) with φ_n values using, evaluated by the three indicated above methods, comparison is adduced. As one can see, the theoretical results obtained according to the Eq. (43) do not correspond to the experimental data either qualitatively or quantitatively.

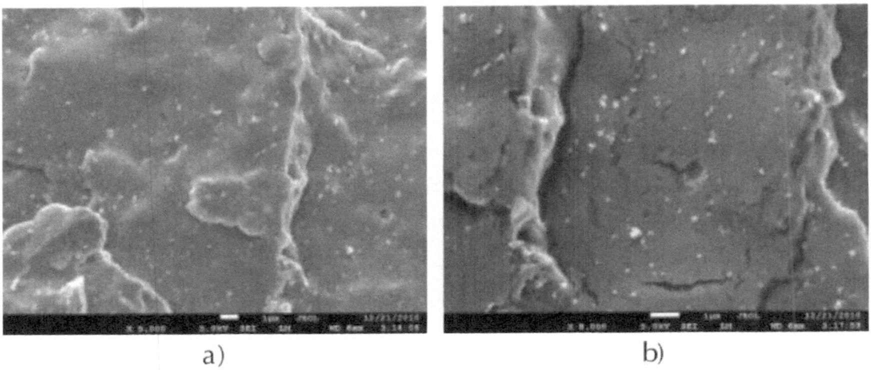

FIGURE 13 The electron micrographs of chips of nanocomposites PP/$CaCO_3$ with nanofiller contents W_n 1 (a) and 4 (*b*) mass. %.

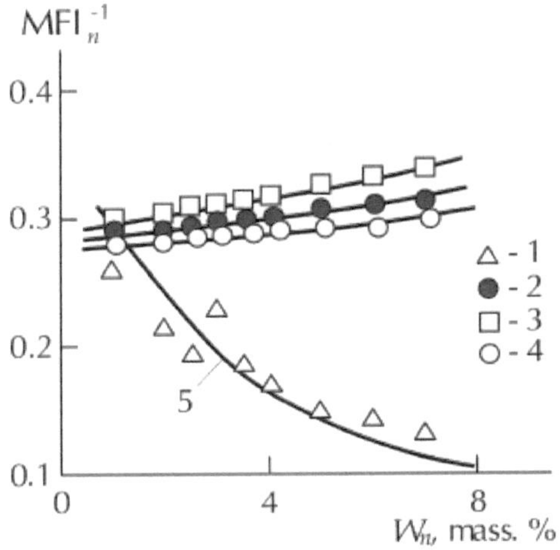

FIGURE 14 The dependences of melt viscosity MFI_n^{-1} on nanofiller mass contents W_n for nanocomposites PP/$CaCO_3$. 1—the experimental data; 2–4—calculation according to the Eq. (3.43) without accounting for (2) and with accounting for (3) nanofiller particles aggregation and at the condition ρ_n=const (4). 5—calculation according to the Eq. (49).

The indicated discrepancy requires the application of principally differing approach at the description of polymer nanocomposites melt viscosity. Such approach can be the fractal analysis, within the framework of which the authors [68] proposed the following relationship for fractal liquid viscosity η estimation:

$$\eta(l) \sim \eta_0 l^{2-d_f}, \tag{49}$$

where l is characteristic linear scale of flow, η_0 is constant, d_f is fractal dimension.

In the considered case the nanoparticles $CaCO_3$ aggregate radius $D_{ag}/2$ should be accepted as l. Since the indicated aggregate surface comes into contact with polymer, then its fractal dimension d_{surf} was chosen as d_f. The indicated dimension can be calculated as follows [56]. The value of nanofiller particles aggregate specific surface S_u was estimated according to the equation [69]:

$$S_u = \frac{6}{\rho_n D_{ag}}, \tag{50}$$

and then the dimension d_{surf} was calculated with the aid of the equation [56]:

$$S_u = 410\left(\frac{D_{ag}}{2}\right)^{d_{surf}-d} \tag{51}$$

As earlier, η value was considered as reciprocal value of MFI_n and constant η_0 in the Eq. (49) was accepted equal to $(MFI_m)^{-1}$. At these conditions and replacement of proportionality sign in the Eq. (49) by equality sign and also using in connection with this the additional constant 6.8 one can calculate theoretical values of MFI_n, if D_{ag} magnitude is expressed in microns. In Fig. 14, the comparison of the received by the indicated mode values $\eta = MFI_n^{-1}$ with the experimental dependence $MFI_n^{-1}(W_n)$ is adduced, from which theory and experiment good correspondence follows (the mean discrepancy of these parameters makes up ~8%).

The Eq. (49) allows to make a number of conclusions. So, at the mentioned above conditions conservation D_{ag} increase, that is, the initial nanoparticles aggregation intensification, results in melt viscosity reduction, whereas d_{surf} enhancement, that is, nanoparticles (aggregates of nanoparticles) surface roughness degree increasing, raises nanocomposites melt viscosity. However, as it follows from the Eqs. (50) and (51), D_{ag} increasing is accompanied by d_{surf} reduction. The indicated factors, critical for nanofillers, are not taken into account for the continuous treatment of melt viscosity for polymer microcomposites (the Eqs. (38), (39) and (42)).

Hence, the results adduced above have shown that the developed for microcomposites rheology description models do not give adequate treatment of melt viscosity for particulate-filled nanocomposites. The indicated nanocomposites rheological properties correct description can be received within the framework of fractal model of viscous liquid flow. It is significant, that such approach differs principally from the used ones at microcomposites description. So, nanofiller particles aggregation reduces both melt viscosity and elasticity modulus of nanocomposites in solid-phase state. For microcomposites melt viscosity enhancement is accompanied by elasticity modulus increase.

The similar effects were observed and at melt viscosity investigation of polymer nanocomposites with other types of inorganic nanofillers (organoclay, carbon nanotubes). At the rheology of these nanocomposites description, preserving general methodology, one should take into account special features of nanofiller the indicated types. So, the authors [70] performed investigation and theoretical description of melt viscosity of nanocomposite poly(butylenes terephthalate)/Na^+-montmorillonite (PBT/MMT) as a function of nanofiller (organoclay) contents.

In Figs. 15 and 16, the dependences of the ratios G_n/G_m and η_n/η_m on MMT mass contents and the comparison of the parameters η_n/η_m and $(1+\varphi_n)$ are adduced for nanocomposites PBT/MMT, respectively. These plots do not correspond again to the Eqs. (38) and (39), as it was described above for the case of nanocomposites PP/$CaCO_3$. Therefore the model [68] (the Eq. (49)) was used for the description of nanocomposites PBT/MMT melt viscosity. In the considered case Na^+-montmorillonite plates packet (tactoid) thickness l_T should be accepted as a linear scale l, since such packet length and width are constant. l_T value can be determined according to the following equation [64]:

$$l_T = \left(N_{pl} - 1\right) d_{001} + N_{pl} d_{pl}, \tag{52}$$

where N_{pl} is Na^+-montmorillonite plates number per packet (tactoid), d_{001} is distance between plates (interlayer spacing), d_{pl} is silicate plate thickness, equal to ~1 nm [56].

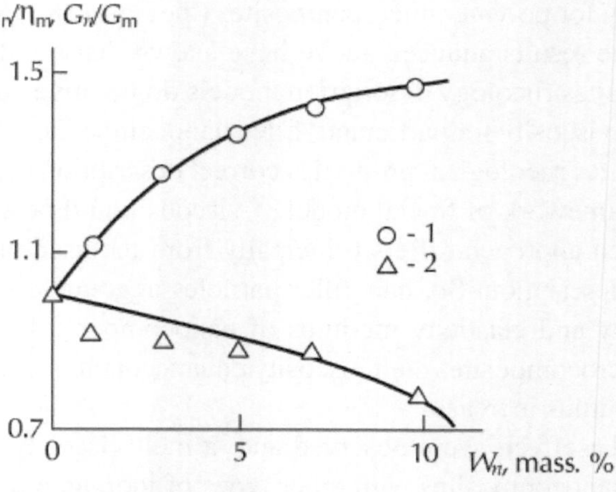

FIGURE 15 The dependences of ratios of shear moduli G_n/G_m (1) and melt viscosities η_n/η_m (2) of nanocomposite G_n, η_n and matrix polymer G_m, η_m on nanofiller mass contents W_n for nanocomposites PBT/MMT.

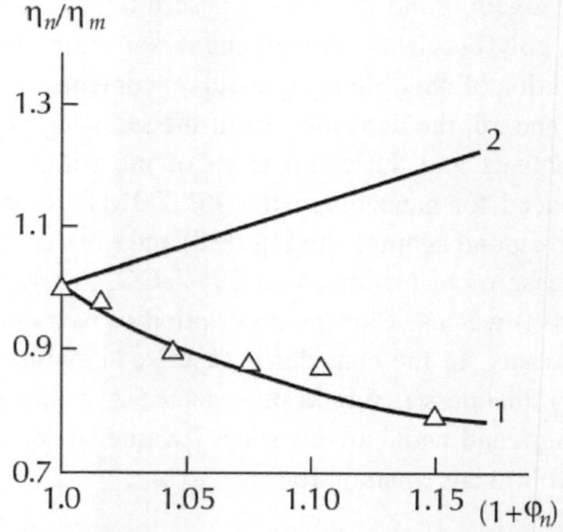

FIGURE 16 The dependence of nanocomposite and matrix polymer melt viscosities ratio η_n/η_m on nanofiller volume contents $(1+\varphi_n)$ (1) for nanocomposites PBT/MMT. The straight line 2 shows relation 1:1.

N_{pl} and d_{001} values are determined by the level of interfacial adhesion organoclay-polymer matrix, which is characterized quantitatively by the parameter b_{ad} [71]:

$$N_{pl} = 24 - 5.7b_{ad}, \tag{53}$$

$$d_{001} = 1.27b_{ad} \tag{54}$$

In its turn, b_{ad} value can be estimated with the aid of the following percolation relationship [56]:

$$\frac{E_n}{E_m} = 1 + 11\left(1.955\varphi_n b_{ad}\right)^{1.7}, \tag{55}$$

where E_n and E_m are elasticity moduli (Young's moduli) of nanocomposite and matrix polymer, respectively.

The Eq. (55) was received for nanocomposites polymer/organoclay with organoclay intercalated structure and its application in the given case proves to be correct by organoclay packets (tactoids) (N_{pl}>1) availability, that is a typical sign of Na^+-montmorillonite intercalated structure [56]. For nanocomposites PBT/MMT N_{pl} value is varied within the limits of 1.5–16.6 within the range of W_n=1–10 mass. % [70].

As dimension d_f in the Eq. (49), as above, the dimension d_{surf} of organoclay plates surface is accepted, since this very surface comes into contact with polymer matrix. d_{surf} can be estimated by two methods. In case of the firstly from them (theoretical one) first the organoclay specific surface S_u is calculated according to the Eq. (50), where D_{ag} value for Na^+-montmorillonite is accepted equal to arithmetical mean of its plate three sizes, and further the dimension d_{surf} can be determined with the aid of the Eq. (51). Calculation according to the described technique gives d_{surf}=2.77 for Na^+-montmorillonite.

The second method uses the experimental techniques (X-raying and low-temperature nitrogen adsorption) for d_{surf} estimation that gives 2.83 and 2.78, respectively [72]. As one can see, the greatest discrepancy between theory and experiment makes up ~ 2% at d_{surf} determination.

However, as it has been shown in Refs. [73, 74], polymer macromolecule cannot reproduce such high roughness of the surface by virtue of its final rigidity, defined by statistical segment final size, and therefore real (effective) surface fractal dimension d_{surf}^{ef} is determined as follows [73]:

$$d_{surf}^{ef} = 5 - d_{surf} \tag{56}$$

Accounting for the made above estimations the Eq. (49) can be rewritten as follows [70]:

$$\eta_{rel} = 1.35\left[\left(N_{pl} - 1\right)d_{001} + N_{pl}d_{pl}\right]^{d_{surf}^{ef} - 3}, \tag{57}$$

where values d_{001} and d_{pl} are given in nm.

In Fig. 17, the comparison of the received by the indicated mode dependence $\eta_{rel}(W_n)$ (curve 1), where η_{rel} is determined as the ratio η_n/η_m, with the experimentally determined η_{rel} values, from which a good correspondence of theory and experiment follows.

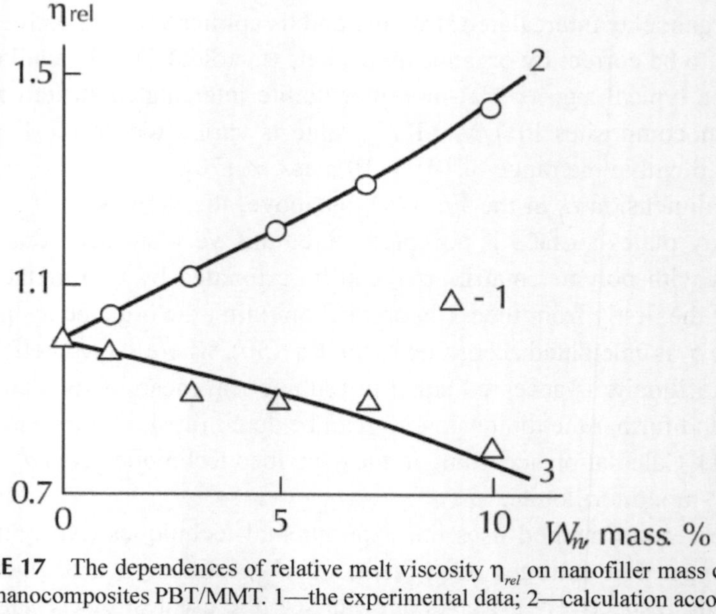

FIGURE 17 The dependences of relative melt viscosity η_{rel} on nanofiller mass contents W_n for nanocomposites PBT/MMT. 1—the experimental data; 2—calculation according to the Eq. (43); 3—calculation according to the Eq. (57).

The Eq. (57) allows making the following conclusion. The increase of both l_T and d_{surf} (if the value d_{surf}^{ef} is used) results in melt viscosity reduction. The indicated factors, critical for organoclay, are not taken into account again in continuous melt viscosity treatment for polymer composites (the Eq. (42)).

Hence, the results stated above confirmed that the models, developed for the microcomposites rheology description, did not give melt viscosity adequate treatment for nanocomposites polymer/organoclay as well. And as earlier, the indicated nanocomposites rheological properties description can be obtained within the framework of a viscous liquid flow fractal model. Na^+-montmorillonite plates aggregation in packets (tactoids) simultaneously reduces both melt viscosity and elasticity modulus in solid-phase state of nanocomposites.

Let us consider in the conclusion of the present section melt viscosity behavior as a function of nanofiller contents for nanocomposites polypropylene/carbon nanotubes (PP/CNT), studied in Refs. [55, 76]. In Figs. 18 and 19 the dependences of the ratios G_n/G_m and η_n/η_m on nanofiller contents and the parameters η_n/η_m and $(1+\varphi_n)$ for nanocomposites PP/CNT, respectively, are adduced. And as and in case of the two considered above nanocomposites classes, the discrepancy between the experimental data and behavior, predicted by the Eqs. (38) and (39), is observed again. In Fig. 20, the dependence of MFI_n on nanofiller mass contents W_n for nanocomposites PP/CNT, calculated according to the Eq. (43), is adduced. As one can see, the qualitative discrepancy between theoretical calculation (curve 1) and experimental data (points) is observed. If the Eq. (43) assumes melt viscosity increasing (MFI_n reduction) at W_n growth, then the experimental data discovers opposite tendency ($MFI_n > MFI_m$).

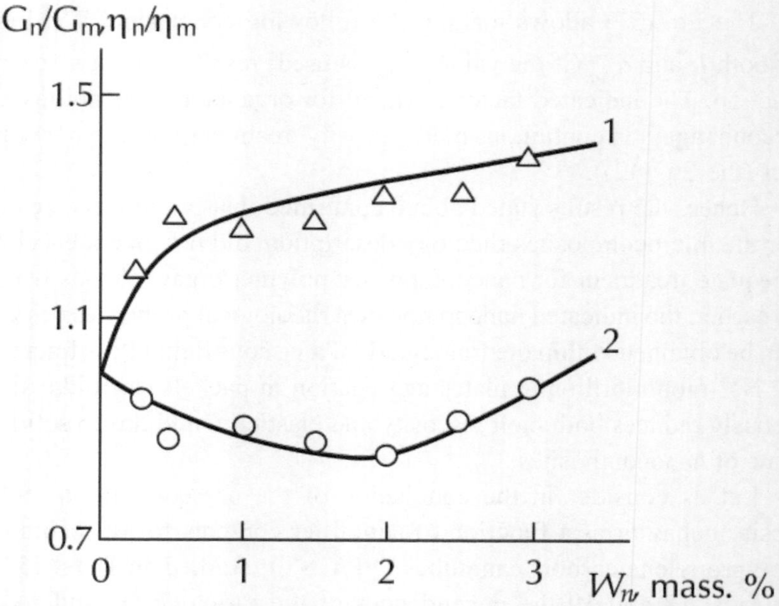

FIGURE 18 The dependences of shear moduli G_n/G_m (1) and melt viscosities η_n/η_m (2) ratios of nanocomposite G_n, η_n and matrix polymer G_m, η_m on nanofiller mass contents W_n for nanocomposites PP/CNT.

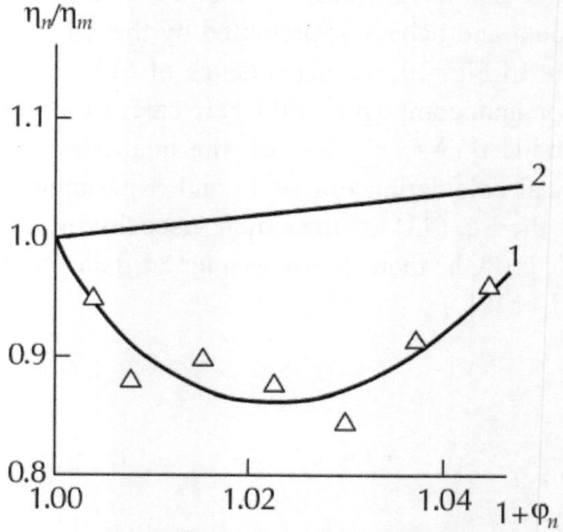

FIGURE 19 The dependence of nanocomposite and matrix polymer melt viscosities ratio η_n/η_m on nanofiller volume contents $(1+\varphi_n)$ (1) for nanocomposites PP/CNT. The straight line 2 shows relation 1:1.

The indicated discrepancy forces to search on other approach for the description of nanocomposites PP/CNT melt viscosity. And as earlier, the fractal model of viscous liquid flow (the Eq. (49)) [68] was chosen as such approach. As it is known through Ref. [77], CNT by virtue of their strong anisotropy (in case of the used carbon nanotubes of mark "Taunite" the ratio of length to diameter is larger than 45) and low transversal stiffness form ring-like structures with radius R_{CNT}, which can be determined with the aid of the equation [78]:

$$\varphi_n = \frac{\pi L_{\text{CNT}} r_{\text{CNT}}^2}{\left(2R_{\text{CNT}}\right)^3}, \tag{58}$$

where L_{CNT} and r_{CNT} are length and exterior radius of carbon nanotubes, respectively.

Further the radius R_{CNT} was accepted as scale l. Since CNT surface comes into contact with polymer matrix, then its fractal dimension d_{surf} was chosen as d_f. d_{surf} value, which can be evaluated according to the Eqs. (50) and (51). The calculation according to the indicated technique gives the value d_{surf}=2.79. As it has been noted above, a macromolecular chain of polymer matrix cannot "reproduce" CNT surface high roughness and therefore it "perceives" the indicated surface as a much smoother one. In such case d_{surf} effective value (d_{surf}^{ef}) should be used, which is determined according to the Eq. (56). And as earlier, η value in the Eq. (49) was considered as MFI$_n$ reciprocal value and the constant η_0 was accepted equal to 1.1 (MFI$_m$)$^{-1}$. At these conditions and the replacement of proportionality sign in the Eq. (49) by equality sign the theoretical values MFI$_n$ can be calculated, if R_{CNT} value is expressed in microns. In Fig. 20, the comparison of the obtained by the indicated mode MFI$_n$ values with the experimental dependence MFI$_n$(W_n) is adduced, from which a good correspondence of theory and experiment follows.

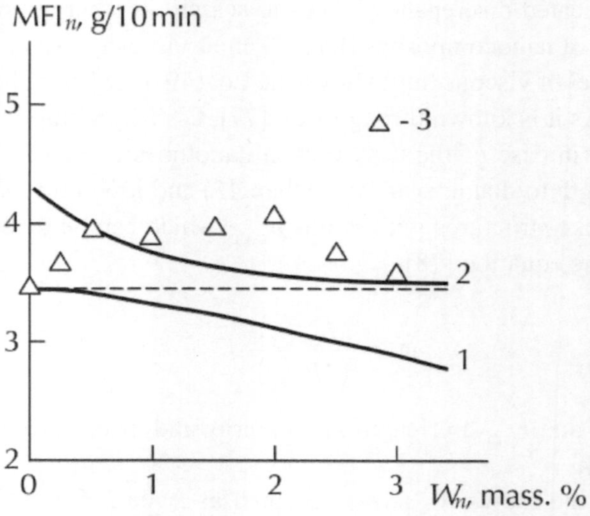

FIGURE 20 The dependences of melt flow index MFI_n on nanofiller mass contents W_n for nanocomposites PP/CNT. 1—calculation according to the Eq. (43); 2—calculation according to the Eq. (49); 3—the experimental data.

The Eq. (49) allows to make a conclusions number. So, at the mentioned above conditions preservations R_{CNT} increase results in nanocomposite melt viscosity reduction. The similar influence d_{surf} enhancement, that is, CNT roughness degree increase, exercises in virtue by using the Eq. (56) of Chapter 2.

As the adduced above data have shown, the polymer nanocomposites with three main types of inorganic nanofiller and also polymer-polymeric nanocomposites melt viscosity cannot be described adequately within the framework of models, developed for the description of microcomposites melt viscosity. This task can be solved successfully within the framework of the fractal model of viscous liquid flow, if in it the used nanofiller special feature is taken into account correctly. Let us note that unlike microcomposites nanofiller cotents enhancement does not result in melt viscosity increase, but, on the contrary, reduces it. It is obvious, that this aspect is very important from the practical point of view.

3.3 THE STRUCTURAL MODEL OF POLYMERS SYNTHESIS REACTIONS IN MELT

At present it has been established that polymers synthesis in solutions is controlled by the macromolecular coil structure, which is a fractal object [79–81] (see also Section 2.7). Moreover the synthesis process, irrespective of the used method (radical polymerization, polycondensation and so on), proceeds according to the irreversible aggregation cluster-cluster mechanism, that is, presents itself small macromolecular coils sets in larger ones. In this case the fractal dimension of macromolecular coils in solutions varies within the limits of 1.50–2.12 [82] that corresponds to the indicated aggregation mechanism. However, at polymers synthesis in melt macromolecular coil surroundings change occurs—instead of solvent molecules now it is surrounded by the same coils. As it has been shown in Ref. [6] this results in fractal dimension of macromolecular coil in melt enhancement approximately up to 2.50. The macromolecular coil structure change should tell on the synthesis process kinetics. The authors [83] performed the quantitative treatment of the assumed changes within the framework of fractal analysis and irreversible aggregation models on the example of polyether polyols polyesterification reaction [84].

In Fig. 21 the kinetic curves conversion degree—reaction duration *Q-t* for two polyols on the basis of ethyleneglycole (PO-1) and propyleneglycole (PO-2) are adduced. As it was to be expected, these curves had autodecelerated character, that is, reaction rate ϑ_r was decreased with time. Such type of kinetic curves is typical for fractal reactions, to which either fractal objects reactions or reactions in fractal spaces are attributed [85]. In case of Euclidean reactions the linear kinetics (ϑ_r=const) is observed. The general Eq. (2.107) was used for the description of fractal reactions kinetics. From this relationship it follows, that the plot $Q(t)$ construction in double logarithmic coordinates allows to determine the exponent value in this relationship and, hence, the fractal dimension Δ_f value. In Fig. 3.22 such dependence for PO-1 is adduced, from which it follows, that it consists of two linear sections, allowing to perform the indicated above estimation. For small t (t 50 min) the linear section slope is higher and Δ_f=2.648 and for t>50 min Δ_f=2.693. Such Δ_f increase or macromolecular coil density enhancement in reaction course is predicted by the irreversible aggregation model for multiple nucleation sites, which corresponds most completely to the conditions of real polymerization reactions [86]. The indicated dimensions correspond to the irreversible aggregation mechanism particle-cluster, that is, unlike polymers synthesis in solutions reaction in melt proceeds by particles (oligomer molecules) adding to a growing macromolecular coil. It is obvious, that Δ_f increase in the Eq. (107) of Chapter 2 in comparison with synthesis in solution should raise essentially reaction duration by virtue of the power form of this relationship that is observed in practice [87]. However, this enhancement is compensated to a certain extent by higher values c_0 and much higher η_0 magnitudes for melt in comparison with the solution, as a result of which synthesis in melt duration remains on an acceptable level [84].

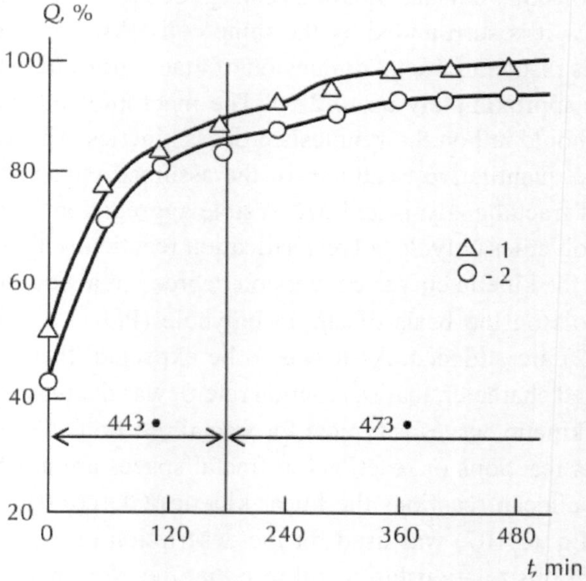

FIGURE 21 The kinetic curves conversion degree-reaction duration Q–t for polyols on the basis of ethyleneglycole (1) and propyleneglycole (2).

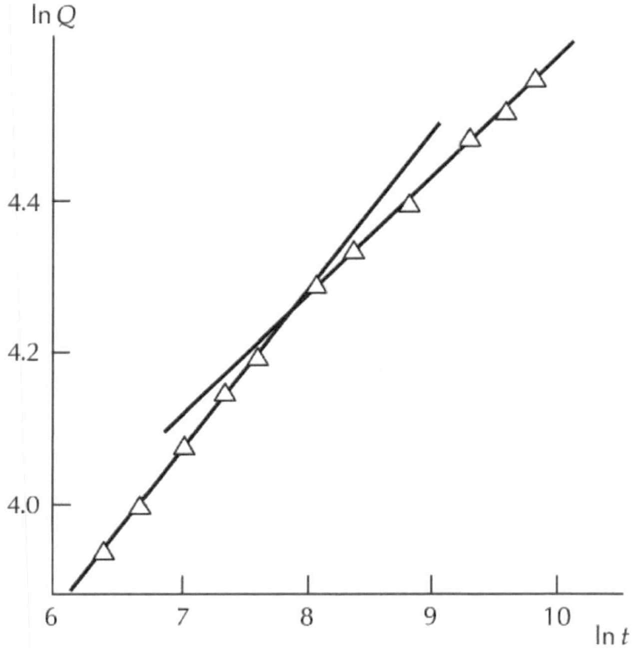

FIGURE 22 The dependence of conversion degree Q on reaction duration t in double logarithmic coordinates for polyol on the basis of ethyleneglycole.

The authors [84] have shown polyesterification reaction rate ϑ_r reduction approximately by an order of magnitude within the range of Q=0.5–0.9. Within the framework of fractal approach ϑ_r value can be received by the Eq. (108) of Chapter 2 differentiation by time t that gives the Eq. (108) of Chapter 2. In Fig. 23 the dependences $\vartheta_r(Q)$, received experimentally [84] and calculated according to the Eq. (108) of Chapter 2, in which the product $c_0\eta_0$ is supposed constant and proportionality coefficient is the received by method of the best theory and experiment correspondence, comparison is adduced. As it follows from the data of Fig. 23, the good correspondence is obtained between the indicated dependences. Let us note that according to the Eq. (108) of Chapter 2 ϑ_r reduction at t (or Q) growth is due to reaction fractal character only [83].

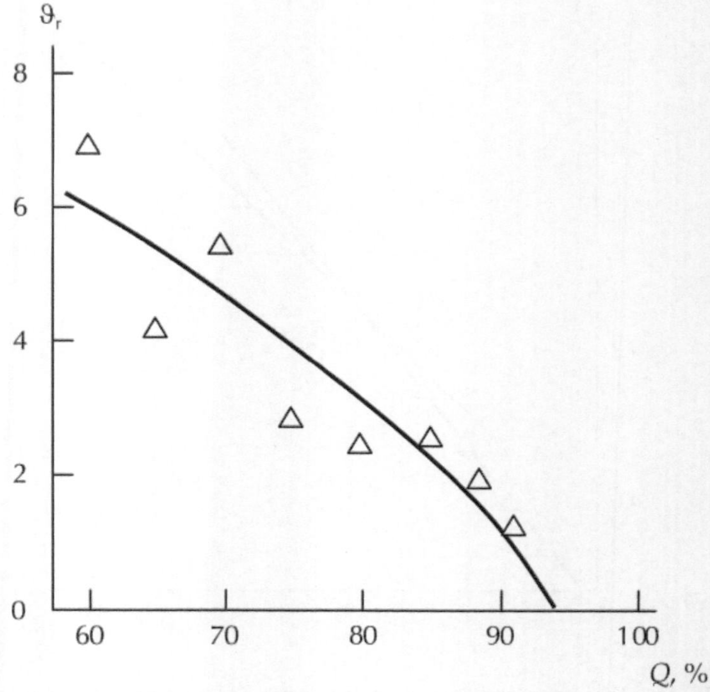

FIGURE 23 The dependences of reaction rate ϑ_r on conversion degree Q, received experimentally (1) and calculated according to the Eq. (108) of Chapter 2 (2), for polyol on the basis of ethyleneglycole.

Another variant of ϑ_r estimation is proposed in Ref. [22]. In this case the Eq. (106) of Chapter 2 is valid for fractal reactions.

In Fig. 24, the dependence $\vartheta_r(t)$ in double logarithmic coordinates is adduced. As it was to be expected, this dependence proves to be linear, that allows one to determine from its slope the value h=0.56. Thus, this result confirms polyesterification reaction proceeding in heterogeneous (fractal) medium. For such mediums (particularly at elevated temperatures) the connectivity is characterized by not spectral dimension d_s, but by its effective value d_s', which takes into account temporal (energetic) disorder availability in the system [22]. d_s' value is linked with heterogeneity exponent h by the Eq. (6), which for h=0.56 gives the value d_s'=0.88 [83].

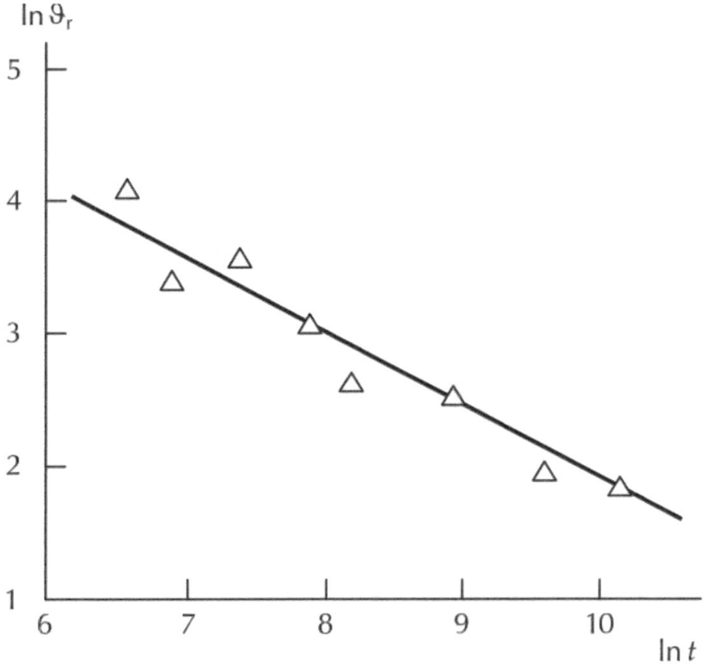

FIGURE 24 The dependence of polyesterification reaction rate ϑ_r on its duration t in double logarithmic coordinates for polyol on the basis of ethyleneglycole.

Argyrakis [24] has shown that at the investigation of chemical reactions on fractal objects the corrections by small clusters in system availability are necessary. Just such corrections require the usage in theoretical calculations of not generally accepted spectral (fracton) dimension d_s [23], but its effective value d_s'. It is obvious, that the availability in real polymers molecular weight distribution by virtue of the indicated above causes requires application of just dimension d_s'. The authors [21] offered the Eq. (11) for the value d_s' determination, calculation according to which for Δ_f=2.648 (T=443 K) gave the value d_s'=0.918 and for Δ_f=2.693 (T=473 K) d_s'=0.954. Both indicated values d_s' correspond well to the estimation according to the Eq. (6) (the discrepancy is less than 8%). This circumstance confirms, that in the synthesized PBT melts macromolecular coils of different sizes are formed, that is, certain molecular weight distribution has place.

Let us note that synthesis temperature enhancement results in some h reduction (from 0.541 up to 0.523). This corresponds to the theorem about subordination (the Eq. (15)). And at last, as it has been shown in Ref. [88], medium character defines the reaction order n:

$$n = \frac{2-h}{1-h} \tag{59}$$

For the obtained above h values the magnitude $n \approx 3$. Let us note that for Euclidean spaces (objects) at $h=0$ only one value of reaction order is possible, namely, $n=2$.

Hence, the results stated above have shown that the structural analysis of polyesterification reaction in melt, using fractal analysis and irreversible aggregation models notions, allows to give precise enough description of this reaction even without applying purely chemical aspects. Let us note that fractal analysis is a more strict mathematical calculus than often used for synthesis kinetics description empirical equations.

KEYWORDS

- **aromatic polyethersulfonoformals**
- **binary hookings**
- **devil's staircase**
- **hetero-chain semirigid-chain polymers**
- **high density polyethylene**
- **Kirkwood-Riseman theory**
- **liquid with fixed structure**
- **rigid-chain polymers**
- **structure absence**
- **structureless liquid**

REFERENCES

1. Kalinchev, E. L.; Sakovtseva, M. B.; *Properties and Processing of Thermoplastics.* Leningrad, Chemistry, **1983,** 288 p.

2. Li, G.; Stoffy, D.; Nevil, K.; *New Linear Polymers*. Moscow, Chemistry, **1972,** 280 p.
3. Sanditov, D. S.; Bartenev, G. M.; *Physical Properties of Disordered Structures*, Novosibirsk, Science, **1982,** 256 p.
4. Dolbin, I. V.; *The Fractal Model of Polymer Melts Structural Stabilization*. Saarbrücken, LAP Lambert Academic Publishing GmbH Company. **2012,** 168 p.
5. Feder, F.; *Fractals*. New York, Plenum Press, **1990,** 248 p.
6. Vilgis, T. A.; Flory theory of polymeric fractals—intersection, saturation and condensation. *Physica A,* **1988,** *153(2),* 341–354.
7. Kozlov, G. V.; Temiraev, K. B.; Malamatov, A. Kh.; Shustov, G. B.; The polymer melts viscosity: fractal analysis and prediction. Bulletin of KBSC RAS, **1999,** *2,* 95–99.
8. Mashukov, N. I.; Mikitaev, A. K.; Gladyshev, G. P.; Belousov, V. N.; Kozlov, G. V.; Molecular weight characteristics of modified LDPE. *Plastics,* **1990,** *11,* 21–23.
9. Kavassalis, T. A.; Noolandi, J. A new theory of entanglements and dynamics in dense polymer systems. Macromolecules, **1988,** *21(9),* 2869–2879.
10. Wu, S.; Chain structure and entanglement. *J. Polymer Sci.: Part B: Polymer Phys.* **1989,** *27(4),* 723–741.
11. Prevorsek, D. C.; DeBona, B. T. On chain entanglements in high—T_g amorphous polymers. *J. Macromol. Sci. Phys.* **1981,** *B19(4),* 1329–1334.
12. Graessley, W. W.; Edwards, S. F.; Entanglement interactions in polymers and the chain contour concentration. Polymer, **1981,** *22(10),* 1329–1334.
13. Kozlov, G. V.; Dolbin, I. V. A fractal form of the Mark-Kuhn-Houwink equation. High-Molecular Compounds. B, **2002,** *44(1),* 115–118.
14. Termonia, Y. A creep compliance simulations study of the viscosity of entanged polymer melts. Macromolecules, **1996,** *29(6),* 2025–2028.
15. Kozlov, G. V.; Zaikov, G. E.; Structure of the Polymer Amorphous State. Utrecht, Boston, Brill Academic Publishers, **2004,** 465 p.
16. Novikov V. U.; Kozlov, G. V.; Fractal analysis of macromolecules. Successes of Chemistry, **2000,** *69(4),* 378–399.
17. Novikov V. U.; Kozlov, G. V.; Polymer structure and properties within the framework of fractal approach. Successes of Chemistry, **2000,** *69(6),* p. 572–599.
18. Kopelman, R.; Excitons dynamics resembling fractal one: geometrical and energetic disorder. In book: Fractals in Physics. Ed. Pietronero, L.; Tosatti, E.; Amsterdam, Oxford, New York, Tokyo, North-Holland, **1986,** 524–527.
19. Afaunov V. V.; Kozlov, G. V.; Mashukov, N. I.; Zaikov, G. E.; Fractal analysis of polyethylene inhibited thermooxidative degradation processes. Journal Applied Chemistry, **2000,** *73(1),* 136–140.
20. Dolbin, I. V.; Kozlov, G. V.; Zaikov, G. E.; The theoretical estimation of effective spectral dimension for polymer melts. Proceedings of International Interdisciplinary seminar "Fractals and Applied Synergetics, FaAS-01." Moscow, Publishers MSOU, **2001,** 41–42.
21. Kozlov, G. V.; Dolbin, I. V.; Zaikov, G. E.; The theoretical estimation of effective spectral dimension for polymer melts. *J. Appl. Polymer Sci.* **2004,** *94(4),* 1353–1356.
22. Klymko, P. W.; Kopelman, R.; Fractal reaction kinetics: exciton fusion on clusters. *J. Phys. Chem.* **1983,** *87(23),* 4565–4567.
23. Alexander, S.; Orbach, R.; Density of states on fractals: "fractons." *J. Phys. Lett.* (Paris), **1982,** *42(17),* p. L625-L631.

24. Argyrakis, P.; Percolation and fractal behavior on disordered lattices. In book: Fractals in Physics. Ed. Pietronero, L.; Tosatti, E.; Amsterdam, Oxford, New York, Tokyo, North-Holland, **1986,** 513–518.
25. Kozlov, G. V.; Gazaev, M. A.; Novikov, V. U.; Mikitaev, A. K.; Simulation of amorphous polymers structure as percolation cluster. Letter to Journal Engineering Physics, **1996,** *22(16),* 31–38.
26. Beloshenko, V. A.; Kozlov, G. V.; Lipatov, Yu. S.; Mechanism of cross-linked polymers glass transition. Physics of Solid Body, **1994,** *36(10),* 2903–2906.
27. Berstein, V. A.; Egorov, V. M.; Differential Scanning Calorimetry in Physics-Chemistry of Polymers. Leningrad, Chemistry, **1990,** 256 p.
28. Lobanov, A. M.; Frenkel, S. Ya. To question on nature of so-called transition "liquid-liquid" in polymer melts. High-Molecular Compounds. A, **1980,** *22(5),* 1045–1057.
29. Sokolov, A. M.; Dimensions and other geometrical critical indices in percolation theory. Successes of Physical Sciences, **1986,** *150(2),* 221–256.
30. Rammal, R.; Toulose, G.; Random walks on fractal structures and percolation clusters. *J. Phys. Lett.* (Paris), **1983,** *44(1),* p. L13-L22.
31. Baranov V. G.; Frenkel, S. Ya.; Brestkin Yu.V.; Dimensionality of different states of linear macromolecule. Reports of Academy of Sciences of SSSR, **1986,** *290(2),* 369–372.
32. Kozlov, G. V.; Shustov, G. B.; Dolbin, I. V.; Zaikov, G. E.; The physical significance of heterogeneity parameter in fractal kinetics of reactions. Bulletion of KBSC RAS, **2002,** *1,* 141–145.
33. Halsey, T. C.; Jensen, M. H.; Kadanoff, L. P.; Procaccia, I.; Shraiman, B. I.; Fractal measures and their singularities: the characterization of strange sets. *Phys. Rev. A*, **1986,** *33(2),* 1141–1151.
34. Bolotov V. N.; Positrons annihilation in fractal mediums. Letters to Journal Engineering Physics, **1995,** *21(10),* 82–84.
35. Nigmatullin, R. R.; Fractional integral and its interpretation. Theoretical and Mathematical Physics, **1992,** *90(3),* 354–367.
36. Meilanov, R. P.; Sveshnikova, D. A.; Shabanov, O. M.; Sorption kinetics in systems with fractal structure. Proceedings of Higher Educational Institutions, North-Caucasus region, natural sciences, **2001,** *1,* 63–66.
37. Dolbin, I. V.; Kozlov, G. V.; The physical significance of reactionary medium heterogeneity for polymer solutions and melts. Reports of Adygean International Academy of Sciences, **2004,** *7(1),* 134–137.
38. Dolbin, I. V.; Kozlov, G. V.; Zaikov, G. E.; The Structural Stabilization of POlymers: Fractal Models. Moscow, Publishers "Academy of Natural Sciences", **2007,** 328 p.
39. Kozlov, G. V.; Zaikov, G. E.; The Structural Stabilization of Polymers: Fractal Models. Leiden, Boston, Brill Academic Publishers, **2006,** 345 p.
40. Emanuel, N. M.; Thermooxidative aging of polymers. High-Molecular Compounds. A, **1985,** *27(7),* 1347–1363.
41. Emanuel, N. M.; Buchachenko, A. L.; The Chemical Physics of Polymers Molecular Fracture and Stabilization. Moscow, Science, **1982,** 359 p.
42. Dolbin, I. V.; Kozlov, G. V.; Mashukov, N. I.; The fractal model of regimes change for thermooxidative degradation of polymer melts. Successes of Modern Natural Sciences, **2002***(4),* 101–102.

43. Kozlov, G. V.; Dolbin, I. V.; Zaikov, G. E.; Structural criterion of change of a kinetic curves type in the process of a thermooxidative degradation. In book: Chemical Reactions: Quantitative Level of Liquid and Solid Phase. Ed. Zaikov, G.; Jimenez, A.; New York, Nova Science Publishers, Inc.; **2004,** 103–114.
44. Kozlov, G. V.; Dolbin, I. V.; Zaikov, G. E.; Structural criterion of change of a kinetic curves type in the process of a thermooxidative degradation. In book: Physical Chemistry of Low and High Molecular Compounds. Ed. Zaikov, G.; Dalinkevich, A.; New York, Nova Science Publishers, Inc.; **2004,** 107–118.
45. Kozlov, G. V.; Shustov, G. B.; Zaikov, G. E.; Burmistr, M. V.; Korenyako V. A.; The influence of polymer melt structure on chemical mechanisms of polyarylatearylenesulfonoxide melt thermooxidative degradation. Problems of Chemistry and Chemical Technology, **2002***(3),* 67–72.
46. Kozlov, G. V.; Shustov, G. B.; Burmistr, M. V.; Korenyako, V. A.; Zaikov, G. E.; The intercommunication of chemical and physical characteristics of heterochain polyethers melts thermooxidative degradation. Problems of Chemistry and Chemical Technology, **2004,** *2,* 95–99.
47. Dolbin, I. V.; Kozlov, G. V.; The structural analysis of polycarbonate thermooxidative degradation within the framework of scaling approaches. Reports of Adygen International Academy of Sciences, **2002,** *6(1),* 72–76.
48. Kozlov, G. V.; Shustov, G. B.; Zaikov, G. E. A polymer melt structure role in the thermooxidative degradation process of heterochain polyethers. Journal of Applied Chemistry, **2002,** *75(3),* 485–487.
49. Balankin, A. S.; Synergetics of Deformable Body. Moscow, Publishers of Ministry of Defence of SSSR, **1991,** 404 p.
50. Shlyapnikov Yu.A.; Kiryushkin, S. G.; Mar'in, A. P.; Antioxidative Stabilization of Polymers. Moscow, Chemistry, **1986,** 256 p.
51. Kozlov, G. V.; Shustov, G. B.; Zaikov, G. E.; The structural aspect of the interrelation of the characteristics of thermal and thermooxidative degradation of heterochain polyethers. In book: Aging of Polymers, Polymer Blends and Polymer Composites. V. 2. Ed. Zaikov, G.; Buchachenko, A.; Ivanov, V.; New York, Nova Science Publishers, Inc.; **2002,** 151–160.
52. Gammet, L.; The Grounds of Physical Organic Chemistry: Rates, Equilibrium and Reaction Mechanisms. Moscow, World, **1972,** 564 p.
53. Chen, Z. -Y.; Deutch, J. M.; Meakin, P.; Translational friction coefficient of diffusion limited aggregates. J.; Chem. Phys.; **1984,** *80(6),* 2982–2983.
54. Kozlov, G. V.; Batyrova, H. M.; Zaikov, G. E.; The structural treatment of a number of effective centers of polymeric chain in the process of the thermooxidative degradation. *J. Appl. Polymer Sci.* **2003,** *89(7),* 1764–1767.
55. Malamatov, A. Kh.; Kazancheva, F. K.; Kozlov, G. V.; Mathematical simulation of melt viscosity within the framework of fractal analysis. Review of Applied and Industrial Mathematics, **2005,** *12(4),* 1032–1033.
56. Mikitaev, A. K.; Kozlov, G. V.; Zaikov, G. E.; Polymer Nanocomposites: Variety of Structural Forms and Applications. New York, Nova Science Publishers, Inc.; **2008,** 319 p.
57. Malamatov, A. Kh.; Kozlov, G. V.; Burya, A. I.; Kudina, E. F.; Mikitaev, A. K.; The melt viscosity of nanocomposites on the basis of high density polyethylene. Bulletin

of National Academy of Sciences of Belarus, series physical-technical sciences, **2007,** *4,* 25–27.
58. Kozlov, G. V.; Afaunova, Z. I.; Zaikov, G. E.; Methods of deseribing oxidation reactions in fractal spaces. Polymer International, **2005,** *54(4),* 1275–1279.
59. Kozlov, G. V.; Shustov, G. B.; Zaikov, G. E. the fractal physics of the polycondensation processes. J.; Balkan Tribological Association, **2003,** *9(4),* p. 467–514.
60. Kozlov, G. V.; Shogenov, V. N.; The fractal model of nanocomposites polypropylene/calcium carbonate melt viscosity. Nanotechnologies. Science and Industry, **2011,** *6,* 39–44.
61. Kozlov, G. V.; Tlenkopachev, M. A.; Zaikov, G. E.; The rheology of particulate-filled polymer nanocomposites. In: Polymer Yearbook–2011. Polymers, Composites and Nanocomposites. Ed. Zaikov, G.; Sirghie, C.; Kozlowski, R.; New York, Nova Science Publishers, Inc.; **2011,** 157–165.
62. Kozlov, G. V.; Tlenkopachev, M. A.; Zaikov, G. E.; Niyazi, F. F.; The rheology of particulate-filled polymer nanocomposites. In book: Unique Properties of Polymers and Composites. Pure and Applied Science Today and Tomorrow. V. 1. Ed. Bubnov, Yu.; Vasnev, V.; Askadskii, A.; Zaikov, G.; New York, Nova Science Publishers, Inc.; **2012,** 261–269.
63. Mills, N. J.; The rheology of filled polymers. *J. Appl. Polymer Sci.* **1971,** 1*5(11),* 2791–2805.
64. Sheng, N.; Boyce, M. C.; Parks, D. M.; Rutledge, G. C.; Abes, J. I.; Cohen, R. E.; Multiscale micromechanical modeling of polymer/clay nanocomposites and the effective clay particle. Polymer, **2004,** *45(2),* 487–506.
65. Sumita, M.; Tsukumo, Y.; Miyasaka, K.; Ishikawa, K.; Tensile yield stress of polypropylene composites filled with ultrafine particles. *J. Mater. Sci.* **1983,** *18(5),* 1758–1764.
66. Honeycombe, R. W. K.; The Plastic Deformation of Metals. Cambridge, Edward Arnold Publishers, Ltd, **1968,** 398 p.
67. Kozlov, G. V.; Sultonov, N. Zh.; Shoranova, L. O.; Mikitaev, A. K.; Aggregation of nanofiller particles in nanocomposites low density polyethylene/calcium carbonate. Science-Capacious Technologies, **2011,** *12(3),* 17–22.
68. Goldstein, R. V.; Mosolov, A. B.; Flow of fractally beaten ice. Reports of Academy of Sciences, **1992,** *324(3),* 576–581.
69. Bobryshev, A. N.; Kozomazov V. N.; Babin, L. O.; Solomatov V. I.; Synergetics of Composite Materials. Lipetsk, NPO ORIUS, **1994,** 154 p.
70. Dzhangurazov, B. Zh.; Kozlov, G. V.; Mikitaev, A. K.; The fractal model of melt viscosity for nanocomposites poly(butylenes terephthalate)/organoclay. Nanotechnologies. Science and Industry, **2011,** *5,* 45–50.
71. Dzhangurazov, B. Zh.; Kozlov, G. V.; Mikitaev, A. K.; The influence of interfacial adhesion level on nanofiller structure in nanocomposites polymer/organoclay. Surface. X-raying, Synchrotron and Neutron Studies, **2011,** *7,* 96–99.
72. Pernyeszi, T.; Dekany, I.; Surface fractal and structural properties of layered clay minerals monitored by small-angle X-ray scatterin and low-temperature nitrogen adsorption experiments. Colloid Polymer Sci.; **2003,** *281(1),* p. 73–78.

73. Van Damme, H.; Levitz, P.; Bergaya, F.; Alcover, J. F.; Gatineau, L.; Fripiat, J. J.; Monolayer adsorption of fractal surfaces: a simple two-dimensional simulation. J.; Chem. Phys.; **1986,** *85(1),* 616–625.
74. Kozlov, G. V.; Shustov, G. B.; The structural analysis of soots reactive ability. Chemical Technology, **2006,** *1,* 24–26.
75. Kozlov, G. V.; Zhirikova, Z. M.; Aloev V. Z.; Zaikov, G. E.; The fractal model of nanocomposites polypropylene/carbon nanotubes melt viscosity. In book: Nanostructured Polymers and Nanochemistry: Research Progress. Ed. Haghi, K.; Kubica, S.; Zaikov, G.; Torun, REKPOL, **2012,** 44–51.
76. Zhirikova, Z. M.; Kozlov, G. V.; Aloev, V. Z.; The fractal model of melt viscosity of nanocomposites polypropylene-carbon nanotubes. Thermophysics of High Temperatures, **2012,** *50(6),* 785–788.
77. Kozlov, G. V.; Yanovskii, Yu. G.; Zhirikova, Z. M.; Aloev, V. Z.; Karnet, Yu. N.; Carbon nanotubes geometry in medium of polymer composite matrices. Mechanics of Composite Materials and Structures, **2012,** *18(1),* 131–153.
78. Bridge, B.; Theoretical modeling of the critical volume fraction for percolation conductivity in fiberloaded conductive polymer composites. J.; Mater. Sci. Lett.; **1989,** *8(2),* 102–103.
79. Kozlov, G. V.; Malkanduev, Yu. A.; Zaikov, G. E.; Fractal analysis of polymer molecular weight distribution: dynamic scaling. *J. Appl. Polymer Sci.* **2003,** *89(9),* 2382–2384.
80. Kozlov, G. V.; Malkanduev, Yu.A.; Zaikov, G. E.; Gelation in the radical polymerization of dimethyl diallyl ammonium chloride. *J. Appl. Polymer Sci.* **2004,** *93(3),* 1394–1396.
81. Kozlov, G. V.; Beeva, D. A.; Zaikov, G. E.; Mikitaev, A. K.; The fractal physics of branched polymers synthesis: polyhydroxyether. J.; Characterization and Development of Novel Materials, **2012,** *4(4),* 357–386.
82. Kozlov, G. V.; Dolbin, I. V.; Zaikov, G. E.; Fractal physical chemistry of polymer solutions. J.; Balkan Tribological Association, **2005,** *11(3),* 335–373.
83. Naphadzokova, L. Kh.; Kozlov, G. V.; Zaikov, G. E.; The structural model of polyesterification reaction in melt. Chemical Physics and Mesoscopy, **2008,** *10(1),* 72–76.
84. Vaidya, U. R.; Nadkarni, V. M.; Polyester polyols from glycolyzed PET waste: effect of glycol type on kinetics of polyesterification. *J. Appl. Polymer Sci.* **1989,** *38(6),* 1179–1190.
85. Kozlov, G. V.; Zaikov, G. E.; The physical significance of reaction rate constant in Euclidean and fractal spaces at the polymers thermooxidative degradation consideration. Theoretical Grounds of Chemical Technology, **2003,** *37(5),* 555–557.
86. Witten, T. A.; Meakin, P.; Diffusion-limited aggregation at multiple growth sites. *Phys. Rev. A*, **1983,** *28(10),* 5632–5642.
87. Korshak V. V.; Vinogradova, S. V.; Nonequilibrium Polycondensation. Moscow, Science, **1972,** 695 p.
88. Anacker, L. W.; Kopelman, R.; Fractal chemical kinetics: simulations and experiments. J.; Chem. Phys.; **1984,** *81(12),* 6402–6403.

INDEX

A

Acceptor-catalytic polycondensation, 207
Accessibility degree, 177–178
Admixtures (flocculation) adsorption, 179
Amorphous glassy polymers, 114
APESF. See Aromatic polyethersulfonoformals (APESF)
Aromatic copolyethersulfoneformals, 117
Aromatic polyamides, 73–74
Aromatic polyethersulfonoformals (APESF), 256, 261
Autoacceleration, 154
Autodeceleration, 135
Avogadro number, 279

B

Block-copolymer PAASO, 262
Boltzmann constant, 208
Budtov's monography, 211
Burgers vector, 116

C

Chain rigidity, 61, 82–83
  enhancement, 214–215
Chemically limited aggregation of type cluster–cluster (CLA Cl–Cl), 10
Chlorobenzene, 141
Cluster–cluster aggregation, 199
Cluster–cluster mechanism, 148
Comonomers functional groups, 143
Complex molecular architecture
  Budtov's monography, 211
  chain rigidity enhancement, 214–215
  dendrimers macromolecule, 233–234
  double logarithmic coordinates construction, 212
  flexible junction, 214
  fractal object density, 221
  Huggins constant, 211–212, 221–222
  intrinsic viscosity values, 216–219
  Kuhn segment length, 236–238
  macromolecule dynamics, 217
  Mark-Kuhn-Houwink equation, 211
  molecular weight and hydro dynamical, 217
  polymerization degree, 226
  polymer molecular weight variation, 227
  polystirene modification, 225
  PSD macromolecule, 218, 234–236
  PSMSO, 227–230
  Stox formula, 226
  "swarm," 222, 224
  tetrahydrofuran and chloroform, 215–216, 223–224
Consumed oxygen amount, 270–271, 275–276
Copolymer chain rigidity reduction, 120

D

Decay rate, 58
Dendrimers macromolecule, 233–234
Devil's staircase, 267
4,4'-Diaminodiphenylmethane (DDM), 147
Diffusion-limited aggregation of type cluster–cluster (DLA Cl–Cl), 10
Diffusive processes intensity, 179
Diluted solution and condensed state polymers
  aromatic copolyethersulfoneformals, 117
  copolymer chain rigidity reduction, 120
  Euclidean space, 120
  formal fragments contents, 121

gelation transition, 119
genetic intercommunication, 125
glass transition temperature, 117, 123–124
intercommunication analysis, 117
macromolecular coil, 117
molecular characteristics, 125, 128
parameters and fractal dimension, 126
polyarylatepolyarylenesulfonoxide, 123
polycondensation different methods, 124
polymer chain flexibility, 120
polymers condensed (block) state, 124
polymers, solid-phase state, 118–119
quantitative model, 119
solid-phase polymers, 117
solvent type and polymer chain structure, 117–118
sulfone fragments contents, 118
terachloroethane, 121–122
thermo-fluctuational clusters formation, 119
Dimethyldiallyl ammonium chloride (DMDAAC), 58–59, 154–156
Dimethyl sulfoxide (DMSO), 59
DMDAAC. See Dimethyldiallyl ammonium chloride (DMDAAC)
DMSO. See Dimethyl sulfoxide (DMSO)
Double logarithmic coordinates, 178
Dynamical scaling description, 207–208

E

Elementary link, 27, 53–54
Euclidean dimension, 14
Euclidean space, 120

F

Flocculation description, 184
Flory constant, 60
Flory exponent, 55
Flory formula, 16
Flory-Fox equation, 60
Flory-Huggins interaction parameter, 77–81

G

Gammet's constant, 277
Gaussian conformation, 75
Glass transition temperature, 63

H

Hetero-chain semirigid-chain polymers, 256
High density polyethylene (HDPE), 256–257, 261
Huggins constant, 65, 67, 211–212, 221–222

I

Intrinsic viscosity values, 216–219

K

Kirkwood-Riseman theory, 277
Kuhn segment
displacements, 69–72
length, 62, 69, 72, 236–238

L

Lattice animals, 17
Leaking coil, 45, 50, 56–57, 78, 83, 85, 88
Linking intrinsic viscosity, 32–33
Liquid with fixed structure, 263
"Looseness" degree, 176
Low-molecular flocculator effectiveness, 181
Lowmolecular solvents. See polymer solutions, fractal physics

M

Macromolecular coil density, 182–183
Macromolecular coil, intercomunication
analytical intercommunication, 68
aromatic polyamides, 73–74
attractive forces, 57
average end-to-end distance, 89–90, 94–95
branching and stucture (see Macromolecule branching)

chain rigidity, 61, 82–83
compensated state, 68
decay rate, 58
DMDAAC, 58–59
DMSO, 59
factors number, function, 87–88
flexible and rigid-chain polymers, 75–77
Flory constant, 60
Flory exponent, 55
Flory-Huggins interaction parameter, 77–81
fractal parameter, 64
freely connected chain, 74
Gaussian conformation, 75
glass transition temperature, 63
gyration radius, 62, 64
Huggins constant, 65, 67
interaction parameter, 57
intercommunication, 62–63
intrinsic viscosities, 64
Kuhn segment displacements, 69–72
Kuhn segment length, 62, 69, 72
macromolecule division, 91
mean magnitude, 74
MW values, 92, 95
PESF intrinsic viscosity, 60
PMMA and PS, 85
polymer chain rigidity, 69
polymerization degree, 69
polymers classification, 77
polymer-solvent interactions, 69, 82
polysaccharides, 58–59
repulsive interactions availability, 59
rodexman, 58–59
Shultz-Blashke equation, 66
solubility parameter, 80
solvent solubility parameters, 85–86
solvent thermo-dynamical affinity, 83
solvent vaporization molar energy, 78
statistical flexibility, 83
stirene-acrylonitril-methacrylate, 87
transition solvent-nonsolvent, 78
Macromolecular coils mobility, 206
Macromolecule branching
branched polymers, 170
branching factor, 157, 159
bromide-containing aromatic copoly-ethersulfones (B-PES), 157
excluded volume factor, 168–169
intensification, 160–161
interactions polymer-solvent, 158–159
intrinsic viscosities, 160
Mark-Kuhn-Houwink equation, 157
maximum dependence, 170
N-methylpyrrolydone, 157, 161
polyarylate F-2, 162
polyhydroxiether (PHE), 171–172
polymerization degree, 164, 171
polyphenylquinoxalines (PPQX), 157, 166, 172–173
trifunctional branching nodes, 164
volume interaction parameter, 159
Macromolecules fractal analysis
aggregation and degradation processes, 6–7
availability, 4
branched polymers formation, 15
calculation, Flory theory, 4–5
CLA Cl–Cl, 10
cluster, model, 9
connectivity degree, 18
deterministic fractals, 2–3
DLA Cl–Cl, 10
electroconducting polymer polypyrrole polymerization, 12
Euclidean dimension, 14
Flory formula, 16
fractals, 13
fraction filling, 19–20
internal concentration, 20
lattice animals, 17
linear fractals, 3–4
macromolecular coil dimension, 2
Monte-Carlo method, 2, 16
particle-cluster (P-Cl), 8
phantom fractal, 14–15
polydimethylsiloxane (PDMS), 8
polymeric statistical fractals, 4, 12
polymers synthesis process, 9
repulsion interactions, compensation, 5
SAW, 2–3

solvent phantom fractal border, 15
"spreading" dimension, 18
Suzuki rule, 5
TSAW, 11–12
UCD, 13
Mark-Kuhn-Houwink equation
application, 28
bulk interactions, 34
cationic water-dissolved polyelectrolytes, 52
coil elements, interactions, 39
copolymer PAA-TMAC, 52
"elementary link," 27
fractal analysis representations, 31–32
linking intrinsic viscosity, 32–33
macromolecular coil
elements, interactions, 31
form factor, 28
fractal dimension, 36–37, 40, 42, 44
swelling coefficient, 34
parameters, 37
polyarylate F-2 solutions, 32
polyarylate PF-2 M, 38
polymer chain chemical structure, 37
polymer density, 54
polymerization process, 41
polymer solvent interactions, 39
proportionality coefficient, 29, 52
rodexman macromolecular coil, 35
SAN-MMA, 39–40, 46
solubility parameters, 41, 46–47
solutions intrinsic viscosity, 27
stift factor dependence, 43
technique community verification, 46–47
TMAC, 52–55
translational diffusivity, 36
two- and three-solubility parameters, 39
Matrix polymer melt viscosities, 290–291
Melt flow index (MFI), 256, 260–261
Melt flow index values, 280
Microcomposite melt viscosity, 282
Molecular weight distribution (MWD)
acceptor-catalytic polycondensation, 207
Boltzmann constant, 208
cluster–cluster aggregation, 199
curve, 203
dynamical scaling description, 207–208
dynamics prediction, 211
macromolecular coils mobility, 206
normalization constant, 200–202
polycondensation methods, 206
polymerization process, 199, 209–210
renormalization, 204
Shiyan's theory, 199
Monte-Carlo method, 2, 16
MWD. See Molecular weight distribution (MWD)

N

Nanocomposite and matrix polymer, 294
Nanofiller particles aggregation, 285
Nanofiller particles density, 283
Nanogel, 149
Nanoreactor, 149, 154

O

Organoclay-polymer matrix, 291

P

Poly(dimethyl dialyl amonium chloride) (PDMDAAC)
accessibility degree, 177–178
admixtures (flocculation) adsorption, 179
average viscous molecular weight, 177–178
concentration, 180
concentration reduction, 177
diffusive processes intensity, 179
double logarithmic coordinates, 178
external "active zone," 183
flocculation description, 184
fractal dimension, 176
high-molecular polymers, 182
irreversible aggregation model, 181–182

"looseness" degree, 176
low-molecular flocculator effectiveness, 181
macromolecular coil density, 182–183
Polyarylates, 262, 275
Polycondensation methods, 206
Polydimethylsiloxane (PDMS), 8
Polyhydroxiether (PHE), 171–172
Polymer chain flexibility, 120
Polymer chain rigidity, 69
Polymeric statistical fractals, 4, 12
Polymerization degree, 69, 164, 171, 226
Polymerization process, 199, 209–210
Polymer melts, fractal analysis
APESF, 256, 261
"binary hookings" network, 257–258
block-copolymer PAASO, 262
consumed oxygen amount, 270–271, 275–276
fractional differentiation and integration, 267–268
Gammet's constant, 277
heterogeneity parameter, 268
heterogeneous mediums, 265–266
high-density polyethylene (HDPE), 256, 261
intrinsic viscosity, dependence, 259
Kirkwood-Riseman theory, 277
macromolecular coil compactness enhancement, 272
macromolecular entanglements network density, 258
melt flow index (MFI), 256, 260–261
percolation cluster formation, 263
"permeable" macromolecular coil, 269
polyarylates, 262, 275
polycondensation mode, 272
polymer nanocomposites melt (see Polymer nanocomposites melt)
reactive sites number, 273
semicrystalline polypropylene, 274–275
"structureless liquid," 263
thermooxidative degradation, 267, 276–277
transition "liquid 1–liquid 2," 263
viscosity, 256–257
Polymer nanocomposites melt
Avogadro number, 279
electron micrographs, 287
equimolar (stoichiometric) components content, 280–281
HDPE and EP interaction rate, 281
HDPE modification, 278–279
inorganic nanofillers, 282, 296
matrix polymer melt viscosities, 290–291
melt flow index values, 280
microcomposite melt viscosity, 282
nanocomposite and matrix polymer, 294
nanofiller particles aggregation, 285
nanofiller particles density, 283
organoclay-polymer matrix, 291
polymer density, 279
PP/CNT, 293, 295
relative melt viscosity, 292
rheological properties, 289
solid-phase composition, 278–279
structure, 279
Polymers condensed (block) state, 124
Polymer solutions, fractal physics
complex molecular architecture (see Complex molecular architecture)
disolution, thermodynamics
amorphous glassy polymers, 114
Burgers vector, 116
disolvation temperature, 114–115
heat effect, 113–114
local order domains, 114
PMMA, 115
"tie" chains, 114
films formation, mechanisms
intrinsic viscosity, 187
macromolecular "binary hooking," 185–186
macromolecular "stiffness section," 189
mechanical properties, 191
network nodes density, 190

PASF film samples, 185, 195–196
prediction technique verification, 192
"seeds" role, 188
shear modulus, 192
structure formation, 186
temperature dependence, 197–198
thermo-dynamical affinity, 187
uniaxial tension, 185
Witten-Sander clusters (WS clusters), 188
Witten-Sander irreversible aggregation model, 188
low molecular solvents, structural parameters
attractive interactions, 111
bisphenol A, 103
condensed state structures, 108
Flory-Huggins interaction parameter, 108–109
fractal dimension, 104
organic, 105
PC macromolecular coil, 107
polyarylatedimethylsiloxane film samples, 108
polymer-solvent interactions, 103
repulsive interactions, 111
solvent cohesion energy density, 105
solvent quality, characteristics, 113
space fraction, 106
spectral (fraction) dimension, 103
structure complication, 106
"swarm," 109–111
systems polymer-solvent, 107
zero-dimensional, 103
macromolecular coil, intercomunication (see Macromolecular coil, intercomunication)
Mark-Kuhn-Houwink equation, fractal variant (see Mark-Kuhn-Houwink equation)
MWD (see Molecular weight distribution (MWD))
structures intercomunication (see Diluted solution and condensed state polymers)
synthesis, macromolecular coil structure influence
autoacceleration, 154
autodeceleration, 135
chlorobenzene, 141
cluster–cluster mechanism, 148
comonomers functional groups, 143
compositional heterogeneity, 141–143
copolymerization reaction duration, 144
copolymers composition, 144–145
4,4'-diaminodiphenylmethane (DDM), 147
dianurethane and dichlorandhydride, 141
dimethyldiallyl ammonium chloride (DMDAAC), 154–156
Euclidean three-dimensional space, 147
fractal dimension, 130
Gordon-Tailor-Wood equation, 143
heterogeneity degree, 135
hexachlorobenzene, 147
high-temperature polycondensation, 137–138
intercommunication, 148
Mark-Kuhn-Houwink equation, 129, 133
mean-viscous molecular weight, 134
microgels self-diffusivity, 151–153
microheterogeneity coefficient, 146

nanochemistry, 149
"nanogel," 149
"nanoreactor" qualities, 149
none quilibrium polycondensation process, 135
organic solvents, 128
polyarylate F-2, 129–130, 132, 140
polyarylates, 134
polycondensation process, 128, 133, 137
polymerization duration, 154
polyurethanearylates, 141–142
reactionary medium initial viscosity, 136
reaction rate, 136
screening repulsive interactions, 138
solvent influence, 128
structural reformation, 154
temperature dependence
catalysis, 96
flocculation, 96
Huggins and Shultz-Blashke equations, 99
intrinsic viscosity, 97–98, 102
isophthalic acid (F-1), 96–97
polyarylate (PAr), 96–97
polymer solubility parameter, 100–101
polymer-solvent interaction, 98
Shultz-Blashke equation, 98
solvent solubility parameter, 102
synthesis processes, 96
triacetate cellulose, 100–101
Polymers, solid-phase state, 118–119
Polymers synthesis, reactions in melt
conversion degree, 300
degree-reaction duration, 298–299
irreversible aggregation mechanism, 298
kinetic curves conversion degree, 297
polycondensation, 297
polyesterification reaction rate, 301
radical polymerization, 297
temperature enhancement, 302
Poly(sodiumoxi) methylsylseskvioxanes (PSMSO), 227–230
Polyphenylquinoxaline, 172–173
Polyphenylquinoxalines (PPQX), 157, 166
Polypropylene/carbon nanotubes (PP/CNT), 293
Polystirene modification, 225
PP/CNT. See Polypropylene/carbon nanotubes (PP/CNT)
PSMSO. See Poly(sodiumoxi) methylsylseskvioxanes (PSMSO)
Ptitsyn-Eizner approximation, 60

R

Repulsive interactions availability, 59
Rigid-chain polymers, 69, 75–77, 97–98, 102, 235–237
Rimman-Liouville fractional derivative, 90
Rodexman, 58–59

S

SAN-MMA. See Stirene with acrylonitrile and methyl methacrylate (SAN-MMA)
SAW. See Self-avoiding walks (SAW)
Self-avoiding walks (SAW), 2–3
Shiyan's theory, 199
Shultz-Blashke equation, 66
Solid-phase polymers, 117
Solvent thermo-dynamical affinity, 83
Solvent type and polymer chain structure, 117–118
Solvent vaporization molar energy, 78
Stirene-acrylonitril-methacrylate, 87
Stirene with acrylonitrile and methyl methacrylate (SAN-MMA), 39–40, 46
Structure absence, 263
Structureless liquid, 263
Sulfone fragments contents, 118
Suzuki rule, 5

T

Terachloroethane, 121–122

Tetrahydrofuran and chloroform, 215–216, 223–224
Thermo-fluctuational clusters formation, 119
Thermooxidative degradation, 267, 276–277
"Tie" chains, 114
TMAC. See Trimethylammonium methacrylate chloride (TMAC)
Transition "liquid 1–liquid 2," 263
Transition solvent-nonsolvent, 78
Trifunctional branching nodes, 164
Trimethylammonium methacrylate chloride (TMAC), 52–55
True self-avoiding walk (TSAW), 11–12
TSAW. See True self-avoiding walk (TSAW)

U

UCD. See Upper critical dimension (UCD)
Upper critical dimension (UCD), 13

W

Witten-Sander clusters (WS clusters), 188
Witten-Sander irreversible aggregation model, 188